Sustainable Landscape Planning and Natural Resources Management

IEREK Interdisciplinary Series for Sustainable Development

Sustainable Landscape Planning and Natural Resources Management (SLNR) is a peer-reviewed book series that presents groundbreaking research through the intersection of two main areas that of which are 'Sustainable Landscape Planning' and 'Natural Resources Management'. The series initially covers a wide range of academic disciplines in sustainability, including landscape ecology, green urbanism, urban regeneration, land restoration and renaturation, energy management, and nature conservation. Moreover, it also incorporates the latest developments and research in engineering topics that contribute to more sustainable cities, such as Geosciences, infrastructure, civil engineering, geology, mining, water, and natural resources conservation and management.

By showcasing methods for discovering, managing, and employing natural resources, the series aims to provide approaches that enable the responsible use of these resources. Ultimately, the SLNR series is committed to advancing the United Nations' Sustainable Development Goals for 2030.

The wide range of topics covered in this series are primarily related to sustainable development, involving durable infrastructure projects, civil engineering, urban regeneration, the built, sustainability, climate change mitigation, energy management, geosciences, geology, and the future of cities, all of which depend on the availability of natural resources and effective wealth management. The series also features conceptual and empirical contributions from various interrelated fields of earth science, with an emphasis on the era of new discovery. It focuses on providing practical solutions to ensure sustainable development and provides new case studies that offer concrete examples of how to manage natural resources effectively.

The series encompasses the best research papers from various international conferences organized by IEREK conferences, also accepts proposals from other international conferences, including interdisciplinary contributions. It exhibits innovative approaches and highlights how they can best support both economic and sustainable development through the responsible use of natural resources and global, local, and individual activities for the welfare of all societies on Earth.

The SLNR series is a valuable source for experts in government, international organizations, consultancies, industry, business sectors, and academics. Published in cooperation with IEREK, the Springer SLNR series displays important new research in various sectors, delivering valuable insights and practical solutions to the complex challenges of sustainable development for the readers.

Jyoti Devi
Editor

Mutation Breeding in Crop Improvement

Principle and Application

Springer

Editor
Jyoti Devi
Department of Agriculture
Amity University
Noida, Uttar Pradesh, India

ISSN 2948-1910 ISSN 2948-1929 (electronic)
Sustainable Landscape Planning and Natural Resources Management
IEREK Interdisciplinary Series for Sustainable Development
ISBN 978-3-032-17910-4 ISBN 978-3-032-17911-1 (eBook)
https://doi.org/10.1007/978-3-032-17911-1

This Springer imprint is published by the registered company Springer Nature Switzerland AG
The registered company address is: Gewerbestrasse 11, 6330 Cham, Switzerland

Preface

Mutations, which are heritable changes in genetic material, play a vital role in creating genetic variability that is essential for crop improvement and development. The principles, methods, and significance of mutant breeding in agriculture are discussed in this book, with a focus on how crucial it is to solving global food security issues. A crucial component of contemporary crop development initiatives is mutation breeding, which entails the deliberate creation of mutations using physical, chemical, or biological mutagens. It makes it possible to develop improved plant cultivars with modified characteristics, such as increased yield, resistance to disease, and tolerance to stress. The book also discusses the history of mutant breeding, emphasizing significant findings and the development of methods from the first physical mutagens, such as X-rays and gamma rays, to sophisticated molecular tools, such as CRISPR Cas9. The mechanics and uses of site-directed and insertion mutagenesis are discussed, as well as the difference between spontaneous and induced mutations. It covers the relative advantages of chemical, physical, and biological mutagens, as well as the ways in which each works and the developments in molecular genetics that have transformed mutant breeding. The successful creation of more than 2000 mutant crop types globally highlights the accomplishments of mutation breeding, which have a major impact on agricultural sustainability and productivity. The potential of mutation breeding to produce climate-resilient crops, enhance nutritional value, and tackle environmental issues is also examined. Mutation breeding is still a vital strategy for developing novel crop cultivars and achieving global food security since it is economical and effective. This book is a great tool for researchers and students to learn how to properly use induced mutations in biological research and crop development. Agricultural scientists, breeders involved in crop development programs, horticulturists, floriculturists, and graduate and postgraduate students will find it interesting and helpful. The end of the book demonstrates how cutting-edge genomic technologies and mutant breeding could be combined in the future to address novel agricultural issues.

Noida, India

Jyoti Devi

Contents

Role of Mutagenesis in Genetic Improvement of Crop Plants

Gayatri Kumari, Kajal Kumari, Sanjeev Kumar, and Nitish Kumar

Abstract

Even though the world is currently going through a rapid phase of agricultural expansion, concerns about food security persist. The demands of an exponentially expanding population will require 70% more food by 2050. We need to improve our current crop varieties' genetic diversity, climate change adaptation, input efficiency, high yield, nutritional value, and ability to adapt to different agro-ecosystems without causing environmental harm if we want to escape this dilemma. Mutation breeding also known as induced mutation is one of the several breeding methods used to enhance the genetic variation among crop species and develop many different varieties of crop plants. Mutation breeding over the past 50 years has become increasingly prevalent. Both chemical and physical mutagens alter the genetic makeup of crops which brings crops with novel characteristics. By the use of mutation induction, there is an enhancement of nutritional quality of crop species and their production increases and the adaptability of the most significant crops in the world, including oilseeds, pulses, millets, wheat, and rice increases. Commercially released mutant cultivars have undoubtedly contributed significantly to the resolution of food and nutritional security issues in numerous nations, as seen by the overall area they cover. By creating mutants through the irradiation of in vitro grown tissues, it is now possible to treat vast populations in a way that was previously impossible. This method speeds up the selection of desired features and increases the gene pool, which enhances traditional breeding.

Keywords

Mutagens · Mutagenesis · Mutation breeding · Crop improvement

G. Kumari · K. Kumari · N. Kumar (✉)
Department of Biotechnology, Central University of South Bihar, Gaya, Bihar, India
e-mail: nitish@cub.ac.in

S. Kumar
School of Agricultural Science, KK University, Bihar Sharif, Nalanda, Bihar, India

1 Introduction

As the human population is constantly increasing the demand of food also increasing. To meet the constantly increasing demand for food, we need to focus on crop improvement techniques, so that all people get proper food and nutrition. Attempts to address the problems of global nutrition and food security must include crop genetic enhancement. Mutation breeding is one of the most prominent methods for the advancement of agriculture (Adamu et al. 2007). Mutation, a natural process which synthesizes new gene variants or alleles, which is heritable change in an organism's genetic characteristic. Mutation is primarily responsible for all genetic alterations in any creature, including plants. The process of permanently altering an organism's genetic makeup is referred to as mutagenesis.

Induced mutations can be used in the mutation breeding technique, to create new crop types with desired traits. Crop plants may develop novel and heritable character variants as a result of mutation. The recombination of alleles on homologous chromosomes and their independent assortment during meiosis improve the variety obtained in this way.

Plant mutagenesis and mutation breeding, for desirable traits in a wide range of food crops, are crucial methods for increasing genetic variability. Induced mutagenesis, for locating important regulatory genes and molecular pathways, is one of the best methods. The agents by which mutation can be created are known as mutagens. There are two types of mutagens that cause induced mutations: physical and

J. Devi (ed.), *Mutation Breeding in Crop Improvement*, Sustainable Landscape Planning and Natural Resources Management,
https://doi.org/10.1007/978-3-032-17911-1_1

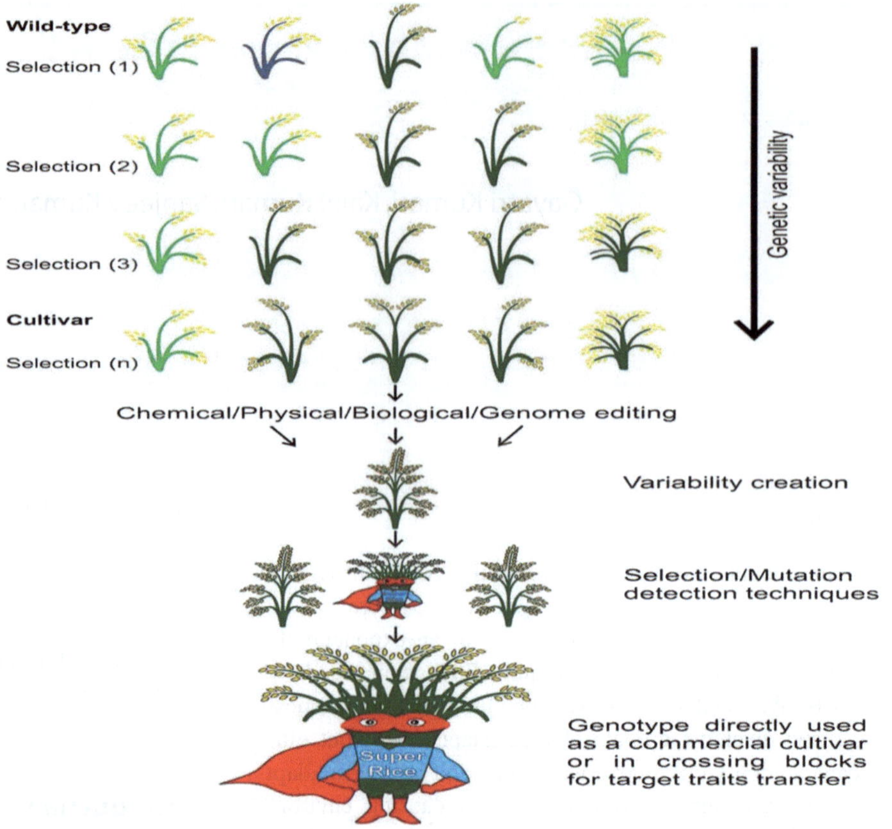

Fig. 1 Method for obtaining mutant variety. Mutagenesis is induced with a chemical or physical seed treatment given to an M_0 population which gives rise to M_1 plants. Seeds from M_1 plants will produce M_2 plants and DNA analysis of M_2 plants will be done to find mutations, then M_3 population will be produced from M_2 plants. (*Source* Viana et al., 2019)

chemical. Physical mutagens cause lesions or chromosomal anomalies in plants, like X-rays, gamma rays, ion beams and fast-moving thermal neutrons cause mutations. Equally, chemical mutagens cause point mutations. In order to select desired mutants' mutations can be induced and genetic variants can be produced by the use of mutagenic agents. The type of mutagen, dosage, and exposure duration all affect the rate, frequency, and efficiency of mutation in plants. To create a new crop variety by mutation, the parent plant is typically treated with a suitable mutagen. The variance created by mutation is used as a catalyst for evolution by the natural selection.

The idea of using induced mutagenesis to improve crops was first proposed at the start of the twentieth century. Over the last eight decades, the practice of mutant breeding has been effectively employed to enhance crops and support the work of conventional plant breeding techniques. Induced mutation is the most effective way to change the genetic makeup of crop plants that can be challenging to introduce by means of cross-breeding and any other breeding techniques. As a result, throughout the past few years, different workers have employed different kind of mutagens for the production of genetic variability in a number of pulse crops, including *Cicer arietinum, Vicia faba, Vigna mungo.* The first disease-resistant barley mutant was discovered as early as 1942. This prompted additional mutagenesis research, which led to the introduction of mutants in a range of crops. Of these types, 370 were legumes and 1468 were cereals. The bulk of grain varieties originated from wheat (197), barley (269), and rice (434). Method for obtaining mutant variety in plant is presented in Fig. 1.

Since the identification of X-ray mutation effects, inducing mutation has previously been acknowledged as a possible method for the improvement of crops. Genetic variety has been steadily declining, which ultimately led to the artificial induction of mutation. Just after Muller demonstrated in 1927 that irradiation by X-ray might significantly increase the mutation rate in Drosophila, Stadler demonstrated in 1928 that X-ray radiation combined with radium caused a significant phenotypic difference in sterility in maize tassels and in barley seedlings. Later, newly constructed nuclear research facilities generated gamma and ionizing radiations, such as neutrons, and beta (β) and alpha(α) particles, which are the commonly employed physical mutagens. During World War II, chemical mutagens that were easier to work with, more accessible, and less damaging were combined with radiation-based treatments. Chemical mutagens, for example, methylnitrosourea (MNU), sodium azide (SA), and ethyl methanesulfonate (EMS) were discovered as a result of this work. Chemical mutagens have become more and more popular because their handling is easy and having high mutation induction rate even used in small quantity. Ethyl

Table 1 Examples of commonly used physical mutagens

Mutagen	Source	Characteristics
X-rays	X-ray machine	Type of electromagnetic radiation which penetrates tissues from a few millimeters to many centimeters
Gamma rays	Nuclear reaction and radioisotopes	Electromagnetic radiation, very penetrating into tissues; Cobalt-60 and Cs-137 are sources
Neutrons	Nuclear reactors or accelerators	They are of different types (fast, slow, thermal); penetrate tissues to many centimeters, source is uranium 235
Beta particles	Accelerators or radioisotopes	These are electrons; ionize; shallowly penetrating; sources include 32P and 14C
Alpha particles	Radioisotopes	A helium nucleus capable of heavy ionization; very shallowly penetrating
Protons	Accelerators or nuclear reactors	Derived from hydrogen nucleus; penetrate tissues up to several centimeters
Ion beam	Particle accelerators	Produced positively charged ions are accelerated at a high speed, deposit high energy on a target

methanesulfonate, or EMS, is an alkylating agent and the most commonly employed chemical mutagen (Jankowicz-Cieslak et al., 2016). When EMS preferentially alkylates purines, especially guanine, it results in a thymine base over a cytosine residue opposite the O-6-ethyl guanine, which causes a random point mutation during replication. The majority of the alterations in EMS-mutated populations are caused by base pair transitions from GC to AT. In vitro cultivated somatic cells or calli, seeds, seedlings, tubers, rhizomes, corms, bulbs, or stolons are examples of parent plant material, anthers, inflorescences, and spikes are other plant components that are utilized in mutagenesis.

Keeping in view the benefits of mutation breeding this chapter will explore the different types of mutagens, their mechanisms of action, and their impact on agricultural advancements, highlighting their potential in sustainable food production.

2 Types of Mutagens

2.1 Physical Mutagens

Physical mutagens may affect DNA directly or indirectly. Direct effect is the direct interaction of the radiation with DNA, while, in indirect effect, DNA interacts with radiation generated by reactive oxygen species. The physical mutagens are divided into two classes: non-ionizing radiations and ionizing radiations. Ionizing radiation includes gamma rays, X-rays, and beta and alpha particles while, ultraviolet (UV) rays are included in non-ionizing radiation. The most commonly used irradiation rays of these are gamma rays (Jain 2010) and the second most commonly used irradiation rays are X-rays after gamma radiation (Beyaz and Yildiz 2017). Ion beam radiation has developed into a potent and distinct mutagen in recent years. In addition to gamma rays, scientists are now conducting experiments with other forms of carcinogenic radiation. One example of the usage of other physical mutagens is the recent use of fast neutrons to cause significant genetic material deletions. Examples of commonly used physical mutagens are presented in Table 1.

2.1.1 X-rays

X-ray were discovered by W. C. Roentgen in 1895. Since then, this has been widely employed for clinical diagnosis, leading to a significant number of studies on the impact of X-rays on living things. Because plants can undergo mutations when exposed to high amounts of X-rays, scientists have been employing this in mutant breeding programs.

2.1.2 Gamma Rays

Gamma rays, a kind of physical mutagens, has emerged as the most widely used physical mutagen since the 1950s. Its widespread availability and versatility in applying to a variety of applications, including food irradiation, mutant breeding, and medical applications, account for its widespread use (Mba and Shu 2012). There are various sources for the commercial generation of gamma rays. Caesium-137 (137Cs) and Cobalt-60 radioisotopes are two examples of such sources. According to Mba et al. (2012), Potassium-40 is the naturally occurring isotope which are present in the surrounding emits gamma rays.

2.1.3 UV Rays

The mutagenic impact of ultraviolet (UV) light was initially identified by Altenberg in 1934. Since then, numerous studies have been conducted to evaluate how UV light affects crop plants. The main natural source of UV radiation is sunlight, which contains UV radiation. UV radiation has wavelength longer than X-rays but shorter wavelength than visible light, placing it in the middle of the electromagnetic spectrum. UV radiation is classified into three types of ranges based on its wavelength: UV-B, UV-A, and UV-C. The stratosphere's ozone layer primarily influences UV-C, which has the maximum energy. Due to its restricted ability to penetrate tissue, UV radiation can only be used in specific applications. But in tissue culture, UV light has developed into a powerful mutagen for mutant breeding.

2.2 Chemical Mutagens

Chemical mutagens work by replacing a base, altering or mispairing a base and damaging a base. Ever since the chemical mutagens were discovered, the identification of hundreds of different types of chemical mutagens have been done, including base analogs, alkylating agents, azide, nitroso compounds, acridine dyes, and more (Kharkwal 2012). They can be categorized into three different types—base analogs, alkylating agents, and intercalating agents. Examples of commonly used chemical mutagens are presented in Table 2.

2.2.1 Base Analogs

Base analogs replace the nitrogenous bases in nucleic acids and closely mimic DNA bases. These alterations may alter gene transcription and DNA replication. 2-Aminopurine and 5-bromouracil are examples of base analogs of adenine and thymine, respectively. Base analogs cause point mutations in the plant genome.

2.2.2 Alkylating Agents

It works by adding alkyl groups to the bases and cause DNA damage. Since alkylating compounds are the most effective chemical mutagens for producing novel plant mutants, they have been utilized as plant chemical mutagens for many years. Methyl methanesulfonate (MMS), ethyl methanesulfonate (EMS), mustard gas, and ethylenimine (EI) are four of the frequently utilized alkylating chemicals that can be used in induced mutagenesis. Alkylating agents can be classified into mono-, bi-, and poly-functional. Bifunctional and poly-functional alkylating chemicals are among them; they can cross-link DNA, which causes DNA breakdown later on. Alkylating agents work well and are comparatively simple to work with in tests. In addition to being a bactericide and pesticide, it is also well known that sodium azide (NaN3) can be extremely mutagenic to both animals and plants.

2.2.3 Intercalating Agents

Intercalating agents intercalates between DNA double strands and cause interference with replication and transcription. Among the chemical mutagens are 4', 6-diamino-2-phenylindole (DAPI), ethidium bromide and acridine dye is regarded as an intercalating agent that can intercalate between neighboring dsDNA base pairs. According to Hashemi Shabestari et al. (2017), intercalating agents are employed for DNA imaging as fluorescent markers. Acridine dyes can be used as fluorescent markers, but they can also stretch the DNA strand, which makes it possible to add or remove nucleotide bases. They cause frameshift mutations most frequently.

Table 2 Examples of commonly used chemical mutagens

Mutagen group	Example	Mode of action
Alkylating agents	Methanesulfonate(MMS), Ethyl methanesulfonate(EMS), 1-methyl-1-nitrosourea (MNU), dimethyl sulfate (DMS), 1-methyl-2-nitro-1-nitrosoguanidine (ENNG), N,N-dimethylnitrous amide (NDMA), N,N-diethylnitrous amide (NDEA)	Add methyl or ethyl groups to bases and, depending on the affected atom, the alkylated base may then degrade to yield an abasic site, which is mutagenic and recombinogenic, or mispair to result in mutations upon DNA replication
Hydroxylamine	Hydroxylamine	Same as alkylating agents
Azide	Sodium azide	Same as alkylating agents
Base analogs	2-Aminopurine, 5-bromouracil, 5-bromodeoxyuridine, maleic hydrazide	These intercalate between DNA bases and causes the distortion of the double helix, DNA polymerase recognizes this stretch as an additional base and inserts an extra base opposite this molecule and causes frameshift mutation
Nitrous acid	Nitrous acid	Remove amino group from bases and causes point mutation, deamination of adenine gives hypoxanthine which can pair with cytosine rather than guanine
Acridines	Acridine orange	These intercalate between DNA bases and causes the distortion of the double helix, DNA polymerase recognizes
		This stretch as an additional base and inserts an extra base opposite this molecule and causes frameshift mutation

3 Mechanism of Action of Mutagens

Mutagens are agents that induce mutation by DNA damage, by replacing or altering bases, causing replication error which leads to permanent changes in the genome.

3.1 Mechanism of Action of Physical Mutagens

3.1.1 Mechanism of Action of Ionizing Radiations

Ionizing radiation primarily interacts with DNA directly to produce strand breakage, may work indirectly by promoting the cell's production of reactive substances like hydroxyl radicals. Due to biological systems' aqueous nature, ionizing radiation's impacts on water produce a wide variety of reactive oxygen species. These species can harm bases and produce various degradation products and adducts.

3.1.2 Mechanism of Action of X-rays

Through direct DNA damage, including breaks and other structural changes, X-rays cause mutations that alter the genetic material. With their high energy, X-rays can enter cells and directly interact with DNA molecules. DNA strand breakage (either single- or double-strand breaks) or other structural alterations, such as cross-linking, may result from this interaction. Errors or mutations may result from these DNA damage-related disruptions to the regular DNA replication process. Changes to the genetic code that may result in negative consequences include base pair substitutions, deletions, and insertions.

3.1.3 Mechanism of Action of Gamma Rays

It causes DNA damage, either directly or indirectly, which results in a variety of mutations, including nucleotide changes, single- or double-strand breaks, and chromosomal abnormalities. DNA molecules may be directly impacted by gamma rays, leading to single-strand or double-strand breaks can cause mutations during DNA replication or repair by upsetting the structure of the molecule. Gamma rays have the ability to alter DNA bases, which can result in improper pairing during replication and mutations (Pathirana, 2011). When gamma rays contact with water molecules, which make up a large portion of cells, they can produce extremely reactive free radicals, such as hydroxyl radicals. Following their interaction with DNA, these free radicals may cause the same kinds of damage as previously mentioned, such as base changes, which can result in mutations. Free radicals have the ability to break DNA as well. DNA damage can result in chromosomal abnormalities on a broader scale.

3.1.4 Mechanism of Action of UV Rays

The primary way that UV light damages DNA and results in mutations is by forming pyrimidine, pyrimidone photoproducts, and cyclobutane pyrimidine dimers, which cause cytosine to thymine transitions. DNA can be damaged in a number of ways by direct interactions with UV light, especially UV-B and UV-A. Additionally, UV light can indirectly harm DNA by producing reactive oxygen species, which can oxidize DNA bases and result in mutations.

3.2 Mechanism of Action of Chemical Mutagens

By changing DNA structure through chemical interactions such as base analogs, deamination, alkylation, and intercalation, chemical mutagens cause mutations that result in incorrect base pairing and mistakes during DNA replication.

3.2.1 Mechanism of Action of Base Analogs

During DNA replication, these chemicals can be integrated into DNA by mimicking the bases of conventional DNA. But because of their marginally differing chemical structures, the bases are not paired correctly. For example 5-bromouracil, a base analog of thymine, can cause a GC to AT transition by pairing with guanine rather than adenine.

3.2.2 Mechanism of Action of Deaminating Agents

By removing amino groups from DNA bases, these substances change the bases' structure and base pairing characteristics. For example, nitrous acid deaminate adenine to hypoxanthine, which pairs with cytosine rather than thymine and cause GC to AT transition.

3.2.3 Mechanism of Action of Alkylating Agents

These substances alter the characteristics of base pairing by adding alkyl groups to DNA bases, which can result in improper base pairing, disruption of DNA repair and replication, DNA strands cross-linking to stop replication.

3.2.4 Mechanism of Action of Intercalating Agents

By inserting themselves between neighboring base pairs in DNA, these planar molecules cause structural disruptions. This may result in DNA base insertions, deletions, or frameshift mutations.

4 Application of Induced Mutagenesis in Crop Improvement

Crop production has increased significantly as a result of induced mutagenesis, according to Kharkwal and Shu, with novel kinds created from these mutants being utilized in various places worldwide. The production of superior crop varieties worldwide has been greatly aided by the genetic variability created by induced mutagenesis with different mutagens in current plant breeding. This has been particularly important in recent decades, producing plants with traits including high yield, early maturity, and resistance to lodging and pests. With an emphasis on the outcomes of agricultural development by mutagenesis, Ahluwalia et al. (2004) assessed the worldwide impact of the varieties generated and released in main crops. The production of enhanced commercial agricultural varieties and the generation of new genetic resources with superior features and extraordinary combinations of traits were the two key objectives of mutation breeding. Although the primary objective of mutant breeding is the production of superior varieties, mutagenesis-generated variability can be used for a number of purposes in plant breeding. In traditional or hybrid breeding operations, for instance, they can be used as parents. The extensive use of induced mutations in plant breeding has led to the development of exceptional types of mutant plants (Kharkwal 2023). These mutants are crucial for creating crops with higher yields, higher quality, increased stress tolerance and yield contributing traits, longer shelf lives, and lower agronomic input requirements. T-DNA insertional mutagenesis has led to a rapid advancement in our understanding of plant biochemistry, physiology, and development. It has been proposed that auxin transport, inhibition, uptake, and signal transduction are affected by auxin mutants like aux1, mp, and lop1. The discovery of cytokinin-resistant mutants, cell division mutants, photomorphogenic mutants (det1, cop), and mutants with high cytokinin levels (amp1) helped to clarify the mechanism of cytokinin action. In Arabidopsis thaliana, cytokinin mutants including ckr1, ein2, cry1, stp1, and zea3 were discovered by Schmulling et al. in 1997. These mutations have clarified the function of genes regulated by cytokinin in a variety of biological processes, including photosynthesis, chloroplast formation, disease resistance, food metabolism, and cell division. The growth hormone gibberellin's mechanism of action was clarified by Chandler and Robertson in 1999 by screening for dwarf le mutants in peas and dwarf mutants in maize. GA deficiency causes several dwarf mutants, including Rht3 in wheat and d8 in maize, to not react to administered GA3. The development of lodging-resistant and highly fertilizer-responsive cultivars has been greatly aided by these dwarf mutants. By delaying senescence and extending floral life, as demonstrated by their transfer to petunia and tomato, these mutants are extremely beneficial and contribute significantly to the prolongation of fruit's shelf life. There are a number of homeotic mutants in Arabidopsis, Antirrhinum, and Petunia that produce faulty blooms. These mutants' isolation has greatly advanced our knowledge of floral development pattern. Homeotic mutants for leafy cotyledons lec have been created by insertional mutagenesis and are faulty in the development of embryos that stay green. Mutations like the fis mutant that regulate seed development are crucial to understanding the apomixes. Yield and attributes linked to yield are significantly influenced by the developmental patterns of crop plants. In the near future, the modification of these patterns will take on a new role in plant breeding. Enhancing the well-adapted cultivars by altering one or two essential traits is the main objective of mutation-based plant breeding (Oladosu et al., 2016). Plant height, seed dehiscence, and disease resistance are among the improved characteristics that boost output and quality attributes like oil profile content, malting quality, and the size and quality of the starch granules. Rice, wheat, maize, and barley grain yields improved because of high-density planting and resistance to loading. Compared to cross-fertilizing species, induced mutation provides more opportu-

Table 3 Role of mutagens in crop improvement

Mutagens	Dose	Remarks	References
EMS	1.5%	Plant height decreases, panicle length increases, panicle height increases	Identification of variability for agronomically important traits in rice mutant families (Luz et al. 2016)
Sodium azide	10 mM–50 mM	Dwarfed mutant and chlorophyll mutant, no. of plants reduced	Mutagenic effects of sodium azide on quality of maize seeds (Eze et al., 2015)
X-ray	50 Gy–300 Gy	Plant height decreases, loses viability	Optimization of X-ray irradiation dose for induced mutation in bread wheat varieties cultivated in Albania (Ylli et al. 2024)
Gamma ray	30, 50, 70 Gy	High yield, starch content increases, amylasecontent increases	Mutation breeding of sweet potato by gamma ray radiation (Shin et al. 2011)
Alpha particles	1, 10, 30, 40 Gy	Germination percentage decreases	Alpha particles and cobalt-60 gamma rays have different biological effects in upland cotton pollen grains (Yue et al., 2012)

nities for self-fertilizing species to improve. Usually, there are two causes for this. Recessive mutations in such populations are difficult to select, incorporate, and sustain. Another issue is plant breeding in cross-fertilizing species, which is more typically a challenge of managing the diversity that already exists than a lack of variability per se. However, a number of successful plant development initiatives were used to address the unpredictability caused by mutagens. The role of mutagens in crop improvement is presented in Table 3.

5 Future Prospectus

Recently, interest in mutation research has increased because induced mutagenesis is increasingly being used in plant molecular biology to isolate and identify genes and study their structure and function. Future crop development initiatives will surely be significantly impacted by this discovery. Very soon, genetic engineers will use new technologies and mutation as tools for plant breeders. Nevertheless, most of the cultivars that have been released to the public thus far were produced by a mix of mutation and direct selection. In the present era for crop improvement, the use of mutant breeding has assumed a new and expansive paradigm because to advancements in in vitro culture and molecular technologies (Mba C, 2013). Heavy ion beam irradiation is now an effective way and common to produce mutation in a wide variety of plant species because of its high frequency and broad range. By creating better and more resilient traits, in vitro mutagenesis has recently increased agricultural productivity and germplasm innovation. To improve crop species, a small number of tissues and calli may be subjected to mutagenesis using in vitro culture procedures. In vitro mutagenesis is commonly utilized these days, several plants, including sugarcane and bananas, have been successfully regenerated with this method. However, it is now possible to produce many plants that are propagated from seeds again using cell suspension cultures, such as barley, wheat, rice, and maize. In the future, It would also be crucial to create in vitro cell selection methods for disease resistance. It might be feasible to combine the most recent methods of chromosomal doubling, regeneration of doubled haploid plants, microspore culture, cell suspension, anther culture, and haploid cell irradiation to produce genotypes with desired traits. Additionally, the produced mutation has showed utility in the production of genetic maps that will aid future marker-assisted molecular plant breeding. Recent years have seen a rise in the use of mutation breeding as a useful technique for crop growth.

Plant production and quality could be rapidly increased by directly using mutation in the creation of molecular maps in structural and functional genomics.

6 Conclusion

For 50 years, induced mutations have been a key component of plant breeding, helping both developed and developing countries increase their food production. Traditional mutation breeding is still used for the benefit of communities, in addition to the use of state-of-the-art labs and modern genomic technologies for mutation induction and discovery. As long as the objectives and selection criteria are well-defined, mutation breeding has shown itself to be useful, flexible, and ready for use on any crop. More than 170 distinct plant species now have 3211 documented mutant variants. Although induced mutagenesis is nearly seven decades old, it has been proven to be effective in unlocking the plant genetic resources' potential and provide the building blocks required for plant breeders to produce the targeted smart crop variants. Mutation breeding-produced crop varieties are greatly enhancing livelihoods and promoting global food and nutritional security. Since traditional breeding techniques have long been employed to reduce genetic variety, one of the most crucial tactics for boosting genetic diversity and variation in crops to overcome bottleneck circumstances is induced mutagenesis.

References

Adamu AK, Aliyu H. Morphogical effects of sodium azide on tomato (Lycopersicon esculentum Mill). Science World Journal. 2007;2(4). https://doi.org/10.4314/swj.v2i4.51755

Ahloowalia BS, Maluszynski M, Nichterlein K. Global impact of mutation-derived varieties. Euphytica. 2004 Feb;135(2):187-204.

Beyaz R, Yildssiz M. The Use of Gamma Irradiation in Plant Mutation. Breeding in Book Plant Engineering. 2017 Nov 17:33-46. DOI: 10.5772/intechopen.69974

Eze JJ, Dambo A. Mutagenic effects of sodium azide on the quality of maize seeds. Journal of advanced laboratory Research in Biology. 2015;6(3):76-82. E-ISSN: 0976-7614

Jain SM. Mutagenesis in crop improvement under the climate change. Romanian biotechnological letters. 2010 Mar 1;15(2):88-106.

Jankowicz-Cieslak J, Till BJ (2016) Chemical mutagenesis of seed and vegetatively propagated plants using EMS. Curr Protoc Plant Biol 1(4):617–635. https://doi.org/10.1002/cppb.20040

Kharkwal MC. A brief history of plant mutagenesis. InPlant mutation breeding and biotechnology 2012 (pp. 21-30). Wallingford UK: CABI. https://doi.org/10.1079/9781780640853.0021

Kharkwal MC (2023) Role of mutation breeding in crop improvement with special reference to Indian subcontinent. In: Mutation breeding for sustainable food production and climate resilience 2023. Springer Nature Singapore, Singapore, pp 355–428. https://doi.org/10.1007/978-981-16-9720-3_13

Luz VK, Silveira SF, Fonseca GM, Groli EL, Figueiredo RG, Baretta D, Kopp MM, Magalhães AM, Maia LC, Oliveira AC (2016) Identification of variability for agronomically important traits in rice mutant families. Bragantia 75(1):41–50. https://doi.org/10.1590/1678-4499.283

Mba C (2013) Induced mutations unleash the potentials of plant genetic resources for food and agriculture. Agronomy 3(1):200–231. https://doi.org/10.3390/agronomy3010200

Mba C, Afza R, Shu QY (2012) Mutagenic radiations: X-rays, ionizing particles and ultraviolet. In: Plant mutation breeding and biotechnology 2012. CABI, Wallingford UK, pp 83–90. https://doi.org/10.1079/9781780640853.0083

Oladosu Y, Rafii MY, Abdullah N, Hussin G, Ramli A, Rahim HA, Miah G, Usman M (2016) Principle and application of plant mutagenesis in crop improvement: a review. Biotechnol Biotechnol Equip 30(1):1–6. https://doi.org/10.1080/13102818.2015.1087333

Pathirana R (2011) Plant mutation breeding in agriculture. CABI Rev 2(2011):1–20. https://doi.org/10.1079/PAVSNNR20116032

Schmülling T, Schäfer S, Romanov G (1997) Cytokinins as regulators of gene expression. Physiol Plant 100(3):505–519. https://doi.org/10.1111/j.1399-3054.1997.tb03055

Shin JM, Kim BK, Seo SG, Jeon SB, Kim JS, Jun BK, Kang SY, Lee JS, Chung MN, Kim SH. Mutation breeding of sweet potato by gamma-ray radiation. African Journal of Agricultural Research. 2011 Mar 18;6(6):1447-54. ISSN 1991-637X ©2011 Academic Journals

Ylli A, Mitrushi F, Elezi F, Ylli F. Optimization of X-ray irradiation dose for induced mutation in bread wheat varieties cultivated in Albania. DOI:10.37392/RAPPROC.2024.13

Viana VE, Pegoraro C, Busanello C, Costa de Oliveira A (2019) Mutagenesis in rice: the basis for breeding a new super plant. Front Plant Sci 10:1326. https://doi.org/10.3389/fpls.2019.01326

Yue J, Wu L, Wu Y, Tang C (2012) Alpha-particles and 60Co γ-rays have different biological effects on upland cotton (Gossypium hirsutum L.) pollen grains. J Agric Sci 4(3). https://doi.org/10.5539/jas.v4n3p145

In Vivo and in Vitro Random Mutagenesis Methods in Crop Plants

Preeti Singh, Shashi Prakash, and Hidayatullah Mir

Abstract

A crucial technique in crop improvement is mutagenesis, which creates genetic variation to produce new plant varieties with enhanced nutritional value, disease resistance, and stress tolerance. By changing the structure of DNA, induced random mutagenesis—achieved by physical (gamma rays, X-rays, UV radiation) and chemical (EMS, MMS, sodium azide) mutagens—speeds up reproduction. Directly exposing seeds, pollen, or entire plants to mutagens is known as "in vivo mutagenesis." Desired traits are then assessed in the field. Using plant tissue culture, in vitro mutagenesis is typically carried out at the cellular level. It offers early and effective variant selection as well as more control over the induction of mutations. Crop improvement in cereals (rice, wheat), legumes (soybean, lentil), oilseeds (rapeseed), fruits (banana, strawberry), and vegetables (tomato, cassava) has been greatly aided by these techniques, resulting in commercially successful mutant varieties with improved agronomic performance. Mutagenesis has drawbacks despite its advantages, such as the randomness of mutagenic events, the rare chance of getting advantageous features, the possibility of genetic instability, and time-consuming mutant screening. However, advances in molecular markers and genome sequencing has led to improvements in mutation detection and enhanced selection efficiency, resulting in a more successful mutation breeding programs. With a focus on its use in modern crop breeding, this chapter elucidates the significance, methods, uses, and limitations of in vivo and in vitro mutagenesis.

Keywords

Random induced mutagenesis · Genetic variability · In vivo mutagenesis · In vitro mutagenesis · Crop improvement

1 Introduction

In the present scenario, crop improvement is critically focused on addressing global food security, adapting to climate change, and ensuring sustainable agricultural productivity. Breeding for better adaptation to shifting environmental conditions, enhanced crop yield and quality, decreased susceptibility to pests and diseases, and long-term agricultural sustainability is greatly possible by a diverse gene pool. While modern biotechnological methods require highly sophisticated and expensive infrastructure, traditional breeding methods to a great extent rely on natural genetic recombination and selection, which can be time-consuming and limited in certain crops. Genetic diversity is one pre-requisite for plant breeders to choose and combine desired traits. In crops with little natural diversity, mutagenesis has shown itself to be a useful technique for generating novel genetic variability. Hugo de Vries introduced the concept of "mutation" at the beginning of the twentieth century. Spontaneous mutation have been one of the powerful forces driving evolution in nature that has led to the emergence of new genera, species, or individuals, which is essentially random and less frequent. Farmers have relied on spontaneous mutations as a source of genetic diversity since the beginning of agriculture. This inherent variability has also been used by plant breeders to choose traits with desired agronomic qualities, such as non- shattering pods or heads, shrunken seed of maize, development of parthenocarpy in bananas and grapes, and the

P. Singh · S. Prakash
Department of Horticulture (Fruit and Fruit Technology), Bihar Agricultural University, Sabour Bhagalpur, Bihar, India

H. Mir (✉)
Regional Horticulture Research Sub-Station, Sher-e-Kashmir University of Agricultural Sciences and Technology of Jammu, Bhaderwah, J&K, India
e-mail: hidayatmay14@yahoo.co.in

J. Devi (ed.), *Mutation Breeding in Crop Improvement*, Sustainable Landscape Planning and Natural Resources Management,
https://doi.org/10.1007/978-3-032-17911-1_2

removal of bitterness in crops like eggplant and cabbage. For a given trait, the natural variability and spontaneous mutation frequency found in plant populations are frequently restricted. Early in the twentieth century, plant breeders started using chemical and physical mutagens to induce mutations artificially in order to overcome this restriction and increase genetic variability in crop improvement initiatives.

2 Induced Mutagenesis: A Catalyst for Expanding Genetic Diversity and Crop Improvement

Induced mutagenesis, with its distinct advantages over conventional and biotechnological breeding techniques, has emerged as a crucial tool in crop improvement. The capacity of induced mutagenesis to produce novel genetic diversity more frequently in less time is one of its major benefits. By bringing in novel mutations that might not arise naturally, it has caused the gene pool to grow. According to Joint FAO/IAEA, 2025, the top crop plants with maximum number of registered mutants in the world are rice, barley, chrysanthemum, wheat soyabean, and maize and the continents of the world developing most number of mutant varieties are Asia, Europe, and North America (Fig. 1a and b). Breeders are able to identify and develop improved cultivars more quickly thanks to the increased variability brought about by the induction of mutation. In addition, induced mutagenesis provides a quicker and more effective way to improve crops than traditional breeding techniques. Mutagenesis can directly cause advantageous mutations in a single generation, while traditional breeding frequently takes several generations to stabilize desired traits. Because of its effectiveness, it is especially useful for creating crops that are more resilient to pests, diseases, and environmental stressors like salinity and drought. The non- transgenic nature of induced mutagenesis is another noteworthy benefit that has extended its use in modern crop improvement programs. Mutagenesis causes natural genetic changes via gene and chromosomal mutations without adding external genetic material, in contrast to genetic engineering, which introduces foreign DNA into a plant's genome. It also provides a more straightforward but efficient way to improve crops. Even though biotechnology makes precise gene manipulation possible, it frequently encounters ethical dilemmas, regulatory obstacles, and low public acceptance. This characteristic gives breeders a more secure and well-liked option for crop improvement by making mutagenesis more acceptable in areas with stringent laws governing genetically modified organisms (GMOs).

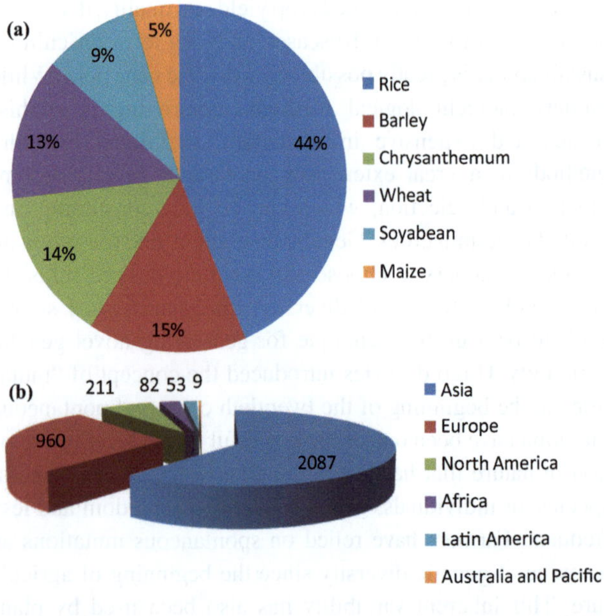

Fig. 1 **a** Top crop plants with maximum number of mutant varieties. *Source* IAEA mutant database. http://mvgs.iaea.org/] **b** Regions with mutant varieties developed in crop plants. *Source* IAEA mutant database. http://mvgs.iaea.org/)

The contribution of induced mutagenesis to improving crop plant with respect to yield, nutritional quality, stress tolerance, etc., is quite noteworthy. Induced mutagenesis has led to notable improvements in crop plants globally. For instance, "Golden Promise" barley, developed in the UK through gamma radiations, exhibits semi-dwarf stature and salt tolerance. In the US, "Star Ruby" grapefruit developed through thermal neutron technique is known for its deep red flesh and enhanced sweetness. Vietnam's "VND 95–20" rice is another noteworthy achievement of random mutagenesis, which offers enhanced yield and disease resistance. Several crop varieties developed through this technique have demonstrated enhanced nutritional profiles, such as increased protein content in barley and elevated β-carotene levels in rice. Such examples underscore the effectiveness of induced mutagenesis in creating resilient and productive crop varieties. Induced mutagenesis continues to be a powerful tool in agricultural research, providing sustainable solutions for improving crop productivity, resilience against biotic and abiotic stress, and nutritional quality. Its ability to complement traditional and modern biotechnological approaches ensures its continued relevance in global efforts to enhance food security. Induced random mutagenesis, both in vivo and in vitro, has become a powerful tool to accelerate crop improvement by generating novel genetic variations in crops with narrow genetic base that can be harnessed for enhancing important agronomic traits such as yield, nutritional quality, disease resistance and abiotic stress tolerance (Ahloowalia et al. 2004; Mba et al. 2010).Induced mutagenesis involves the application of physical (e.g., gamma rays, X-rays, fast neutrons, etc.) or chemical mutagens (e.g., ethyl methane sulfonate (EMS), sodium azide, etc.) to create genetic alterations in plant genomes (Tables 1 and 2). The type of mutagen and its optimum doze/concentration is standardized in order

Table 1 Physical mutagens used in in vivo and in vitro mutagenesis

Physical mutagen	Mode of action	Application (In vitro/In vivo)	References
Gamma Rays (^{60}Co, ^{137}Cs)	Causes DNA strand breaks, base damage, and chromosomal aberrations	Both (used in seed treatment, in vitro callus irradiation, protoplasts, and meristem cultures)	Shu et al. (2012); Jain (2012); Perveen et al. (2023a); Perveen et al. (2025)
Ultraviolet (UV) Radiation (UV-A, UV-B, UV-C)	Causes pyrimidine dimers, leading to replication errors	In vitro (used in plant cell and tissue cultures)	Bado et al. (2023)
Fast neutron radiation	Induces double-strand breaks and large deletions in DNA	Both (used in mutation breeding and gene knockout studies)	Shu et al. (2011); Till et al. (2018)
Laser irradiation	Induces oxidative stress and DNA strand breaks	In vitro (used in somatic embryogenesis and protoplast culture)	Bhagwat and Duncan (1998)
Microwave radiation	Causes protein denaturation, cell membrane disruption, and oxidative stress	In vivo (used for seed and pollen treatment)	Ahloowalia et al (2004)
Thermal stress (Cold and heat shock)	Causes protein misfolding, chromatin remodeling, and epigenetic modifications	In vitro (used in tissue cultures for stress-induced mutations)	Shu et al. (2011)

Table 2 Chemical mutagens used in in vivo and in vitro mutagenesis

Chemical mutagen	Mode of action	Application (In vitro/In vivo)	References
Ethyl Methanesulfonate (EMS)	Alkylates guanine at the O6 position, leading to G/C → A/T transitions	Both (used in cell suspensions, callus, and seeds)	Bado et al. (2023); Till et al. (2018)
Methyl Methanesulfonate (MMS)	Alkylates DNA, leading to basemispairing and strand breaks	Both (effective in tissue cultures and seed treatment)	Shu et al. (2011)
Sodium Azide (NaN_3)	Induces point mutations by generating azide metabolites that react with DNA	Both (widely used in cereals for seed mutagenesis)	Ahloowalia et al. (2004)
Hydroxylamine (HA)	Hydroxylates cytosine, leading to C → T transitions	In vitro (used in plant cell cultures)	Bhagwat and Duncan (1998)
Nitrous acid (HNO_2)	Deaminates adenine, cytosine, and guanine, leading to base substitutions	Both (applied to seeds and cultured cells)	Bado et al. (2023)
Acridine orange	Intercalates into DNA, causing frameshift mutations	In vitro (used in plant tissue cultures)	Jain (2012)
Base Analogs (5-Bromouracil, 2-Aminopurine)	Incorporate into DNA instead of normal bases, leading to incorrect base pairing	In vitro (used for genetic modification in protoplasts)	Shu et al. (2011)
Proflavine	Intercalates into DNA and induces frameshift mutations	In vitro (mainly in bacterial and plant cell cultures)	Patade and Suprasanna (2009)
Mustard gas (Sulfur mustard)	Alkylates DNA, leading to cross-linking and strand breaks	In vivo (used historically but highly toxic)	Bhagwat and Duncan (1998)

to achieve maximum frequency of desirable mutations. This process can lead to beneficial mutations that are not naturally available in conventional breeding pools, thereby expanding genetic variability and enabling the development of superior cultivars (Shu et al. 2011).

3 Types of Induced Mutagenesis

3.1 In Vivo Mutagenesis

This approach involves exposing whole plants, seeds, pollen, or vegetative propagules to mutagenic agents under natural physiological conditions. It has been widely employed in cereals, legumes, and horticultural crops to improve important economic traits. Examples include high-yielding mutant rice, wheat, maize, sorghum, and barley varieties that have significantly contributed to global food production (Parry et al. 2009). The major advantage of in vivo mutagenesis is its ability to produce stable genetic alterations with minimal soma clonal variation, ensuring that desirable traits are heritable and reproducible across generations (Shu et al. 2012). This technique is crucial for plant breeding, especially in species that are difficult to regenerate through tissue culture (Shu et al. 2011). The effectiveness of in vivo mutagenesis depends on the type of mutagen used, the dosage, and the genetic background of the plant species. It has been widely employed in cereal crops such as rice, wheat, and barley, as well as in legumes and horticultural plants (Mba et al. 2010). With the advancement of molecular biology techniques, mutation breeding is now complemented by genome sequencing and marker-assisted selection, making the identification and utilization of beneficial mutants more efficient (Parry et al. 2009).

3.2 In Vitro Mutagenesis

Unlike in vivo mutagenesis, this technique involves subjecting plant cells, tissues, or callus cultures to mutagens under controlled laboratory conditions. It is particularly useful for species that are difficult to propagate conventionally or require precise genetic modifications. In vitro mutagenesis has been instrumental in improving crops such as bananas, where tissue culture techniques combined with gamma irradiation have led to the development of Fusarium wilt-resistant clones (Mba et al. 2010). Additionally, it has enabled the production of salt-tolerant rice mutants with improved ion homeostasis and osmotic balance (Shu et al. 2011).

4 General Procedure for Induction in Vivo and in Vitro Random Mutagenesis in Crop Plants

The process of in vivo and in vitro mutation breeding primarily involves three key steps with some differences. The first step is the induction of random mutations within the plant genome using physical and chemical mutagen. This is followed by the second step, which involves screening the resulting mutants to identify those with desirable traits. Lastly, the third step focuses on selecting mutants that exhibit the targeted characteristics. Both in vivo and in vitro approaches of induced mutagenesis have contributed significantly to crop improvement by enhancing yield, quality, disease resistance, and stress tolerance. Understanding the procedural differences between in vivo and in vitro random mutagenesis is essential for selecting the most suitable method based on crop type, available resources, and desired outcomes. Although the process of induction of mutation and screening of mutants varies between seed and vegetatively propagated crops under in vivo and in vitro mutation conditions, there are few fundamental principles that applies in all mutation studies or programs, which are as follows:

a. The mode of reproduction and commercial propagation method of crop plant must be known, as it affects the methodology to be adopted for mutagenic treatment and selection.
b. A comprehensive understanding of genetic basis of trait of interest and ploidy level of crop is critical. For instance, modification of monogenic trait is quite easier than polygenic traits.
c. Choice of mutagen (physical or chemical), its doze, exposure time, infrastructure facility and screening techniques is vital for the successful selection and development of mutants.

4.1 Procedure for in Vivo Mutagenesis in Seed-Propagated Plants

In vivo mutagenesis in seed-propagated plants involves several key steps to ensure effective mutation induction and selection of desirable traits (Fig. 2). The general procedure includes:

4.1.1 Selection of Plant Material

The success, reliability and uniform response of induced in vivo mutagenesis in seed propagated crops depends on the quality and genetic stability of the selected seeds. The seeds should be of high quality, uniform shape and size, and should be free from all kind of physical defects, diseases, and contaminants (Ahloowalia et al. 2004). The seeds should have high vigor and germination rates, ideally above 85%.The seeds selected for treatment should be genetically pure and must belong

4.1.2 Pre-treatment of Seeds

Before exposure to mutagens, seeds are pre-treated in order to improve their permeability and ensure uniform absorption of mutagen treatment. The pre-treatment of seeds includes soaking of seeds in distilled water for 6–12 h at room temperature. The hydration process softens the seed coat, making it more permeable to mutagens and prevents excessive damage by buffering internal cellular structures. Some protective agents such as sodium thiosulfate ($Na_2S_2O_3$) and antioxidants like ascorbic acid (Jain 2012) aids in reducing the oxidative stress and lethality caused during mutagenesis.

4.1.3 Mutagen Treatment

The selected seeds are subjected to either chemical mutagens or physical mutagens to induce mutations. Chemical mutagens are widely used due to their ability to cause point mutations by modifying DNA bases. Commonly used chemical mutagens includes: EMS (Ethyl Methanesulfonate), Sodium Azide (SA), and Nitrosomethylurea (NMU). The seeds are submerged and stirred in prepared mutagen solution of known strength. The treatment is generally conducted at 20–25°Cfor a particular time period in order to optimize the mutagen activity, as lower or higher temperatures may reduce effec-

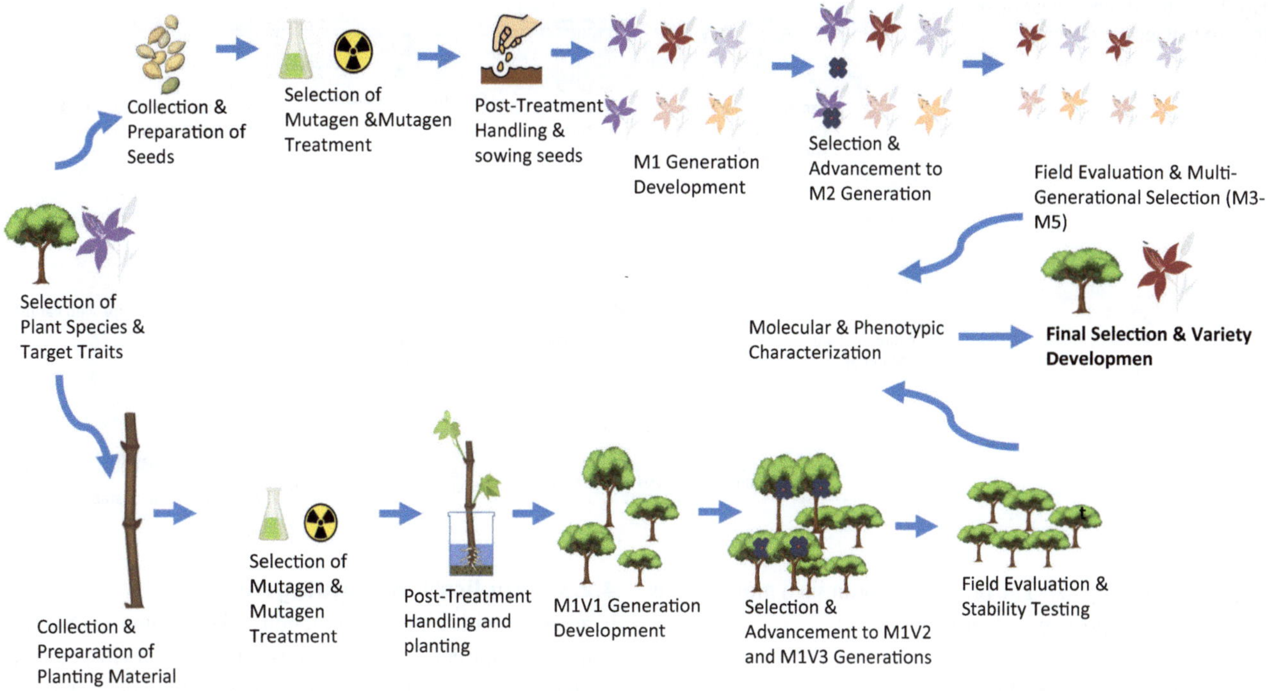

Fig. 2 Pictorial representation of in vivo mutagenesis in seed and vegetatively propagated crop plants

tiveness or increase toxicity. After the treatment, the seeds are rinsed under running tap water for 4–6 h to remove residual mutagen. In case of physical mutagens such as radiation induce mutations by causing DNA breaks and chromosomal rearrangements. Common radiation sources are Gamma rays (Cobalt-60 Source), X-rays, and fast neutrons. The seeds are irradiated with desired radiations carefully adjusted based on plant species and mutagenic sensitivity in a controlled radiation chamber. The seeds are then kept in darkness for 24 h to reduce oxidative stress (Jain 2012). This allows damaged DNA to stabilize before sowing.

4.1.4 Post-treatment Handling of Seeds

After mutagenesis treatment, proper handling ensures seed viability and mutation retention. The seeds are dried in shade at room temperature to prevent overheating. Drying ensures storage longevity without moisture-induced degradation. Treated seeds are stored in cool and dry conditions to maintain viability. Proper labeling is done to record mutagen type, dose, and treatment duration.

4.1.5 Field Sowing and Evaluation

Once mutagenized seeds (M1 generation) are dried, they are sown in the field to evaluate the mutation effectiveness for various parameters such as germination rate, seedling vigor and other morphological traits like height, leaf shape, flowering time. High planting densities are recommended to compensate for possible mutagen-induced lethality or sterility (Datta 2023). Desirable mutants (e.g., plant stature, yield, quality, disease resistance, drought and salt tolerance) are selected.

4.1.6 Selection and Advancement of Mutants

Harvested seeds from M1 plants (M2 generation) are grown and screened for phenotypic variations. Stable and desirable mutations are selected and advanced through multiple generations (M3-M5 Generations) (Shu et al. 2012). Marker-assisted selection and biochemical assays may be used to confirm desirable mutations.

4.1.7 Evaluation and Commercialization

Selected mutants are subjected to agronomic trials for yield performance, stress tolerance, and disease resistance. Thereafter, the approved lines are released as new cultivars or integrated into breeding programs (Mba et al. 2010).

4.2 Procedure for in Vivo Mutagenesis in Vegetatively Propagated Plants

Mutagenesis in vegetatively propagated crops involves exposing plant materials such as tubers, bulbs, shoot tips, nodal segments, or suckers to physical or chemical mutagens to induce desirable genetic changes. Since these crops do not undergo sexual reproduction, mutations are directly fixed in clonal progeny, making this method highly valuable for crop

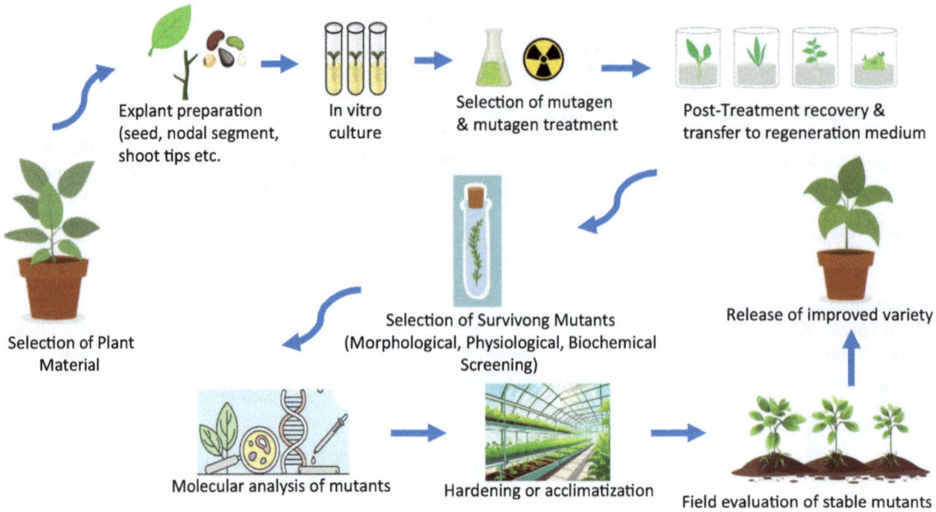

Fig. 3 Pictorial representation of in vitro mutagenesis in crop plants

improvement (Fig. 3). The procedure for in vivo mutagenesis of vegetatively propagated crops includes:

4.2.1 Selection of Plant Material

The choice of plant material is crucial for effective mutation induction and subsequent selection of desirable traits. The planting material selected should be free from viral, bacterial, and fungal infections to ensure reliable mutant screening (Shu et al. 2011). Inclusion of tissue culture-derived materials is one of the reliable methods to eliminate pre-existing diseases and genetic variability (Mba et al. 2010). Clonal plant parts should be genetically uniform to facilitate the detection of induced mutations (Shu et al. 2011). Variability in plant material can complicate distinguishing induced mutations from natural genetic differences (Datta 2023). Different types of planting material can be used, such as tubers (e.g., potatoes, yams), bulbs (e.g., onions, garlic), suckers (e.g., banana, pineapple), and shoot tips and nodal segments.

4.2.2 Pre-treatment Handling

To enhance mutagen absorption and improve mutation efficiency, selected plant parts undergo specific pre-conditioning treatments such as hydration or dessication. Hydration or water soaking of planting material increases the permeability of plant tissues, facilitating uniform mutagen uptake (Mba et al. 2010). Tubers or bulbs are soaked for 6–12 h before treatment. Exposure to mild abiotic stress (e.g., temperature fluctuation, desiccation) can enhance mutagen efficiency by activating stress-response pathways (Datta 2023). In some cases, moderate drying before chemical treatment improves absorption and reduces excessive tissue damage (Shu et al. 2012).

4.2.3 Application of Mutagens

Irradiation with physical mutagens like gamma or X-ray is commonly applied to whole plants, tubers, or in vitro cultured explants. Gamma rays are generally used for inducing useful mutations in banana, potato, and sugarcane (Shu et al. 2012). Mutagenic chemicals (e.g., EMS, colchicine) can be applied via soaking, spraying, or injection. Shoot tips or nodal segments are often immersed in mutagenic solutions for controlled durations (Datta 2023).

4.2.4 Post-treatment Recovery

After mutagenic exposure, plant materials undergo a recovery phase to stabilize mutations and assess viability. Treated plant parts are transferred to nutrient-rich media or planted in soil under controlled conditions (Parry et al. 2009). Early screening of putative mutants ensures viable plant recovery. Some explants may be cultured in vitro to facilitate mutation fixation (Shu et al. 2012). Treated plant parts are rinsed and transferred to growth media or field conditions for recovery. Initial screening is performed to assess viability and growth response (Parry et al. 2009).

4.2.5 Generation of Mutants

The first vegetative cycle (V1) may show physiological stress and chimerism. Successive clonal generations (V2, V3, etc.) are used to stabilize mutations (Shu et al.2012).Chimeric tissues are either selected or resolved through further propagation.

4.2.6 Selection and Evaluation of Mutants

Morphological and agronomic screening of mutant lines in field trials. Molecular and biochemical analyses to confirm stable mutations.

4.2.7 Commercialization and Breeding Integration

Superior mutants are multiplied and evaluated in multilocation trials. Selected clones are released as new varieties or used in further breeding programs (Mba et al. 2010).

4.3 Procedure for in Vitro Random Mutagenesis

In vitro mutagenesis is a technique used to induce genetic mutations in plant cells, tissues, or organs under controlled laboratory conditions. It is a powerful tool to for creating genetic variation in crop plants. Unlike site-directed mutagenesis, which aims at targeted changes in nucleotide sequence of a gene at a specified location, random mutagenesis induces mutations throughout the genome, enabling the selection of desirable traits. The goal is to produce beneficial traits such as disease resistance, improved nutritional profiles, improved yield, or biotic and abiotic stress tolerance. Tissue culture makes induced mutation more efficient by allowing the handling of large populations and by increasing mutation induction efficiency, possibility of mutant recovery and speediness of cloning selected variants. The availability of suitable selection methods could improve its effectiveness and potential applications. The molecular markers available today may provide tools to assist in induction of mutation protocols by investigating genetic variation within populations as well as early detection of mutants with desired traits. This procedure involves several key steps, from selecting plant material to screening for desirable mutants. Here's a detailed procedure for conducting in vitro mutagenesis in plants:

4.3.1 Selection of Plant Material

the first step of in vitro mutagenesis is the selection of appropriate plant material. Explant should be derived from certified virus free source as this initial choice strengthens the complete protocol, including useful mutant identification and requirements for its release as a cultivar. The plant material used must have the ability to regenerate and express desirable traits after mutation. The cultures must be free of any microbial contamination as the microbes may get stimulated and proliferate on exposure to radiation thus weakening the plant. Meristem cultures, callus cultures, and embryogenic cultures are most commonly used cultures for in vitro mutagenesis. Meristem cultures are ideal for minimizing genetic variation and ensuring stability, while callus cultures offer versatility and ease of manipulation for inducing mutations. Embryogenic cultures, on the other hand, are particularly valuable for crops that are difficult to regenerate and offer a direct pathway for the formation of embryos. Careful choice of appropriate tissue culture system, potential for generating mutants with desirable traits for crop improvement can be maximized. Mutagen treatment should be applied to well-established cultures having consistent proliferation rate through 5–6 subcultures.

4.3.2 Exposure to Mutagens

The plant tissues are exposed to mutagenic agents under controlled conditions. Physical or chemical mutagens can be used. The exposure may vary in intensity and duration, depending on the type of mutagen used.

4.3.3 Application of Chemical Mutagen

Chemical mutagens induce mutations by altering the DNA sequence through various mechanisms, including alkylation, base analog incorporation, and oxidative damage. These changes result in base substitutions, frame shifts, or chromosomal aberrations, which can lead to phenotypic variations in plants. As compared with physical mutagens, chemicals may give rise to relatively more gene mutations rather than to chromosomal changes. Chemical mutagens are divided into several groups depending upon its mode of action. Steps involved in application of chemical mutagen.

4.3.4 Preparation of Mutagenic Solution

The mutagenic chemical (EMS, MMS, or a nitroso compound) is dissolved in a solvent like water or ethanol to create a solution at the desired concentration. The concentration of the mutagen varies depending on the crop species and the desired mutation rate.

4.3.5 Immersion of Plant Tissue

Sterilized plant tissues, such as explants or callus cultures, are immersed in the mutagenic solution for a fixed duration. For example, EMS is often used at concentrations of 0.1–0.3 M for 2–4 h, depending on the plant species.

4.3.6 Post-treatment Rinsing

After the treatment, the plant tissues are thoroughly rinsed with sterile water to remove any remaining mutagen, preventing toxicity.

4.3.7 Culture and Regeneration

Following the mutagenic treatment, the tissues are cultured on suitable growth media to promote regeneration. The regenerated plants are then screened for desired traits.

4.3.8 Application of Physical Mutagen

In this technique, physical agents like electromagnetic radiation such as gamma rays, X rays, and UV light and particle radiation, such as fast and thermal neutrons, beta and alpha particles are used to induce mutations in biological material. Physical mutagens such as gamma rays, X-rays, UV radiation, and ion beam irradiation have been widely used in plant breeding programs to develop high-yielding, stress-tolerant, and disease-resistant crop varieties.

4.3.9 Gamma Radiation

Gamma radiation is typically generated using radioactive isotopes like Cobalt-60 or Cesium-137. The plant material is exposed to gamma rays at controlled doses. The radiation dose is a critical factor in determining the extent of mutation. For most plant tissues, doses between 10 and 100 Gray (Gy) are typically used. Lower doses generally induce fewer mutations, while higher doses may cause cell death or undesirable mutations. Explants such as callus or meristem cultures are placed in a lead-shielded chamber, and the radiation is applied for a set duration. For example, exposure times may range from 10 to 60 min depending on the radiation intensity and dose (Bado et al. 2023).

4.3.10 UV Radiation

UV radiation is typically generated using UV lamps that emit UV-C light (wavelength of 100–280 nm). UV light induces mutations by causing thymine dimers in DNA. UV radiation is generally less penetrating than gamma radiation, so it is typically used on surface-sterilized seeds or thin tissues like leaf explants. Exposure times usually range from 10 to 30 min under UV-C light, with the distance between the plant material and the UV source influencing the dose. Plant material is placed under the UV light for a specified time. For example, leaf explants might be exposed to UV-C light for 20 min at a distance of 10 cm from the light source (Bado et al. 2023).

4.3.11 Regeneration of Mutagenized Tissues

After mutagenesis, the treated tissues are transferred to regeneration media to initiate the growth of new shoots and roots. The regeneration process includes Transfer the mutagenized explants onto fresh solid media (MS or other suitable media) to promote callus formation and shoot development and root formation

4.3.12 Selection of Mutant Plants

Once the regenerated plants have developed, selection criteria must be applied to identify the mutants with desirable traits. This can include various visual, biochemical and molecular methods. If the mutation causes a visible phenotypic change (e.g., color, leaf morphology), the mutants can be selected visually. In some cases, specific biochemical assays or marker genes can be used to detect particular mutations or enhanced traits (e.g., increased resistance to herbicides or pathogens). Molecular markers (such as SSRs, SNPs, or QTL analysis) can be used to verify mutations at the DNA level.

4.3.13 Evaluation of Mutants

Once mutant plants are selected, evaluation is done for several parameters which includes phenotypic, genotypic, and stability testing analysis. Phenotypic analysis assesses traits such as growth rate, yield, disease resistance, and stress tolerance. Genotypic analysis involves conducting molecular analyses to confirm the genetic basis of the observed phenotype. Whereas, in stability testing, the selected mutants are grown in different environmental conditions over several generations to ensure the mutation is stable and heritable.

4.3.14 Field Evaluation

The final step is to field test the mutagenized plants under natural conditions to confirm the superiority of the selected mutants. If the plants exhibit enhanced traits, they can be further developed and potentially released for commercial cultivation.

5 Application of Induced Random Mutagenesis in Crop Improvement

(a) **Enhancement of Yield and Quality**: Mutation breeding has successfully improved yield traits in staple crops like rice, wheat, maize, barley, and soybeans. This innovation, primarily through irradiation and chemical mutagenesis has led to the development of high-yielding and lodging-resistant wheat and rice cultivars, significantly boosting global food production. In rice, induced mutagenesis has contributed to the development of high-yielding and short-duration varieties, helping to improve food security, especially in regions with shorter growing seasons (Shu et al. 2011). Similarly, in barley, the introduction of mutant varieties with enhanced tillering capacity and higher grain weight has led to improved productivity (Ahloowalia et al. 2004). Mutation breeding has also been instrumental in improving soybean by developing varieties with increased seed protein and oil content, making them more nutritionally valuable (Mba et al. 2010). Additionally, mutants with enhanced pod-setting ability and resistance to pod shattering have been introduced in rapeseed, soyabean, and lupin, ensuring better harvest efficiency and reduced post-harvest losses (Datta 2023).

Random induced in vitro mutagenesis has emerged as a valuable tool for achieving yield goals by generating novel genetic variations in a small period of time that have contributed to higher productivity and better adaptability. For instance, in soybean, EMS-induced mutations have led to the development of high-yielding lines with improved seed weight and an increased number of pods per plant, significantly enhancing overall crop productivity

Table 3 Random mutagenesis in crop plants for quality and nutrition

Crop	Method	Mutagen	Highlight	References
Chrysanthemum (*Chrysanthemum morifolium*)	In vitro	Gamma irradiation (25 Gy)	Yellow flower mutants	Ahloowalia et al. (2004)
Soybean (*Glycine max*)	In vivo	Gamma irradiation and EMS	Oil quality improvement	Yathaputanon et al. (2009)
	In vivo	Gamma irradiation	Improvement of protein quality	Yathaputanon et al. (2009)
	In vivo	EMS	Low phytic acid content	Yathaputanon et al. (2009)
Maize (*Zea mays*)	In vivo	Gamma irradiation, EMS and NMU	Improvement of protein quality	Tomlekova (2010)
Cassava (*Manihot esculenta*)	In vivo	Gamma irradiation	High-amylose content	Ceballos et al. (2008)
Gerbera (*Gerbera jamesonii*)	In vitro	Gamma irradiations (1.5–30.0 Gy) EMS [0.1–1.0% (v/v)]	Flower color	Ghani et al. (2013)
Strawberry (*Fragaria* × *ananassa)*	In vitro	UV and Gamma irradiation	Anthocyanin-rich mutants with improved fruit colorand shelf life	Ryu et al. (2020)
Papaya (*Carica papaya*)	In vitro	Gamma Irradiation	Resistant to *papaya ringspot virus*(PRSV)	Patade and Suprasanna (2009)
Pomegranate (Punica granatum)	In vitro	Gamma Irradiation	Enhanced fruit size and resistance to fungal infections	Ahloowalia et al. (2004)

(Mudibu et al. 2012). Similarly, in rapeseed, a combination of in vitro mutagenesis and somaclonal variation has resulted in oilseed varieties with higher oil content and greater environmental adaptability, making them more viable for commercial cultivation (Ahloowalia et al. 2004). Additionally, cassava, a vital staple crop in many tropical regions, has benefitted from gamma radiation-induced mutations, leading to the development of varieties with shorter maturity periods and higher starch content. These improvements not only enhance food security but also provide economic benefits by supporting industrial applications of cassava-based products. Collectively, these advancements underscore the role of in vitro mutagenesis in optimizing crop performance, ensuring sustainable agricultural productivity, and meeting the growing demands for food and industrial raw materials.

(b) **Nutritional Improvement**: Random induced mutagenesis has been widely applied to enhance the nutritional quality and remove anti-nutritional factors from crop plants. In rice, gamma irradiation has been used to develop high provitamin-A mutants, contributing to biofortification efforts. Similarly, induced mutations in wheat have led to increased folate content, addressing micronutrient deficiencies. Similarly, iron- and zinc-enriched wheat and barley varieties have been developed through mutagenesis, offering better dietary solutions to micronutrient deficiencies (Mba et al. 2010). In legumes, EMS-induced mutations have successfully enhanced iron and zinc accumulation in lentils and chickpeas, improving their nutritional value for human consumption. In groundnuts, mutagenesis has been used to develop varieties with higher oleic acid content, promoting better cardiovascular health. Protein quality improvement is another key application. In maize, mutation breeding has been used to enhance lysine and tryptophan content, addressing protein-energy malnutrition. Similarly, induced mutations in barley have improved β-glucan content, beneficial for gut health and reducing cholesterol levels. A notable example of removal of anti-nutritional factor in common beans, *i.e.,* phytic acid (hinders absorption of minerals), through mutation is low phytic acid mutant. With advancements in genomic tools, mutant lines with enhanced nutritional profiles are being characterized and integrated into breeding programs, making mutagenesis a valuable tool for developing biofortified crops. A detailed account for induced mutagenesis for improvement of nutritional quality and other important quality traits is mentioned in Table 3.

(c) **Flowering Time and Maturity**: Random mutagenesis has also played a crucial role in modifying flowering time and maturity in crops, helping optimize yield and adaptability to diverse agro-climatic conditions. By inducing mutations in key regulatory genes, breeders have developed early and late maturing crop varieties suited for specific agricultural needs. Early flowering or maturing mutants are valuable in regions with short growing seasons, drought-prone environments and for catching early markets. For example, EMS-induced *Hd1* mutants in rice matured 10–15 days earlier, while gamma-irradiated *VRN-H1* barley exhibited reduced vernalization requirements (Shu et al., 2012). Similarly, EMS-mutagenized sunflower mutants showed early blooming, improving drought resistance. Late flowering or maturing mutants are beneficial for enhanced biomass accumulation and yield stability. Fast neutron-induced wheat mutants with altered *VRN1* had delayed flowering, while gamma-mutagenized tomato with *SP5G* mutations exhibited prolonged fruiting periods (Shu et al. 2012). These mutants enhance climate resilience, support staggered cropping, and optimize hybrid breeding. Random mutagenesis, combined with modern molecular techniques, remains a powerful tool for developing climate-resilient crops and ensuring food security.

(d) **Abiotic and Biotic Stress Resistance:** Climate change and environmental stresses, such as drought, salinity, and extreme temperatures, pose significant challenges to crop productivity. Mutation breeding has helped develop stress-tolerant varieties, ensuring stable yields under adverse conditions. For instance, salt-tolerant rice mutants with improved ion homeostasis and osmotic balance have been developed, enabling cultivation in saline-prone areas (Shu et al. 2011). In maize, drought-resistant mutants with enhanced root architecture and water-use efficiency have been selected through mutagenesis, making them more suitable for semi-arid regions (Shu et al. 2012). Similarly, in sorghum, heat-tolerant mutants with improved photosynthetic efficiency have been introduced, contributing to food security in hot and dry climates (Datta 2023).

Induced in vivo and in vitro mutagenesis has been successfully employed to develop disease and pest-resistant cultivars in several crops. For example, groundnut mutants resistant to fungal infections such as rust and late leaf spot have been developed through gamma irradiation (Motagi et al. 2022). Similarly, in wheat, resistance to fungal diseases like powdery mildew and rust has been achieved through targeted mutagenesis, reducing dependence on chemical fungicides (Parry et al. 2009). In bananas, induced mutations have resulted in cultivars resistant to Panama disease (*Fusarium* wilt), which is a major threat to global banana production. The development of resistant clones through gamma irradiation has played a key role in sustaining banana yields in disease-prone regions (Mba et al. 2010). Mutations in rice have been induced to confer resistance to bacterial blight and sheath blight. A detailed account of induced mutagenesis for resistance/tolerance against biotic and abiotic stress is mentioned in Tables 4 and 5.

Table 4 Random mutagenesis in crop plants against biotic stress

Crop	Method	Mutagen	Highlight	References
Mungbean (*Vignaradiata*)	In vivo	Gamma irradiation and EMS	Resistance to yellow mosaic India virus	Anuragi et al. (2022)
Rice (*Oryza sativa*)	In vivo	Gamma irradiation and EMS	Resistance to blast, powdery mildew, hopper and bacterial leaf stripe	Kharkwal (2012)
Date palm (*Phoenixdactylifera*)	In vitro	Gamma irradiation	Tolerance to bayoud disease	Jain (2012)
Banana (*Musa* spp.)	In vitro	300 mM EMS	Tolerance to Fusarium wilt-resistant lines	Chen et al. (2012)
Papaya (*Carica papaya*)	In vitro	Gamma irradiation	Resistant to papaya ring spot virus(PRSV)	Patade and Suprasanna (2009)
Potato cv. Alanso (*Solanum tuberosum*)	In vitro	Gamma irradiations (22 Gy and109 Gy)	Resistance against Black scurf	Kara and Arici (2019)

Table 5 Random mutagenesis in crop plants against abiotic stress

Crop	Method	Mutagen	Highlight	References
Wheat (*Triticum aestivum*)	In vitro	2 mM NaN_3	Increased drought tolerance	Ahloowalia and Maluszynski (2001)
Rice (*Oryza sativa*)	In vitro	EMS (0.1%–1.5%)	Reduced plantheight	Shu et al. (2011)
Cotton (*Gossypium* spp.)	In vivo	Gamma irradiation	High fiber quality	Haq (2009)
Maize (*Zea mays*)	In vivo	Gamma irradiation, EMS and NMU	Acidity and drought tolerance	Tomlekova (2010)
Sugarcane (*Saccharum officinarum*)	In vitro	Gamma irradiation (0–50 Gy)	Tolerance to salt	Patade and Suprasanna (2009)
Mango (*Mangifera indica* L.)	In vivo	Gamma irradiation (35 Gy)	Increased salinity tolerance	Perveen et al. (2023b, 2024)
Mango (*Mangifera indica* L.)	In vivo	Gamma irradiation (35 Gy)	Dwarf statured plants	Perveen et al. (2022, 2025)

6 Challenges and Limitations Associated with Induced Random Mutagenesis

Random mutagenesis, whether performed in vivo or in vitro, is a fundamental approach for generating genetic diversity in microorganisms, plants, and other biological systems. However, both methods come with distinct challenges and limitations that can affect their efficiency, reproducibility, and practical applications. In vivo approaches allow mutagenesis in the context of a living organism but suffer from unpredictability and ethical concerns. In vitro methods provide controlled mutagenesis but lack the full physiological context of living systems. Advances in directed evolution and genome-editing technologies, such as CRISPR-Cas9, offer alternative approaches with greater precision and efficiency. Below are the key challenges and limitations linked with random in vivo and in vitro mutagenesis:

6.1 Lack of Target Specificity

One of the primary limitations of random mutagenesis, whether in vivo or in vitro, is the inability to control mutation sites. Mutations occur throughout the genome in an unpredictable manner, making it difficult to target specific genes or regulatory elements. In vivo mutagenesis is influenced by cellular repair mechanisms, whereas in vitro methods introduce mutations in an isolated setting, often lacking genomic context.

6.2 Mutation Efficiency and DNA Repair Mechanisms

In vivo mutagenesis efficiency is often reduced by the host organism's DNA repair mechanisms, which can correct induced mutations, thereby limiting their frequency. In contrast, in vitro methods such as error-prone PCR can introduce mutations at higher rates, but their success depends on proper amplification and integration into functional genetic systems (Cirino et al. 2003).

6.3 Risk of Lethal and Deleterious Mutations

Both in vivo and in vitro mutagenesis are prone to generating lethal or highly deleterious mutations that can hinder viability. In vivo mutations may affect essential genes, leading to cell death. Similarly, in vitro mutagenesis may introduce changes that disrupt protein folding or function, rendering the mutated sequences nonfunctional (Araya and Fowler 2011).

6.4 Difficulty in Screening and Selection

Identifying beneficial mutations from a large mutant pool is a major challenge in both methods. In vivo systems require high-throughput screening to differentiate advantageous mutations from neutral or harmful ones. In vitro mutagenesis also demands efficient selection techniques, such as functional assays, to determine whether mutations have desirable effects on proteins or metabolic pathways.

6.5 Unpredictability in Phenotypic Expression

The relationship between genotype and phenotype is complex, and random mutagenesis often leads to unexpected epistatic interactions. In vivo mutations must function within an intact cellular system, making their phenotypic effects difficult to predict. In vitro mutations, while controlled, may not always translate effectively when expressed in living organisms due to differences in regulatory mechanisms and post-translational modifications.

6.6 Limited Genetic Context and Functional Constraints

In vivo mutagenesis occurs within the full genomic context, meaning that interactions with other genomic elements can influence mutation outcomes. However, this can also introduce unpredictability. In contrast, in vitro mutagenesis typically focuses on isolated genes or proteins, lacking the broader genetic context that might affect functionality in a whole-organism system.

6.7 Technical Complexity and Cost

Both methods require specialized techniques and resources. In vivo mutagenesis often involves mutagenic agents or transposon-based strategies, which can be challenging to control. In vitro mutagenesis, particularly approaches like site-directed mutagenesis or DNA shuffling, requires expensive reagents and molecular biology tools, adding to the cost and complexity (Ling and Robinson 2001).

6.8 Ethical and Regulatory Concerns

The use of random mutagenesis in genetic engineering and breeding programs raises ethical and regulatory concerns. In vivo applications, especially in higher eukaryotes, must comply with biosafety and ethical guidelines (Shewry et al. 2001).

Similarly, in vitro mutagenesis in genetic engineering faces scrutiny regarding unintended effects on ecosystems and food safety.

7 Future Prospects

Random induced mutagenesis, both in vivo and in vitro, remains a crucial tool for developing novel crop varieties with enhanced traits such as stress tolerance, disease resistance, and improved yield. Future research will focus on increasing precision, efficiency, and integration with advanced breeding and genome-editing technologies. One key area of advancement is the optimization of mutagenic agents to enhance mutation efficiency while minimizing deleterious effects. The development of targeted chemical and physical mutagens, coupled with improved delivery mechanisms, will enable higher mutational specificity in both in vivo and in vitro applications. High-throughput mutant screening and phenotyping will be a priority, leveraging machine learning and artificial intelligence to accelerate the identification of beneficial traits. Advances in omics technologies, such as genomics, transcriptomics, and metabolomics, will facilitate better characterization of mutants at the molecular level.

The integration of mutagenesis with genome-editing tools like CRISPR-Cas9 will help refine desirable mutations, bridging the gap between random and targeted genetic modifications. Additionally, synthetic biology approaches will allow for the directed evolution of genes, optimizing traits beyond what traditional breeding can achieve. Finally, there will be an emphasis on climate-resilient crop development, using mutagenesis to engineer plants that can withstand extreme environmental conditions. Regulatory frameworks will also evolve to support the safe and sustainable deployment of mutagenized crops. By combining traditional mutagenesis with modern biotechnological innovations, future research will unlock new genetic potentials, ensuring global food security and agricultural sustainability.

References

Ahloowalia B, Maluszynski M, Nichterlein K (2004) Global impact of mutation-derivedvarieties.Euphytica 135:187–204. https://doi.org/10.1023/B:EUPH.0000014914.85465.4f

Ahloowalia BS, Maluszynski, M (2001) Induced mutations–A new paradigm in plant breeding. Euphytica 118(2):167–173

Anuragi H, Yadav R, Sheoran R (2022) Gamma-rays and EMS induced resistance to mungbean yellow mosaic India virus in mungbean [*Vigna radiata* (L.) R. Wilczek] and its validation using linked molecular markers. Int J Radiat Biol 98(1):69–81. https://doi.org/10.1080/09553002.2022.1998710

Araya CL, Fowler DM (2011) Deep mutational scanning: assessing protein function on a massive scale. Trends Biotechnol 29(9):435–442. https://doi.org/10.1016/j.tibtech.2011.04.003

Bado S, Forster BP, Maghuly F (2023) Physical and chemicals mutagenesis in plant breeding. In: Mutation breeding for sustainable food production and climate resilience. Springer Nature Singapore, Singapore, pp 57–97. https://doi.org/10.1007/978-981-16-9720-3_3

Bhagwat B, Duncan EJ (1998) Mutation breeding of banana cv. Highgate (Musa acuminata, AAA Group) for tolerance to *Fusarium oxysporum*f.sp. cubense using chemical mutagens. Sci Hortic 73(1):11–22. https://doi.org/10.1016/S0304-4238(97)00141-6

Ceballos H, Sánchez T, Denyer K, Tofiño AP, Rosero EA, Dufour D, Fahy B (2008) Induction and identification of a small-granule, high-amylose mutant in cassava (Manihot esculenta Crantz). J Agric Food Chem 56(16):7215–7222. https://doi.org/10.1021/jf800603p

Chen YF, Chen W, Huang X et al (2012) *Fusarium* wilt-resistant lines of Brazil banana (*Musa spp.*, AAA) obtained by EMS-induced mutation in a micro-cross-section cultural system. Plant Pathol 1:1–8. https://doi.org/10.1111/j.1365-3059.2012.02620.x

Cirino PC, Mayer KM, Umeno D (2003) Generating mutant libraries using error-prone PCR. Methods Mol Biol 231:3–9.https://doi.org/10.1385/1-59259-395-X:3

Datta SK (2023) Induced mutation breeding. Springer, Singapore. https://doi.org/10.1007/978-981-19-9489-0

Ghani M, Kumar S, Thakur M (2013) Induction of novel variants through physical and chemical mutagenesis in Barbeton daisy (*Gerbera jamesonii* Hook.). J Hortic Sci Biotechnol 88:585–590.https://doi.org/10.1080/14620316.2013.11513010

Haq MA (2009) Development of mutant varieties of crop plants at NIAB and the impact on agricultural production in Pakistan. Induced plant mutations in the genomic era. Food Agri Organ United Nations 61

Jain SM (2012) In vitro mutagenesis for improving date palm (*Phoenix dactylifera* L.). Emir J Food Agric 9:400–407. https://ejfa.me/index.php/journal/article/view/927.

Kara A, Arici SE (2019) Determination of gamma rays efficiency against *Rhizoctoniasolani* in potatoes. Open Chem 17:254–259. https://doi.org/10.3109/09553002.2010.484478

Kharkwal, MC (2012) A brief history of plant mutagenesis. In Plant mutation breeding and biotechnology. Wallingford UK:CABI. 21–30

Ling MM, Robinson DK (2001) Approaches to DNA mutagenesis: an overview. Anal Biochem 288(1):24–34.https://doi.org/10.1006/abio.1997.2428

Mba C, Afza R, Bado S, Jain SM (2010) Induced mutagenesis in plants using physical and chemical agents. Plant Cell Cult: Ess Methods 20:111–130. https://doi.org/10.1002/9780470686522.ch7

Motagi BN, Bhat RS, Pujer S, Nayak SN, Pasupaleti J, Pandey MK, Gowda MC (2022) Genetic enhancement of groundnut: current status and future prospects. Accel Plant Breed 4(Oil Crops):63–110. https://doi.org/10.1007/978-3-030-81107-5_3

Mudibu J, Nkongolo KK, Kalonji-Mbuyi A, Kizungu R V (2012) Effect of gamma irradiation on morpho-agronomic characteristics of soybeans (Glycine max L.). American J Plant Sci 3(3):331–337

Nakagawa H (2009) Induced mutations in plant breeding and biological researches in Japan. Crops 242(188):48. https://www.fao.org/4/i0956e/i0956e.pdf

Patade VY, Suprasanna P (2009) An in vitro radiation-induced mutagenesis-selection system for salinity tolerance in sugarcane. Sugar Tech 11(3):246–251. https://doi.org/10.1007/s12355-009-0042-4

Parry MA, Madgwick PJ, Bayon C, Tearall K, Hernandez-Lopez A, Baudo M, Rakszegi M, Hamada W, Al-Yassin A, Ouabbou H, Labhilili M (2009) Mutation discovery for crop improvement. J Exp Bot 60(10):2817–2825. https://doi.org/10.1093/jxb/erp189

Perveen N, Dinesh MR, Sankaran M, Shivashankara KS, Venugopalan R (2022) Characterization and evaluation of putative mutant populations of polyembryonic mango genotype Nekkare for dwarfing rootstock traits. J Hortic Sci 17(2):261–271.https://doi.org/10.24154/jhs.v17i2.1456

Perveen N, Dinesh MR, Sankaran M, Venugopalan R (2023a) Mutagenic-sensitivity and variability in the putative mutants of polyembryonic mango genotypes. Indian J Hortic 80(1):3–9.https://doi.org/10.58993/ijh/2023.80.1.1

Perveen N, Dinesh MR, Sankaran M, Ravishankar KV, Krishna Jee HG, Hanur VS, Alamri S, Kesawat MS, Irfan M (2023b) Comparative transcriptome analysis provides novel insights into molecular response of salt-tolerant and sensitive polyembryonic mango genotypes to salinity stress at seedling stage. Front Plant Sci 14:1152485. https://doi.org/10.3389/fpls.2023.1152485

Perveen N, Dinesh MR, Sankaran M, Lakkireddy V, Shivashankara KS, Venugopal R, Mir H (2024) Phenolics signatures in response to salinity stress provide novel insights into physiological basis of salt tolerance in mango (*Mangifera indica* L.). J Pl Growth Regul 43(12):4866–4885. https://doi.org/10.1007/s00344-024-11440-6

Perveen N, Dinesh MR, Sankaran M, Bindu KH, Shivashankara KS, Ravishankar KV, Kanade NM, Kalal P, Mir H (2025) Radiation induced mutations alter morpho- biochemical, anatomical and molecular responses in polyembryonic mango genotype Bappakkai. Plant Physiol Rep 30:284–295. https://doi.org/10.1007/s40502-025-00860-5

Ryu J, Kim WJ, Kim SH, Lee KS, Jo HJ, Kim EY, Ha BK (2020) Characterization of genetic variation and antioxidant properties in strawberry (Fragaria× ananassa Duch.) mutant genotypes. Genet Resour Crop Evol 67:1457–1471. https://doi.org/10.1007/s10722-020-00918-3

Shewry PR, Tatham AS, Halford NG (2001) Genetic modification and plant food allergens: risks and benefits. J Chromatogr B: Biomed Sci Appl 756(1–2):327–335.https://doi.org/10.1016/S0378-4347(01)00093-7

Shu QY, Forster BP, Nakagawa H (2011) Plant mutation breeding and biotechnology. CABI Publishing. https://www.fao.org/4/i0956e/i0956e.pdf

Shu QY, Forster BP, Nakagawa H (2012) Plant mutation breeding and biotechnology. CABI Publishing. https://doi.org/10.1017/S001447971300001X

Tomlekova NB (2010) Induced mutagenesis for crop improvement in Bulgaria. http://www-pub.iaea.org/MTCD/publications/PDF/Newsletters/PMR-02-02.pdf

Till BJ, Datta S, Jankowicz-Cieslak J (2018) TILLING: the next generation. Plant genet mol biol 139–160

Yathaputanon C, Kumsueb B, Malipan A, Srisombun S, Bunyarut J (2009) Protein content in high-protein soybean mutants in Thailand. In: Induced plant mutations in the genomics era. Proceedings of an international joint FAO/IAEA symposium. Food and Agriculture Organization of the United Nations, Rome, p 195Ā196. https://www.fao.org/4/i0956e/i0956e.pdf

CRISPR/Cas9 and TALEN-Mediated Genome Engineering for Crop Improvement

Nusrat Perveen, Javid Iqbal Mir, and Hidayatullah Mir

Abstract

Genome editing offers unprecedented opportunities for developing crops with enhanced traits, including improved yield, disease resistance, drought tolerance, and nutritional content. CRISPR/Cas9 and TALEN are most commonly used systems for site-directed mutagenesis. CRISPR/Cas9, a system derived from bacterial defense mechanisms, provides a highly efficient, targeted approach for gene editing that resulting from single-base pair changes or larger structural modifications in plant genomes. On the other hand, TALENs, which utilize custom-designed DNA-binding proteins, offer an alternative method for achieving similar precision in gene modification. This chapter provides a comprehensive overview of the principles, mechanisms, and applications of CRISPR/Cas9 and TALEN in crop improvement. Additionally we also discuss the comparative advantages and limitations of each system, including their efficiency, specificity, ease of use and the regulatory and ethical concerns associated with genome-edited crops.

Keywords

CRISPR/Cas9 · TALEN · Genome editing · Site-directed nuclease

N. Perveen · J. I. Mir
Central Institute of Temperate Horticulture (ICAR-CITH), Srinagar, J&K, India

H. Mir (✉)
Regional Horticulture Research Sub-Station, Sher-e-Kashmir University of Agricultural Sciences and Technology of Jammu, Bhaderwah, J&K, India
e-mail: hidayatmay14@yahoo.co.in

1 Introduction

Crop improvement is the cornerstone of agricultural sustainability, aimed at enhancing the yield, resilience to stress, and nutritional content of crops for addressing global food security. The traditional methods of breeding, like introduction, selection and hybridization have made significant strides in crop improvement but these techniques are constraints by the availability of genetic diversity and are time-intensive. Induced mutations have also been used to improve plant traits like architecture (Perveen et al. 2022, 2025), salinity tolerance (Perveen et al. 2023, 2024). However, its successful application is often constrained by the unpredictability of results and untargeted changes in genome. Being able to understand the genome and edit it to suit the needs of humankind has always been the most coveted dreams of biologists. Thanks to the emergence of genome-editing systems, this dream has come to near realization. The concept of genome editing, a technology in which DNA/RNA could be changed at a specific genomic position was first proposed in 1980. Reprogrammable sequence-specific nucleases (SSNs) form the basis for genome-editing techniques. Homology-directed repair (HDR), which enables precise modification of the target gene using a foreign donor DNA with homologous regions matching the target locus as a repairing template, and non-homologous end joining (NHEJ), which typically results in loss of function mutations are major repair mechanism working in plants. NHEJ system can operate via Ku-dependent pathway where KU70/80 proteins bind to the target site at the end of double-strand breaks and seal the gap using ligase IV enzyme and its co-factor and Ku-independent where the repairing of DSBs leads to generation of single- stranded overhang which are sealed using DNA ligase and DNA polymerase. Site-Directed Nucleases technique includes SDN-1 which involves. SDN endogenous NHEJ repair mechanism for DSB repair, SDN-2which

J. Devi (ed.), *Mutation Breeding in Crop Improvement*, Sustainable Landscape Planning and Natural Resources Management,
https://doi.org/10.1007/978-3-032-17911-1_3

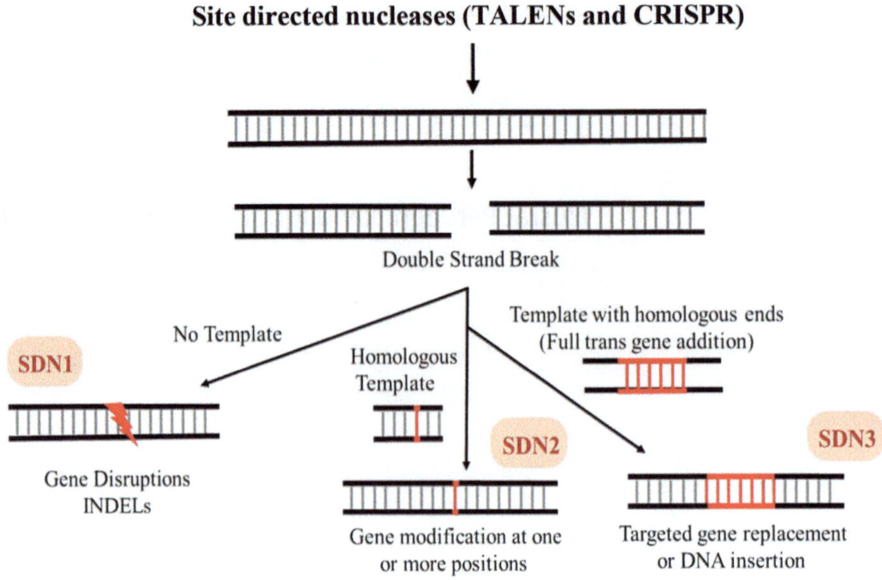

Fig. 1 Representation of SDN1, SDN2, and SDN3 techniques

uses HDR mechanism for specific modification of target site leading to gain of function mutation and SDN3 which uses HDR pathway leading to insertion of new gene. Base and prime editing techniques can result in base alteration and are classified as SDN1. Representation of SDN1, SDN2, and SDN3 techniques are illustrated in Fig. 1.

In the recent years, Site-Directed Nucleases (SDNs) have made genome editing (GE), more precise, faster, and efficient. Meganucleases, ZFNs, and TALENs are protein-dependent DNA cleavage system considered a 1st generation GE tools. Meganucleases contain 20–40 bp long DNA sequence recognition signatures. The major issue in using them is the difficulty in modulating their target specificity and hence they are not generally used in genome-editing programs. Zinc-finger proteins (ZFP) are first engineered endonuclease consisting of artificial bipartite enzymes joined by linker peptide, while TALENs are an improvement over ZFNs. Artificially engineering TAL arrays specific to the target genomic region is easier than constructing ZFNs and its binding efficiency is above 90%. Central domain of this protein contains tandemly repeated amino acid sequence which is almost identical except for amino acids 12 and 13 which defines the DNA-binding specificity. The DNA-binding domain of TAL effectors fuses with catalytic domain of *Fok*I endonuclease to bring about cleavage of target DNA segment. Although TALENs better than ZFNs, in terms of efficiency, it is less suitable for multiplex genome editing.

2 CRISPR: An Adaptive Immune System

Bacteria have been defending themselves from viruses by several defense mechanisms and one such unique defense machinery was found to be an adaptive immunity mechanism consisting of CRISPR and CRISPR associate protein 9 described first in 1987 *in Escherichia coli.* The development of CRISPR/Cas9 gene-editing system earned Charpentier and Doudna the coveted Nobel Prize in Chemistry in 2020. Following interaction with bacteriophages, the CRISPR locus incorporates a brief spacer region (of viral origin plasmid) between the short (30–40 bp) repeated palindromic sequences. This serves as a chronical archive of previous interactions with these bacteriophages. These spacer sequences are particularly responsible for providing specific immunity against phages or plasmids having complementary sequences by targeting them sequentially. Cas proteins adjacent to CRISPR locus contain domains encoding nucleases, polymerase, helicase, and different RNA-binding proteins. Among the various types of CRISPR/Cas system known (Table 1), the most adopted classification divides it into three classes. Type I system consists of six subtypes each having different number of cas genes and Cas3 as the signature proteins Type I-A system contains Cas3 protein as a part of Cascade complex, which has helicase as well as DNase domains needed for cleavage of target genome sequence.

Type II system is divided into three subtypes with Cas9, the RNA dependent endonuclease, as the signature protein along with Cas1, Cas2, and sometimes a fourth protein. Cas9 comprises of two nuclease domain, viz., RuvC which cleaves displaced strand in the crRNA–target DNA complex and HNH which cleaves target strands. Cas9 helps in adaptation, processing of crRNA, and crRNA-tracrRNA-assisted cleavage of the target DNA. Further, the Type III system comprises of Cas10 and can target both DNA and/or RNA. A new system, CRISPR–Cas system IV has been proposed which is guided by protein–DNA interactions, lacks Cas1 and Cas2 proteins and exclusively targets RNA.

Table 1 Variations of the CRISPR system

S. No.	Variations	Features	Effects	Advantages	Disadvantages
1	CRISPR/Cas9	WT Cas9; sgRNA	Double-strand break at the target site	Versatile; effective; stable; easy accessibility	Off-target; PAM limited; different modified alleles
2	CRISPR/Cas9 Nickase	Mutant Cas9 H840A or D10A; sgRNA	Single-strand break	Convenient; efficient; flexible; precise, scalable; robust	PAM limited; 2 sgRNA for Knock Out
3	CRISPRi	dCas9; repressor peptide; sgRNA	Block transcription elongation or knockdown transcripts	Inducible; reversible; low off-target effects;	PAM limited; off-target effects at bidirectional promoters
4	CRISPRa	dCas9; activator peptide; sgRNA	Increase transcription	Low toxicity	PAM limited; complicated to deliver the multiple components

2.1 CRISPR-Mediated Immune Response

CRISPR-mediated immune response occurs in the following steps:

Adaptation: Spacer acquisition acts as a memory card or database of previous encounters with viruses and can be divided into two types, namely, naïve, when cell encounters the virus for the first time and primed when there is repetitive invasion of the virus. Primed spacer acquisition increases the resistance of cell against the invading viruses as multiple spacer sequences makes it difficult for the invading DNA to use mutations to escape the CRISPR–Cas machinery as multiple mutations would be needed to achieve the same. The protospacer adjacent motif (PAM), a short sequence next target sequence that helps distinguish between self and foreign and aids in targeting the invader while shielding its own genome from degradation, was discovered through analysis of the viral sequence that the bacterial CRISPR–Cas system targets.

Expression: The CRISPR locus is transcribed and with the help of Cas ribonucleases the generated RNA is processed to form CRISPR ribonucleoprotein (crRNP) complex. This first creates a lengthy primary transcript with a number of structures that resemble hairpins. This is referred to as pre crRNP. Depending on the subtype, the pre-crRNA is subsequently broken down into smaller units by various Cas proteins. A different process depending on host RNase III and tracrRNA is used for the production of crRNA in type II, a trans-encoded short RNA with complementary bases to pre-crRNA that helps the two couple up with the Cas9 enzyme.

Furthermore, in the type II system, the crRNA-tracrRNA stays connected to the Cas9 enzyme while the crRNA is cut by an unidentified nuclease.

Interference: During interference, crRNA attached with Cas protein searches the protospacer, lands on it and initiates target degradation with the help of specific cas nucleases. Interference in CRISPR systems type I and type II requires the presence of a protosapcer adjacent motif (PAM) motive and complete complementarity between the protospacer and crRNA. Because the PAM sequence aids in both self- and non-self-identification, it is crucial for avoiding the system from attacking its own CRISPR locus. While tracrRNA linked to Cas9 and crRNA performs target recognition and degradation in type I and type III systems, Cas9 is sufficient for causing interference in type II systems. The Cas9 protein contains distinct lobes for target identification and nuclease function.

2.2 CRISPR-Mediated Genome Editing

The *Streptococcus pyogenes* type II CRISPR system is the source of the CRISPR/Cas system. It has genes that encode precursor crRNA, trans-activating crRNA, and Cas9 nuclease. A tiny nuclease with two nuclease domains and a protospacer adjacent motif (PAM)-binding domain is linked to the recognition lobe of the bilobed protein Cas9. The PAM sequence around the target DNA, Cas9-DNA contacts, and the corresponding conformational changes are necessary for target recognition. Cas9-sgRNA cascade scans for 5-NGG-3 sequence in the flanking region of PAM and the required guide RNA complementary in the DNA. Once it locates the complimentary sequence in the seed region, the resultant base pairing results in sgRNA–DNA heteroduplex followed by R-loop formation. At this stage, the two nuclease domains (HNH and RuvC-like) cleave the target resulting in formation of a DSB. CRISPR-based GE includes sgRNA design and fabrication, sgRNA validation, CRISPR–Cas9 construct delivery into plants, and sequencing to identify the intended alteration. Furthermore, compared to the plasmid-mediated approach, which can take more than three months, direct delivery of CRISPR–Cas9 ribonucleoproteins (RNPs) into apple and grapevine protoplasts has been reported to take only two to three weeks in order to achieve effective DNA-free targeted mutations.

2.3 Beyond Cas9

Cas9 possess two nuclease domains, viz., RuvC (a crossover junction endo-deoxyribonuclease) and HNH (containing histidine and asparagine residues). Off-target effect and large size are the major drawbacks of Cas9. Orthologues of SpCas9, StCas9 (*S. thermophilus*), and SaCas9 (*S. aureus*), respectively, requires longer PAMs and are more specific. Recently, *CRISPR*–Cas12a possesses RNase III activity for pre-crRNA processing along with nuclease activity and needs T-rich PAM sequences needs 42-nt crRNA guide region as opposed to approx. 100bp needed by Cas9 and cleaves DNA to generate staggered ends with 5' overhangs and larger indels in contrast to blunt end DSBs and small indels generated by Cas9. Cas13a (previously C2c2) derived from *Leptotrichia shahii* (bacteria) catalyzes maturation of crRNA as well as RNA-guided single-stranded RNA degradation. Cas13a induces single strand breaks with relatively less off-target effect and needs a protospacer flanking site instead of PAM. Further, base editing can modify individual DNA bases without DSBs formation or requirement of donor template. Base editors and prime editors are more modern modification of the CRISPR/Cas system that can be used to create all 12 types of base substitutions, targeted insertions, and deletions without requiring donor templates or producing DSBs. Prime editing introduces precise alterations in crop genomes which can be used for functional genomics, biotic–abiotic-stress tolerance and improving crop yield.

3 TALENs: Protein-Based DNA Modification System

TALE proteins were initially reported from phyto-pathogenic bacteria *Xanthomonas*. The repetitive domains of amino acids that make up the TALE protein each recognize a distinct base. A configurable number of 33–35 amino acids long tandem repeats is the key to TALEs precision and binding selectivity. Fusion of TALEs to the *FokI* nuclease, a type IIS restriction endonuclease present in *Flavobacterium okeanokoites*, results in the generation of TALENs. A DSB can be produced at a particular genomic location using TALENs. TALE is unique in its ability to identify a single-base pair deletion at the variable repeat and differentiate between distinct methylated nucleotides.

3.1 Genome Editing Using TALEN

DNA nucleotide recognition simply requires the usage of a TALE repeat with the appropriate RVD, TALEN construction is simpler than those of its predecessors, such as ZFNs.

Different TALEN designing tools include TALE-NT, idTALEPROGNOS, TALENoffer, E-TALEN, Mojo Hand, and ChopChop. Development of various assembly methods mainly on the basis of Golden Gate cloning lead to increased adoption of TALENs in genome editing in plants which was earlier constrained by the difficulty of constructing TALE through polymerase chain reaction (PCR) owing to the repetitive sequence of TALE. A vector-based technique for TALENs assembly, the Golden Gate assembly methodology uses type IIS restriction enzymes to produce several sticky ends for a particular ligation. This method combines digestion and ligation in one step and is incredibly thorough, economical, and successful. Sequencing is used to verify that the TALE/TALEN assembly is accurate. Mutations in eukaryotes are incorporated using a more sophisticated "Platinum Gate System." Rice mutants were produced when Liu et al. (2025) created ZQTALEN, which targeted the endogenous Nramp5 gene in rice.

Delivery of TALENs can be done via *Agrobacterium*-mediated transformation and has been utilized for delivery of TALEN in different crops like *Arabidopsis*, rice, maize, barley, potato, cabbage, and tomato. Particle bombardment or biolistic delivery is another standard delivery system which can be used in plant species for delivery of TALEN and TALEN proteins were delivered successfully to wheat embryos using particle bombardment method. Comparison between CRISPR/Cas9 and TALEN platforms and Classification of genome-editing techniques are tabulated in Tables 2 and 3, respectively.

4 CRISPR/Cas9 and TALEN-Mediated Genome Editing for Crop Improvement: Case Studies

Within a few years of its inception in 2013, several new gene edited crop germplasm resources were developed using CRISPR technique. Moreover, the CRISPR-mediated genome editing of a waxy corn and browning resistant mushroom was approved by the US FDA exempting the stringent regulations imposed on GM crop. CRISPR-based precision genome-editing tools are presented in Table 4. Schematic demonstration of CRISPR/Cas9 and TALEN-mediated genome editing are illustrated in Fig. 2.

4.1 Improving Resistance in Plants Against Biotic Stress

Availability of whole genome sequence information of the host plant and information about the target gene are the pre-requisites for successfully applying GE tools for crop improvement. Genome of several crops has been sequenced till date and many important genes have been characterized. The understanding of molecular details of plant's innate

Table 2 Comparison between CRISPR/Cas9 and TALEN platforms

S. No.	Particulars	TALEN	CRISPR/Cas9
01	Stands for:	Transcription activator-like effector nuclease	Clustered regularly interspaced short palindromic repeat
02	Recognition type	Site-specific DNA–Protein interaction	Site-specific RNA–Protein interaction
03	Recognition site	A repeated highly conserved 33–34 amino acid sequence with divergent 12th and 13th amino acids of TALE protein known as Repeat Variable Diresidue (RVD)	Single-strand guide RNA
04	Endonuclease	Fok1 nuclease: A bacterial-type IIS restriction endonuclease found in *Flavobacterium okeanokoites,* consisting of an N-terminal DNA-binding domain and a non-sequence-specific DNA cleavage domain at the C-terminal	Cas9 nuclease: CRISPR-associated protein 9 nuclease typically found in *Streptococcus pyogenes* consisting of target DNA-binding HNH-like nuclease domain and a non-sequence-specific RuvC-like nuclease domain
05	Target sequence size	Typically 14–20 bp per TALEN monomer, 28–40 bp per TALEN pair	Typically 20 bp guide sequence and PAM sequence
06	Specificity	Tolerates a small number of positional mismatches	Tolerates positional/multiple consecutive mismatches
07	Targeting limitations	5′ targeted nitrogenous base must be Thymine (T) for each TALEN monomer	Targeted site must precede a PAM (5'-NGG-3') sequence
08	Difficulties of engineering	Difficult: Requires complex molecular cloning methods	Easy: Can be done using standard cloning procedures and oligo synthesis
09	Difficulties of delivering	Difficult: due to the large size of functional components	Moderately difficult: commonly used SpCas9 is large and may cause packaging problems for viral vectors such as AAV, but smaller orthologs exist
10	Off-target effect	Low	High
11	DNA methylation sensitive	Yes	No
12	Mitochondrial genome engineering	Easy	Complicated
13	Precision of genome editing	High	Moderate

Table 3 Classification of genome-editing techniques

Particulars	SDN1	SDN2	SDN3
DNA repair template	Absent	Present; a short DNA sequence identical to the targeted DSB DNA sequence except for one or a few nucleotide changes	Present; DNA repair template contains new DNA sequence (e.g., gene)
DSB repair mechanism	NHEJ	HDR	HDR
Outcome	Targeted, non-specific genetic deletion mutation; screening and selection of the targeted change allows for identification of the desired genomic mutational outcome	A targeted and predetermined point mutation in the desired gene of interest	Integration of specific DNA sequence into the plant genome; insertion of cisgenic, intragenic, or transgenic expression cassettes at a selected genome location
Regulation	Non-GMO; Institutional BioSafety Committee (IBSC) under the Environment Protection Act	Non-GMO; Institutional BioSafety Committee (IBSC) under the Environment Protection Act	GMO; Genetic Engineering Appraisal Committee (GEAC)

immunity through intensive genetic studies has led to identification of specific targets for control of pests and diseases. Host susceptibility (S) genes are important targets for GE for improving biotic-stress tolerance.

Two major strategies that can be used to increase resistance in plants against biotic stress like virus include designing specific sgRNAs to target the virus genome and/or modification of genes involved in the anti-viral pathways. Mutations in eIF4E induced by CRISPR/Cas can change the way viruses interact with plants and mediate resistance to potyviruses. To create homozygous cucumber plants resistant to PRSV, pumpkin mosaic virus, and cucumber vein yellow virus,

Table 4 CRISPR-based precision genome-editing tools

Genome editing tools	Particulars
CRISPR/Cas12a	1. Smaller protein as compared to Cas9 recognizes a "TTTV" PAM sequence and cleaves DNA at a site distal to the target sequence, forming larger deletions 2. Generates cohesive ends with overhangs thus improving the efficiency of HDR-mediated DNA insertions 3. Can be employed for both knockout, knock-ins, and easier multiplexing
RNA editors	1. Utilizes Cas13 to target RNA in plants, such as plant-adapted virus resistance, RNA knockdown, alternative splicing modulation, and RNA base editing 2. Cas13 is a programmable RNA-targeting system for sequence-specific recognition and cleavage
Base editors	1. Integrate a nickase Cas9 (nCas9) with an adenine or cytosine deaminase to catalyze A-to-G or C-to-T substitutions 2. Causes single-base transition substitutions without the need of donor templates and induces DNA double-strand breaks 3. Types: a. Cytosine base editors which convert C:G to T:A b. Adenine base editors which convert A:T to G:C c. Base editors are widely used in plant GE to achieve targeted base modifications 4. Limitation: Can only induce transition substitutions, requires strict sequence suitability, PAM availability requirements and relatively low editing efficiencies
Prime editing	1. It uses nickase Cas9 (nCas9) fused with a reverse transcriptase and guided by a prime editing guide RNA (pegRNA) 2. The pegRNA contains sgRNA for target recognition and a reverse transcription template encoding the desired edit. The nCas9 upon binding induces a nick in the DNA, allowing the pegRNA to initiate reverse transcriptio.n, integrating the desired edit into the genome 3. It enables all 12 types of base substitutions, as well as small DNA insertions and deletions 4. It is not constrained by PAM sequence availability or specific sequence requirements 5. Limitation: Associated with lower editing efficiencies
Prime root	1. It is a novel tool enabling the precise integration of large DNA segments into plant genomes 2. It utilizes a prime editor coupled with recombinase to facilitate accurate large DNA insertion 3. It initially utilizes prime editing to introduce a recombination site into the genome. This is followed by formation of two similar recombination sites on a donor vector by tyrosine recombinase producing an intermediate donor with the desired DNA insert next to a corresponding recombination site. Finally, the tyrosine recombinase integrates the desired DNA insert into the newly established recombination site within the genome
Transposase-assisted target-site integration (TATSI)	1. This recent technology enables targeted DNA insertion in plants by utilizing transposable elements. It utilizes rice Pong transposase protein fused with Cas9 or Cas12a precisely excise and insert DNA into a desired target site 2. within the genome
Cas-Exo	1. Cas-Exo utilizes Cas9 or Cas12a fused with a 5′-exonuclease to enable precise and scar-free insertion of large DNA segments 2. It significantly improves the efficiency of HDR in plants 3. Working: Once the Cas nuclease induces a double-strand break, the 5′ exonuclease creates a longer free 3′ end by degrading the 5′ sequence of the cut strands 4. Can be used for getting stable and heritable knock-ins targeted insertion of large DNA sequence

CRISPR technology was utilized to target the N' and C' terminals of eIF4E. *AtJAZ2* in tomatoes was edited using CRISPR/Cas9, producing mutants that prevent bacterial entry through stomata and conferring disease resistance.*CsLOB1 (LATERAL ORGAN BOUNDARIES 1)* is essential for encouraging pathogen proliferation and pustule production (Hu et al. 2014). Citrus canker development is caused by *PthA4,* which binds to the effector binding element (*EBEPthA4*) in the *CsLOB1* promoter and stimulates CsLOB1 expression (Hu et al. 2014).

Peng et al. (2017) used CRISPR/Cas9 method for modification of the susceptibility gene *CsLOB1* promoter in citrus for generation of citrus canker resistant mutants. Although TALEN- mediated knockout of three TaMLO homeologues gave wheat a robust defense against powdery mildew (PM), it also had negative effects on growth and yield, which limited its application in plant breeding. A knockout mutant with deletion in *TaMLO-B1* was later created by CRISPR/Cas9-mediated genome editing, exhibiting significant resistance against PM without yield penalties.

Also, CRISPR/Cas9 knockout of all *TaEDR1*homeologues provided moderate resistance to powdery mildew besides enhancing stripe rust resistance in wheat without affecting growth and yield parameters making it suitable for breeding. CRISPR/Cas9, has been used to develop high-quality tomato lines resistant to various diseases (Shawky et al. 2024). In

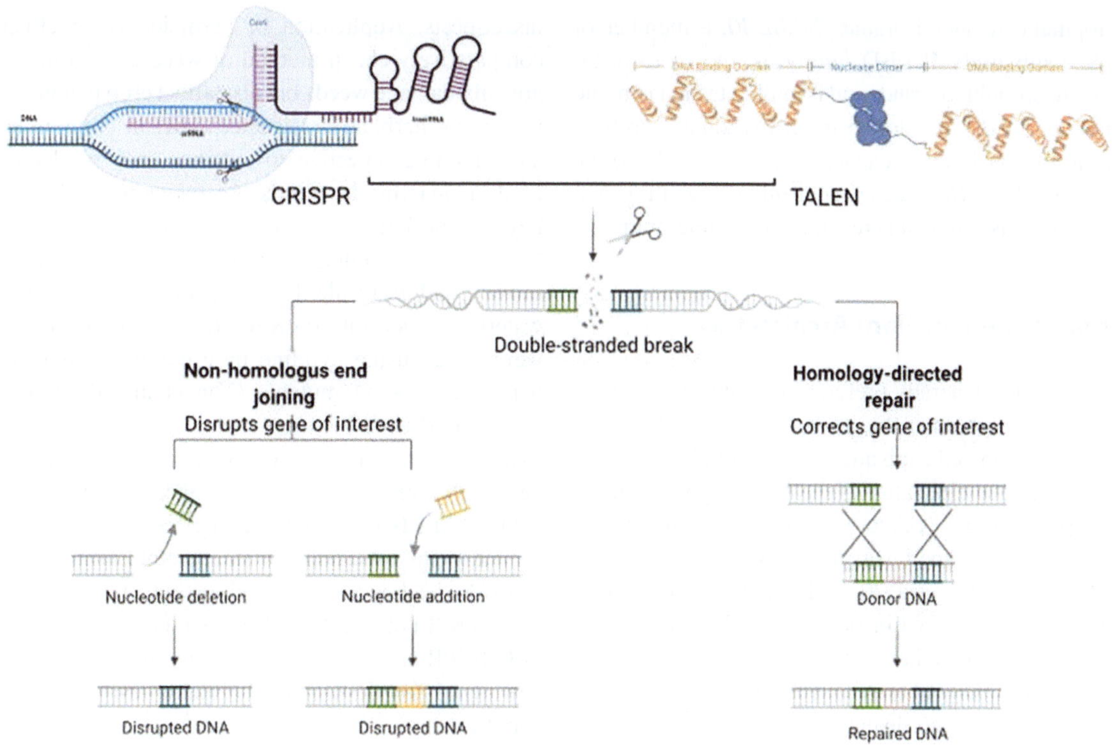

Fig. 2 Schematic demonstration of CRISPR/Cas9 and TALEN-mediated genome editing

tomato, downy mildew (DM) is one of the major fungal diseases causing notable yield loss. *AtDMR6* belonging to the *2-oxoglutarate oxygenase* Fe(II)-dependent super-family is crucial for salicylic acid homeostasis. Over-expression of *AtDMR6*can result in reduced DM susceptibility. Tomatoes with knocked *DMR6* homologues were more resistant to *Pseudomonas syringae.* One of the biggest threats to apple agriculture is fire blight, which is brought on by *Erwinia amylovora.* In order to increase the apple's resistance to fire blight, CRISPR/Cas9 ribonucleoproteins were delivered directly to the apple protoplast, targeting and mutating three genes: *DIPM-1, DIPM-2,* and *DIPM-4.* The banana streak virus, a double-stranded DNA badnavirus, affects bananas, one of the main fruit crops. Seventy-five percent of altered events showed resistance to banana streak virus in the banana variety GonjaManjaya using CRISPR technology (Tripathi et al. 2024).

4.2 Improvement in Abiotic-Stress Resistance

One of the main abiotic stressors causing yield loss worldwide is drought. Abscisic acid (ABA), a phytohormone, is essential for modulating stomatal closure, stress-related gene expression, and transpirational water loss during drought. An essential plasma membrane *H+ATPase* for creating a proton gradient to boost stomatal conductance is the gene encoding Arabidopsis *OPEN STOMATA 2 (OST2).*

Using CRISPR system *ost2* mutants were generated with greater than 32% extraordinary efficiency exhibiting greater stomatal closure with reduced rate of transpirational loss thus increasing drought tolerance. Variants for *ARGOS8 (AUXIN-REGULATED GENE INVOLVED IN ORGAN SIZE 8),* revealed over-expression of *ARGOS8* to decrease ethylene sensitivity and were found to be more drought tolerant and to yield more in the field than the wild type (Shi et al. 2017).

Salinity and drought tolerant rice mutants were generated by targeting *OsDERF1, OsERF922,* and *OsRMC*. Knocking out of such genes can result in increased stress tolerance. Knock out of S gene in tomato *SlAGAMOUS-LIKE 6* (*SlAGL6*) resulted in heat tolerance and improved fruit set under heat stress. Mutants with enhanced *AVP1* expression resulted in improved drought tolerance. *ABRE* (ABA-responsive element)-dependent ABA signaling is transcriptionally activated by *ABA-responsive element-binding protein 1 (AREB1).*

To increase the expression of *AREB1*, CRISPR/dCas9 combined with histone acetyltransferase was introduced. Plants that were activated showed higher chlorophyll content, faster stomatal closure under water deficit, and increased drought tolerance.

Genes particular to plants that include the *LATERAL ORGAN BOUNDARIES DOMAIN (LBD)* are essential for the development of lateral organs. *SlLBD40*, a member of the tomato's subfamily II LBD family, is known to negatively regulate drought tolerance and is implicated in jasmonic acid signaling. *SlLBD40* deletion improves tomato drought tolerance and water-holding capacity (Biswas et al. 2021). Knocking out of *OsRR22* gene, controlling signaling and cytokinin metabolism in plants resulted in salt tolerant rice.

4.3 Improvement in Plant Architecture

Dwarf varieties are generally preferred for intensive cultivation as they are more amenable to be used in high density planting and mechanized cultivation system. Gibberellic acid (GA) involved in cell elongation is the most important determinant of plant height. GA biosynthesis gene in banana *Ma*$GA_{20}OX_2$ was knocked-out using CRISPR to obtain semi-dwarf mutant lines (Shao et al. 2020). Loss-of-function alleles of *GA20ox2* were generated in rice, using CRISPR/Cas9 to produce dwarf mutants (Han et al. 2019). Loss-of-function mutations of GA biosynthesis enzyme *ZmGA20ox3* were generated in maize mutants displaying semi-dwarfing phenotype and enhanced drought resistance (Amadu et al. 2025). *Phytoene desaturase* (PDS) is crucial for converting phytoene into zeta-carotene. When the PDS gene is disrupted, the biosynthesis of gibberellin, carotenoid, and chlorophyll is hindered, resulting in albino and dwarf phenotypes. Although the amino acid sequences of *TFL1* and *FLOWERING LOCUS T (FT)* are substantially conserved, their roles are different. While *TFL1* is a floral repressor and its silence accelerates flowering, *FT* encourages flowering. Apple genome has two *TFL1* homologous genes which are expressed in vegetative tissues (Mimida et al. 2009) a single copy of *MdPDS* gene. Knock out of *PDS* and *TFL1* gene in apple resulted in mutants with albino and dwarf phenotype with earlier flowering (Charrier et al. 2019). Strigolactones (SLs) are crucial for determining plant architecture and inhibiting shoot branching. CRISPR/Cas9-mediated mutation in four *GmSPL9* gene altered plant architecture in soybean (Bao et al. 2019). *PhFT* and *PhTFL1* genes are widely found in plants and in Arabidopsis, they repress flowering by promoting vegetative growth (Baumann et al. 2015). CRISPR-mediated knock out of two paralogous of *PhFT* and four *PhTFL1* paralogous genes in petunia resulted in significant phenotypic alterations in petunia leading to early flowering time. *PhTFL1*mutants exhibited shorter internodes probably by down-regulating *PhGAI* (Abdulla et al. 2024).

4.4 Improvement in Herbicide Resistance

Weeds are the most ubiquitous of all crop pests presenting multitude of problems including harboring pathogens and insect-pests. Application of herbicide is an effective and comparatively cheap method of weed control however they not only damage weeds but also affect crop plants. Therefore, improving herbicide resistance (HR) in crops is an important breeding objective to enhance crop production while easily managing the weeds. *Acetolactate synthaseALS*is the key enzyme for targeting *ALS*-inhibiting herbicides. CRISPR-based targeted mutagenesis of *acetolactate synthase ALS* in watermelon, resulted in generation of herbicide-resistant watermelon germplasm. Key *ALS* cites in tomato and potato were edited using cytidine base editing (CBE) to produce herbicide-resistant mutants (Yan et al. 2018).Wang et al. (2021) used CRISPR/Cas9-mediated genome editing in rice to develop a novel herbicide tolerance allele (*G628W*) with a very high level of herbicide tolerance in mutants. In another study, TALEN-based HR in rice resulted in generation of double point mutations in rice acetolactate synthase gene (*OsALS*) leading to the development of herbicide-resistant rice lines (Li et al. 2016). These studies suggest the potential of CRISPR and TALEN technology in specifically targeting key genes for developing herbicide-resistant/tolerant crop varieties.

4.5 Improvement in Fruit Quality and Shelf Life

Lycopene pigment is responsible for red color in tomato and is known to have a plethora of therapeutic effects. Lycopene gets converted toα- and β-carotene during ripening. In tomato, Li et al. (2018) used CRISPR to knock out different genes involved in converting lycopene to carotenes, viz., *SGR1*, *LCY-E*, *BLC*, *LCY-B1*, and *LCY-B2*resulting in increased concentration of lycopene in the ripe fruits. *Phytoene synthase1* (*PSY1*) is involved in lycopene synthesis during tomato fruit ripening (Giorio et al. 2008). CRISPR/Cas9-mediated knock out of *PSY1*gene resulted in yellow fruit (D'Ambrosio et al. 2018). *Pectate lyases* (*PL*)is responsible for cell wall degradation and fruit softening. CRISPR/Cas9-mediated mutation of tomato *PL* gene increases the fruit firmness and extends the shelf life of tomato fruit. The Sicilian Rouge tomatoes, genetically edited using CRISPR/Cas9 technology to increase the amounts of γ-aminobutyric acid (*GABA*), a potential nutraceutical compound known to reduce blood pressure and promote relaxation has been introduced to the market in Japan by Tokyo-based Sanatech Seed. Pericarp browning caused due to pericarp desiccation and oxidation of phenolic compounds by polyphenol oxidase (*PPO*) and/or peroxidase (*POD*) is a major post-harvest problem in litchi restricting the development of litchi industry. Litchi *PPO* gene (*LcPPO*, JF926153) is a single-copy gene in the litchi genome. Inhibiting *PPO* gene expression can reduce *PPO* activity and inhibit pericarp browning. Knock out of *LcPPO* gene was attempted in "Feizixiao" litchi to demonstrate successful application of CRISPR/Cas9 technology for breeding browning resistant litchi (Wang et al. 2023).

Enzymatic and non-enzymatic browning are the main causes of post-harvest quality loss in potatoes. While non-enzymatic browning, also referred to as cold-induced sweetening (CIS), happens in potatoes stored in cold temperatures, where stored sugars are reduced, enzymatic browning is caused by the polyphenol oxidase (PPO) enzyme converting phenols into quinones. *Agrobacterium*-mediated transient expression approach was utilized for effective delivery of TALENs to target two endogenous genes starch branching enzyme (*SBE1*) and acid invertase (*INV2*) into two potato cultivars, Russet Burbank and Shepody by *Agrobacterium*-mediated infiltration into leaves of potato plants leading to an increased degree of starch branching and CIS (Ma et al. 2017). Furthermore, TALEN-mediated knock out of *PPO* in potato resulted in reduced tuber browning (Dev et al. 2021), while a CRISPR/Cas9-based mutations in *StPPO2* gene resulted in production of potato regenerants with reduced *PPO* activity and enzymatic browning. Aromatic rice varieties are very popular among Asian communities. TALEN was first time used for creating fragrant rice by targeting the *OsBADH2* gene. Novel alleles of *OsBADH2*were generated through CRISPR/Cas9 technology which converted an unscented rice variety, *ASD16*, into aromatic rice (Ashok Kumar et al. 2020). L-idonate dehydrogenase gene (*IdnDH*) plays a major role in the biosynthesis of tartaric acid (TA). It is highly important as it promotes ripening, provides resistance against microorganisms and determines the taste of grape wine. Knock out of *IdnDH* gene in suspension cells of grape variety "Chardonnay" resulted in no biosynthesis and accumulation of TA. The detected mutations were either 1-bp insertions or 1–3 nucleotide deletions with no obvious off-target effect thus proving the efficiency of CRISPR/Cas9 system for precise genome editing in grapes (Ren et al. 2016).

Ethylene plays main role in ripening and shelf life determination in climacteric fruits. Targeting ethylene biosynthesis or signaling pathways are important methods for increasing shelf life of climacteric fruits. CRISPR/Cas9 can be used to slow down fruit ripening process by genetically altering genes that are directly or indirectly involved in regulating ripening. The rate limiting enzymes involved in ethylene biosynthesis include 1-aminocyclopropane-1-carboxylic acid (*ACC*) synthase (*ACS*) and ACC oxidase (*ACO*) genes. *ACO* converts *ACC* into ethylene. *MA-ACO1* has been reported to be the major *ACO* involved in banana fruit ripening. Hu et al. (2021) developed a *MaACO1* disrupted mutants in *Musa acuminata* (AAA group, cv. Brazilian) using CRISPR/Cas9 system to study its effect on banana ripening. Delayed ripening with yellow or green fruit with no speckles were obtained even after 60 days of harvest in *MaACO1*-disrupted banana fruits suggesting that it is an ideal target gene to edit for improved shelf life in banana (Hu et al. 2021). In another study an attempt was made to improve the shelf life of Japanese luxury melon (*Cucumis melo* var. reticulatus, "Harukei 3") by targeting *CmACO1* gene (which is a key gene for shelf life in melons). CRISPR/Cas9-based mutations in *CmACO1* gene resulted in mutant phenotypes with reduced ethylene production and increased fruit firmness beside a delay in fermentation of fresh fruit. The results indicate that apart from improving shelf life of melon, CRISPR technology can help reducing post-harvest losses in fruit and contribute to food security. *RIPENING INHIBITOR* gene (*RIN*) belonging to the MADS-box gene family is one of the major regulators of ethylene biosynthesis in tomato and is involved in the ripening, fruit firmness, color, and flavor development. Knock-out of *RIN* using CRISPR caused partial ripening of tomato fruits (Ito et al.2015).

CRISPR/Cas9-mediated knockout mutants of *AP2a*, *FUL1*, *FUL2*, and *NOR* genes were developed in tomato to understand their role in fruit ripening. The nor mutant produced by CRISPR/Cas9 exhibited earlier ripening and orange ripe phenotypes. The *ful1* or *ful2* double mutant showed a severely blocked ripening phenotype while the *ful1* and *ful2* single mutants exhibited normal ripening phenotypes, suggesting redundant function of *FUL1* and *FUL2* in the fruit ripening process. Epigenetic modifications like DNA cytosine methylation are known to regulate gene expression, promote genome stability, and respond to external environmental stress.

5 Regulatory and Ethical Concerns Associated with Genome-Edited Crops

Regulatory and ethical considerations surrounding crops edited using CRISPR and TALEN technologies are complex, involving scientific, legal, and societal dimensions. While the EU has been implementing process-based regulation for genome-edited crops and has labeled GE crops as genetically modified organisms (GMOs), other nations, such as the United States, deregulate SDN1 and SDN2 crops. Ethical concerns mainly focus on biosafety, environmental impact, and socio-economic implications. It is argued that CRISPR and TALEN allow precise and specifically targeted modifications thus reducing unintended mutations compared to conventional breeding or older genetic modification techniques. However, still the concerns remain for potential off-target effects and ecological risks, such as gene flow to wild relatives and unintended consequences on biodiversity. Another ethical debate involves the intellectual property rights, as large agribusinesses often patent genome-editing technologies, potentially limiting access for small farmers and public research institutions. Additionally, consumer acceptance remains a challenge, as public perception of genome-edited crops varies across cultures and regions, often influenced by misinformation and lack of transparency from biotechnology firms. Many scientists advocate for a risk-based regulatory approach that assesses genome-edited crops based on their final traits rather than the method used to develop them, a stance supported by countries like Canada and Japan. Ethical considerations also extend to food security and sustainability, as genome-editing could enhance crop resilience against climate change, pests, and diseases, potentially reducing reliance on chemical pesticides and fertilizers. However, concerns persist regarding corporate control over agricultural biotechnology and the need for equitable distribution of its benefits, particularly in developing countries. Transparency, public engagement, and interdisciplinary policymaking are crucial to addressing these ethical and regulatory challenges while ensuring that genome-editing technologies contribute to sustainable and equitable food production systems.

6 Conclusion

CRISPR/Cas9 and TALEN technologies have revolutionized genome engineering for crop improvement by enabling precise and efficient genetic modifications. These tools offer unprecedented accuracy in targeted mutagenesis, facilitating the development of crops with enhanced yield, stress tolerance, disease resistance, and improved nutritional profiles. While TALENs provide specificity and versatility, CRISPR/Cas9 has gained widespread adoption due to its simplicity, efficiency, and adaptability to various plant species. The past decade has witnessed immense rise in the use of CRISPR and TALEN-based genome-editing crop improvement for producing mutants with increased tolerance to biotic, abiotic stress, improved nutritional quality and shelf life. Despite these advancements, challenges such as off-target effects, regulatory hurdles, and public acceptance remain critical considerations for the broader application of genome editing in agriculture. Ongoing research and technological refinements, including base editing and prime editing, continue to enhance the precision and efficiency of these methods. As regulatory frameworks evolve and societal perceptions shift, genome engineering holds immense potential to address global food security challenges and drive sustainable agricultural practices.

References

Abdulla MF, Mostafa K, Kavas M (2024) CRISPR/Cas9-mediated mutagenesis of *FT/TFL1* in petunia improves plant architecture and early flowering. Plant Mol Biol 114:69. https://doi.org/10.1007/s11103-024-01454-9

Amadu MK, Beyene Y, Chaikam V, Tongoona PB, Danquah EY, Ifie BE, Burgueno J, Prasanna BM Gowda M (2025) Genome-wide association mapping and genomic prediction analyses reveal the genetic architecture of grain yield and agronomic traits under drought and optimum conditions in maize. BMC Pl Bio 25(1):135. https://doi.org/10.1186/s12870-025-06135-3

Ashok kumar S, Jaganathan D, Ramanathan V, Rahman H, Palaniswamy R, Kambale R, Muthurajan R (2020) Creation of novel alleles of fragrance gene *OsBADH2* in rice through CRISPR/Cas9 mediated gene editing. PLoS ONE 15:e0237018. https://doi.org/10.1371/journal.pone.0237018

Bao A, Chen H, Chen L et al (2019) CRISPR/Cas9-mediated targeted mutagenesis of *GmSPL9* genes alters plant architecture in soybean. BMC Plant Biol 19:131. https://doi.org/10.1186/s12870-019-1746-6

Baumann K, Venail J, Berbel A, Domenech MJ, Money T, Conti L, Bradley D (2015) Changing the spatial pattern of *TFL1* expression reveals its key role in the shoot meristem in controlling Arabidopsis flowering architecture. J Exp Bot 66(15):4769–4780. https://doi.org/10.1093/jxb/erv247

Biswas D, Saha SC, Dey A (2021) CRISPR-Cas genome-editing tool in plant abiotic stress-tolerance. Plant Gene 26:100286. https://doi.org/10.1016/j.plgene.2021.100286

Charrier A, Vergne E, Dousset N, Richer A, Petiteau A, Chevreau E (2019) Efficient targeted mutagenesis in apple and first time edition of pear using the CRISPR-Cas9 system. Front Plant Sci 10:401–412. https://doi.org/10.3389/fpls.2019.00040

D'Ambrosio C, Stigliani AL, Giorio G (2018) CRISPR/Cas9 editing of carotenoid genes in tomato. Transgenic Res 27(4):367–378. https://doi.org/10.1007/s11248-018-0079-9

Dev SS, Joseph J, D'Rosario LL, (2021) Prospects for Genome Editing of Solanum tuberosum: a promising crop for starvation problem. p 137

Giorio G, Stigliani AL, D'Ambrosio C (2008) Phytoene synthase genes in tomato (*Solanumlycopersicum* L.) new data on the structures, the deduced amino acid sequences and the expression patterns. The FEBS J 275(3):527–535. https://doi.org/10.1111/j.1742-4658.2007.06219.x

Han Y, Teng K, Nawaz G, Feng X, Usman B, Wang X, Luo L, Zhao N, Liu Y, Li R (2019) Generation of semi-dwarf rice (*Oryza sativa* L.) lines by CRISPR/Cas9-directed mutagenesis of *OsGA20ox2* and proteomic analysis of unveiled changes caused by mutations. Biotech 9:387. https://doi.org/10.1007/s13205-019-1919-x

Hu Y, Zhang JL, Jia HG, Sosso D, Li T, Frommer WB, Yang B et al (2014) *Lateral organ boundaries 1* is a disease susceptibility gene for citrus bacterial canker disease. PNAS 111:E521–E529. https://doi.org/10.1073/pnas.131327111

Hu C, Sheng O, Deng G, He W, Dong T, Yang Q, Dou T, Li C, Gao H, Liu S, Yi G (2021) CRISPR/Cas9-mediated genome editing of *MaACO1* (aminocyclopropane-1- carboxylate oxidase 1) promotes the shelf life of banana fruit. Pl Biotech J 19(4):654. https://doi.org/10.1111/pbi.13534

Ito Y, Nishizawa-Yokoi A, Endo M, Mikami M, Toki S (2015) CRISPR/Cas9-mediated mutagenesis of the RIN locus that regulates tomato fruit ripening. Biochem Biophysical Res Comm 467(1):76–82. https://doi.org/10.1016/j.bbrc.2015.09.117

Li T, Liu B, Chen CY, Yang B (2016) TALEN-mediated homologous recombination produces site-directed DNA base change and herbicide-resistant rice. J Gen Genom 43(5):297–305. https://doi.org/10.1016/j.jgg.2016.03.005

Li XD, Wang YN, Chen S et al (2018) Lycopene is enriched in tomato fruit by CRISPR/Cas9-mediated multiplex genome editing. Front Plant Sci 9:559. https://doi.org/10.3389/fpls.2018.0059

Liu Q, Lin L, He S, Yu J, Xie C, Gai C, Han Y, Liu C, Huang F, Chen D, Song Y (2025) A simple and efficient TALEN system for genome editing in plants. Pl Mol Bio 115(1):25. https://doi.org/10.1007/s11103-025-01551-3

Ma J, Xiang H, Donnelly DJ, Meng FR, Xu H, Durnford D, Li XQ (2017) Genome editing in potato plants by *Agrobacterium*-mediated transient expression of transcription activator-like effector nucleases. Plant Biotechnol Rep 11:249–258. https://doi.org/10.1007/s11816-017-0448-5

Mimida N, Kotoda N, Ueada T, Igarashi M, Hatsuyama Y, Iwanami H et al (2009) Four *TFL1/CEN*-like genes on distinct linkage groups show different expression patterns to regulate vegetative and reproductive development in apple (*Malus x domestica* Borkh.). Plant Cell Physiol 50:394–412. https://doi.org/10.1093/pcp/pcp001

Peng A, Chen S, Lei T, Xu L, He Y, Wu L, Yao L Zou X (2017) Engineering canker- resistant plants through CRISPR/Cas9-targeted editing of the susceptibility gene *CsLOB1* promoter in citrus. Pl Biotech J 15(12):1509–1519. https://doi.org/10.1111/pbi.12733

Perveen N, Dinesh MR, Sankaran M, Shivashankara KS,Venugopalan R (2022) Characterization and evaluation of putative mutant populations of polyembryonic mango genotype Nekkare for dwarfing rootstock traits. J Hortic Sci 17(2):261–271.https://doi.org/10.24154/jhs.v17i2.1456

Perveen N, Dinesh MR, Sankaran M, Ravishankar KV, Krishna jee HG, Hanur VS, Alamri S, Kesawat MS, Irfan M (2023b) Comparative transcriptome analysis provides novel insights into molecular response of salt-tolerant and sensitive polyembryonic mango genotypes to salinity stress at seedling stage. Front Plant Sci 14:1152485. https://doi.org/10.3389/fpls.2023.1152485

Perveen N, Dinesh MR, Sankaran M, Lakkireddy V, Shivashankara KS, Venugopal R, Mir H (2024). Phenolics signatures in response to salinity stress provide novel insights into physiological basis of salt tolerance in mango (*Mangifera indica* L.). J Pl Growth Regul 43(12):4866–4885. https://doi.org/10.1007/s00344-024-11440-6

Perveen N, Dinesh MR, Sankaran M. Bindu KH, Shivashankara KS, Ravishankar KV, Kanade NM, Kalal P, Mir H (2025) Radiation induced mutations alter morpho-biochemical, anatomical and molecular responses in polyembryonic mango genotype Bappakkai. Plant Physiol Rep 30:284–295. https://doi.org/10.1007/s40502-025-00860-5

Ren C, Liu X, Zhang Z, Wang Y, Duan W, Li S, Liang Z, (2016) CRISPR/Cas9-mediated efficient targeted mutagenesis in Chardonnay (Vitis vinifera L.). Sci Rep 6(1):32289

Shao X, Wu S, Dou T, Zhu H, Hu C, Huo H, et al. (2020) Using CRISPR/Cas9 genome editing system to create *MaGA20ox2* gene-modified semi-dwarf banana. Pl Biotechnol J 18:17–9. https://doi.org/10.1111/pbi.13216

Shawky A, Hatawsh A, Al-Saadi N, Farzan R, Eltawy N, Francis M, Abousamra S, Ismail YY, Attia K, Fakhouri AS, Abdelrahman M (2024) Revolutionizing tomato cultivation: CRISPR/Cas9 mediated biotic stress resistance. Plants 13(16):2269. https://doi.org/10.3390/plants13162269

Shi J, Gao H, Wang H et al (2017) *ARGOS 8* variants generated by CRISPR-Cas9 improve maize grain yield under field drought stress conditions. Plant Biotechnol J 15(2):207–216. https://doi.org/10.1111/pbi.12603

Tripathi JN, Tripathi L (2024) Using CRISPR-Cas9 genome editing to enhance disease resistanceinbanana. CABI Rev 19(1). https://doi.org/10.1079/cabireviews.2024.0043

Wang S, Wang G, Li H, Li F, Wang J (2023) *Agrobacterium tumefaciens*-mediated transformation of embryogenic callus and CRISPR/Cas9-mediated genome editing in 'Feizixiao' litchi. Horticult Pl J 9(5):947–957. https://doi.org/10.1016/j.hpj.2023.01.011

Wang F, Xu Y, Li W, Chen Z, Wang J, Fan F, Tao Y, Jiang Y, Zhu QH Yang J (2021) Creating a novel herbicide-tolerance *OsALS* allele using CRISPR/Cas9-mediated gene editing. Crop J 9(2):305–312. https://doi.org/10.1016/j.cj.2020.06.001

Yan Y, Liu Q, Zang X, Yuan S, Bat-Erdene U, Nguyen C, Gan J, Zhou J, Jacobsen SE, Tang Y (2018) Resistance-gene-directed discovery of a natural-product herbicide with a new mode of action. Nature 559:415–418. https://doi.org/10.1038/s41586-018-0319-4

Induced Mutations for Enhancing Nutrition Quality and Production of Food

Anirban Maji, Subhasis Samanta, Shouvik Gorai, Soham Hazra, Afridi Mondal, Md. Imtiyaz, and Anindita Mondal

Abstract

Mutation means sudden heritable changes of an organism and this change is continuing spontaneously in the living world with a very low frequency both in oligogenic and polygenic traits. However, it is mainly found in oligogenic traits. In crop improvement programme, mutation breeding techniques are widely applied where lack of variation present among the entire gene pool. The variation can be created either by using physical or chemical mutagen to quick develop large variable population within short period of time. Mutation breeding is an excellent application of agricultural crops which of cost-effective, environment-friendly without losing its quality traits. Enormous varieties have been released through this breeding technique in cereals, pulses, oilseeds, vegetables, fibre crops, fruit crops and ornamentals in India. In future, this breeding technique will play a crucial role in yield, quality and disease-resistant, improved physiological traits for climate resilient and sustainable agriculture ensuring the food security.

Keywords

Variation · Mutation breeding · Application · Crop improvement

A. Maji (✉) · S. Samanta · S. Gorai · A. Mondal · Md. Imtiyaz · A. Mondal
Department of Genetics and Plant Breeding, Bidhan Chandra Krishi Viswavidyalaya, Mohanpur, Nadia, West Bengal 741252, India
e-mail: anigpbr@gmail.com

S. Hazra
Department of Agriculture, Brainware University, Barasat, West Bengal, India

1 Introduction

Plant breeding relies heavily on genetic variation, and introducing new variations is essential for incorporating novel traits into breeding programmes. Mutation breeding, also known as 'variation breeding', involves exposing seeds to chemical mutagens, radiation or enzymes to create mutants with beneficial characteristics. These mutants, known as mutagenic plants, can then be used for breeding with other cultivars. Mutants, whether spontaneous or induced, with beneficial alterations in specific traits have contributed to gene discovery, mapping, functional genomics and breeding. The use of physical and chemical mutagens to induce genetic variation offers an alternative approach for generating the new genetic diversity required to breed varieties with desirable traits, as induced mutations significantly enhance the extent of genetic variation. Applied mutagenesis replicates the natural process of spontaneous mutation, maintaining the original genetic structure without introducing external genes, unlike genetically modified organisms. Due to their long-standing record of safety, organisms produced via physical and chemical mutagenesis are not regulated as GMOs in the EU.

Mutation triggers random genetic variations, giving rise to mutant plants with novel and advantageous traits. Mutation breeding enables the rapid creation of new genetic diversity without major alterations to the plant's overall genetic structure, providing opportunities to screen and select mutants with enhanced agro-economic characteristics. However, these techniques result in 100 or even 1000 of random mutations, whose impacts and outcomes are often uncertain. When paired with advanced breeding methods such as speed breeding and single seed descent selection, induced mutagenesis can produce desired cultivars within a reduced breeding timeframe. The rise of modern technologies has made directed mutagenesis possible, allowing for accurate gene manipulation, including targeted gene knockouts and the integration or

J. Devi (ed.), *Mutation Breeding in Crop Improvement*, Sustainable Landscape Planning and Natural Resources Management,
https://doi.org/10.1007/978-3-032-17911-1_4

modification of particular DNA sequences via homologous recombination. Plants grown in extreme conditions such as zero gravity and cosmic radiation experience notable physiological, physical and genetic transformations. Cosmic radiation has the ability to modify genes, and space crop initiatives have gained worldwide interest for providing important biological insights (Wolverton and Kiss 2009). Major crops and their number of mutant varieties in world and India is illustrated in Figs. 1 and 2. This book chapter seeks to offer a thorough overview of the contribution of mutation breeding to advancing crop improvement. It also explores the basics of mutation breeding, including the different types of mutations and mutagens, as well as its applications, such as directed mutagenesis and space mutagenesis.

2 Types of Mutation

2.1 Spontaneous Mutation

Spontaneous mutations occur naturally at a very low frequency, between 10^{-5} and 10^{-8}, as a result of ongoing adaptation and evolutionary processes. This frequency is too low to produce substantial genetic variation in crops, which is crucial for improving desirable traits. These mutations arise rarely and unpredictably, affecting both the timing of their occurrence and the particular genes they influence. While many mutations can be detrimental, hindering an organism's adaptability or even causing fatal consequences, they can occa-

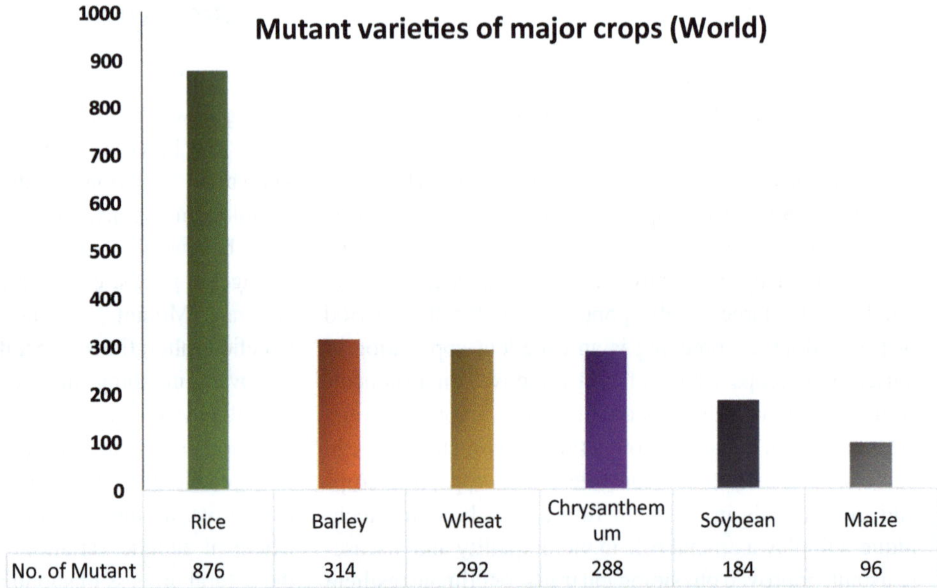

Fig. 1 Major crops and their number of mutant varieties in world. *Source* FAO/IAEA-MVD (2022)

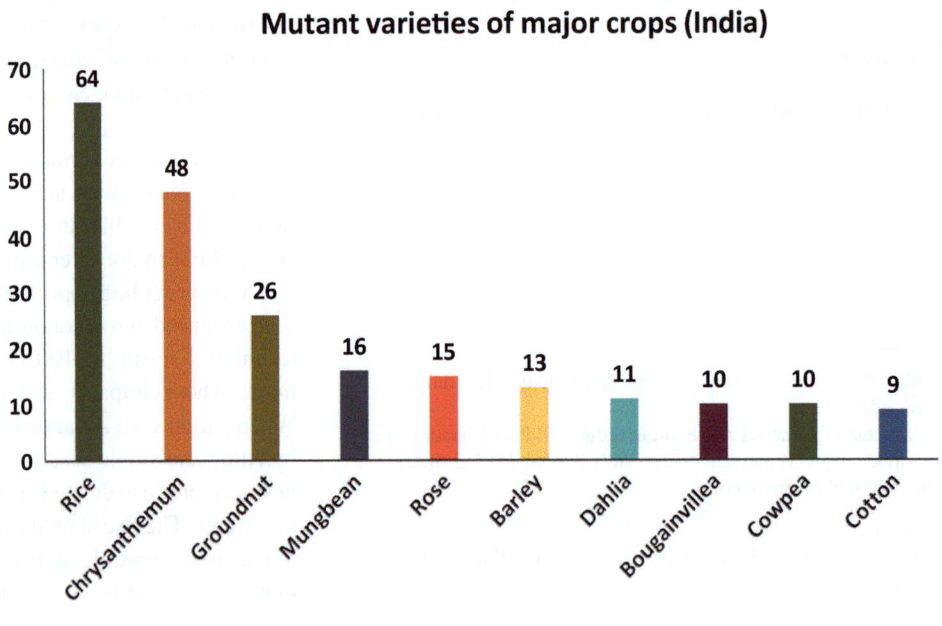

Fig. 2 Major crops and their number of mutant varieties in India. *Source* FAO/IAEA-MVD (2022)

sionally result in the formation of advantageous recombinant genotypes in future generations.

2.2 Induced Mutation

Induced mutation greatly speeds up the genetic change process compared to natural spontaneous mutations. By introducing genetic variability, induced mutagenesis enables breeders to create new traits, aiding in the development of improved crop varieties, advancing gene function studies and pushing forward genetic research. Traditionally, planting materials are exposed to chemical and physical mutagens to induce mutations, resulting in the creation of novel and enhanced crop varieties. Mutagenesis can be applied to various planting materials, including mature plants, especially seedlings, and in vitro cultured cells, offering a flexible approach to genetic enhancement.

2.3 Mutagenic Agents

Irradiation and the use of chemical mutagens are the main techniques for inducing mutations in plants. X-rays and gamma radiation induce a combination of substantial chromosome deletions and point mutations, such as single base substitutions or deletions. In contrast, widely used chemical mutagens include NaN_3, EMS and MNU mainly induce single base substitutions, typically transitioning from G/C to A/T. Traditionally, mutations caused by physical mutagens lead to deletions of up to 803 base pairs, along with a variety of uncharacterized secondary mutations (Shan et al. 2015). The advantage of chemical mutagens is their capacity to create mutant populations with high mutation densities, facilitating the identification of specific mutations within the population.

Gamma rays, a random physical mutagen, have been widely and effectively utilized in applied mutagenesis, with many traits anticipated to be modified by gamma radiation. Ethyl Methanesulphonate (EMS) is regarded as one of the most important chemical mutagens in applied mutagenesis. In terms of mutagenic efficiency, chemical mutagens are as effective as, or even more efficient than, physical mutagens in terms of mutagenic efficiency. Warghat et al. (2011) investigated the study on the effects of sodium azide and gamma radiation on germination rates, survival rates, morphological variations and chlorophyll mutations in musk okra (*Abelmoschus moschatus* L.) concluded that gamma radiation was more effective in producing beneficial mutants. Gamma radiation has been utilized in applied mutagenesis for several crops, including okra, tomato (Im et al. 2018), rice (Pillai et al. 1993), cluster bean (Dube et al. 2011) and bitter gourd. Gamma irradiation leads to chromosome deletions, translocations and abnormalities and is favoured as a mutagenic agent due to its higher penetration of plant tissues, reproducibility and increased mutation frequency. Although gamma radiation can adversely affect seed germination and seedling vigour in the M1 generation, there have also been reports of positive effects on seed germination and seedling development (Beyaz 2020).

At the turn of the century, a significant shift occurred in the use of mutants with the development of efficient TILLING (Targeted Induced Local Lesions in Genomes) techniques. Prior to this, mutation breeding was solely based on forward genetics, where the main approach involved phenotype screening to identify desirable traits in mutant populations. However, TILLING brought forward reverse genetics by allowing the identification of mutations in particular, well-characterized genes. The advent of TILLING has coincided with wider advancements in molecular knowledge of the genetic foundations of crop traits and the development of efficient, next-generation DNA sequencing technologies. In essence, this technology allows for the identification of mutations in any chosen gene within a crop plant's genome, provided the gene's DNA sequence is known and a suitable mutant population exists. Other directed mutagenesis tools include CRISPR–Cas9, TALEN, Mutmap and Mutmap+. Clustered Regularly Interspersed Short Palindromic Repeats (CRISPR) and CRISPR-associated protein 9 (Cas9) offer substantial potential to enhance and fine-tune various traits, such as quality, yield, disease resistance, resilience to environmental stresses and nutritional improvements in different crops. CRISPR–Cas9 allows precise gene editing by inserting, removing or modifying specific DNA sequences. Its simplicity and high efficiency in inducing mutations have made it an invaluable tool for generating new germplasm resources. The capacity for precise gene editing with CRISPR–Cas9 is made possible by access to whole-genome sequencing data and a deep knowledge of gene functions associated with important traits. Transcription Activator-Like Effector Nuclease (TALEN) is an effective genome editing tool employed to introduce beneficial traits into crops. Initially, TALENs gained recognition for their capability to bind specific DNA sequences and enhance gene expression, which could result in hypertrophy (abnormal cell growth), helping plants endure and defend against pathogens. The creation of TALENs involves fusing the FokI restriction endonuclease domain (FokI-R) with a specially engineered TALEN DNA-binding domain (DBD). To achieve precise targeting, two TALEN subunits bind to the target DNA sequence in opposite orientations. TALEN has been employed to improve blight resistance in rice by targeting and inducing mutations in the OsSWEET14 gene. In soybean, TALENs were used to target the FAD2 gene specifically, aiming to improve oil quality by modifying its expression (Haun et al. 2014). MutMap and MutMap+ are methods employed to pinpoint the genomic regions that have undergone changes within a mutant population. In the MutMap approach, a homozygous M3 mutant is crossed with the parent, and the F_2 mutant population generated is sequenced and compared against the reference sequence. MutMap+ is an enhanced version of MutMap, where the M3 mutant bulk is compared with both the M3 parent bulk and a reference sequence for more accurate analysis. SNP variations are identified in both methods and used for further genetic improvements. Both techniques can be used for traits controlled by monogenic recessive genes and have proven useful in a range of crops, including maize, soybean, cucumber and others. The MutMap method detects nucleotide changes induced by both physical and chemical mutagens.

Space breeding represents an innovative intersection of space technology and agriculture. This advanced science harnesses the distinct conditions of the space environment, including particle radiation, microgravity, weak magnetism and high vacuum, to stimulate genetic enhancements in agricultural organisms. By taking advantage of these extraordinary space conditions, which are difficult to replicate on Earth, space breeding generates unique variations in mutagenic materials and leads to the creation of new germplasm resources. The main factors driving plant mutations in space include high-intensity ionizing radiation, microgravity, cosmic magnetic fields (galactic dynamo), ultra-vacuum and other physical stresses.

3 Application of Mutation Breeding Enhancing Food and Nutritional Quality

3.1 Cereals

Mutation breeding has significantly contributed to cereal crop improvement by introducing valuable genetic diversity, particularly targeting higher yield potential, enhanced resilience against environmental stresses, improved grain quality and essential agronomic features. Employing physical and chemical mutagenic agents, especially gamma radiation and ethyl methane sulphonate (EMS), has been instrumental in creating superior cereal varieties possessing beneficial traits such as shorter plant stature, adjusted maturity timing, seed shatter resistance and enhanced disease resistance. This approach predominantly targets the genetic modification of well-established varieties through the alteration of one or two key traits, leading to improvements in yield parameters and quality attributes, including starch composition, malting properties and nutritional value.

In rice, mutation breeding has made substantial contributions to the development of cultivars exhibiting high productivity and enhanced tolerance to environmental stresses. The creation of semi-dwarf genotypes marked a major advancement, offering increased lodging resistance and enabling the application of higher nitrogen levels. One of the most prominent rice mutants, 'Calrose 76', was released in the USA, whereas 'Basmati 370' in Pakistan contributed to enhanced production. In India, a series of rice mutants developed through gamma radiation under the 'PNR' designation exhibited traits such as high yielding, reduced plant stature and early maturity. China's 'Zhefu 802', a widely cultivated mutant variety, expanded to more than 10.6 million ha in a decade. Thailand's aromatic Indica mutants, 'RD6' and 'RD15', developed via gamma irradiation, were extensively cultivated, with the latter covering 0.2 million ha.

A landmark advance in rice mutation breeding was the generation of thermosensitive genic male-sterile (TGMS) lines in both Japonica and Indica backgrounds, a breakthrough that enabled efficient hybrid-seed production via the streamlined two-line system. Similar mutants were created via gamma radiation in the Indica variety '26 Zhaizao' in China. Globally, 876 mutant rice varieties have been developed, including 64 from India. Bangladesh has released 8 rice mutants, including 'Binasail' and 'Binadhan-6', now cultivated on 795,000 ha. In Vietnam, over 55 mutant varieties represent about 15% of the national rice production, covering over 1 million ha (Ahloowalia et al. 2004). In Pakistan, the Nuclear Institute for Agriculture and Biology developed high-performing varieties such as 'Shada', yielding 7 t/ha on 60,000 ha and generating USD 21 million, and 'Shua-92', with yields of 8.5 t/ha across 160,000 ha, contributing USD 223 million (Bughio et al. 2007).

Gamma irradiation continues to be the most extensively employed physical mutagen in rice, accounting for 92% of all induced mutants (FAO/IAEA 2022). It generates a wide spectrum of mutations, ranging from small nucleotide deletions (1–16 bp) to large deletions (up to 129.7 kb) and significant chromosomal rearrangements, including inversions spanning 1.2–3.2 Mb (Morita et al. 2009). Reported mutation frequencies vary from 7.5×10^{-6} to 9.8×10^{-6} (Li et al. 2016). Newer methods such as ion beam radiation (IBR) and fast-neutron irradiation have broadened the scope of rice mutagenesis. Fast neutrons are capable of inducing large-scale genome alterations, impacting 58% of annotated rice genes (Li et al. 2017).

In maize, mutation breeding has relied on agents such as gamma rays, EMS and sodium azide to generate variability in traits including plant architecture, kernel quality and stress adaptability. Studies by Khawaret al. (2010) demonstrated that increasing gamma doses decreased shoot and root lengths, with complete inhibition of germination beyond 2 kGy. Conversely. Marcu et al. (2013) also reported dose-dependent declines in seed vigour and germination rates. While 92 mutant maize varieties have been released globally, India has not yet officially registered any. In maize embryogenic calli, combined treatment with gamma rays and sodium azide has yielded drought-tolerant mutants. EMS-based mutagenesis has facilitated large-scale mutant libraries using exome capture and next-generation sequencing, identifying over 195,000 CG > TA substitutions and more than 4,600 InDels across 32,069 genes (Lu et al. 2018). An EMS-mutagenized mutant panel in tropical maize comprising 2,050 independent M_2 families was developed to support gene discovery (Tran et al. 2020).

For wheat, sodium azide has proven to be an effective mutagen. Dubey et al. (2017) reported mutants with improved yield and enhanced disease resistance. Out of 292 registered wheat mutants worldwide, only 4 originate from India. Gamma irradiation has aided drought tolerance enhancement in Kenyan wheat, where exposure to 150–250 Gy led to the release of the cultivar Njoro BW1 (Njau et al. 2005). Chromosomal translocations induced through radiation successfully introduced resistance to the Hessian fly (*Mayetiola destructor*) from rye into wheat (Friebe et al. 1991). Further, gamma irradiation was employed to confer resistance to Pucciniagraminis f. sp. tritici race Ug99 in Kenyan wheat, resulting in the varieties EldoNgano 1 and EldoMavuno 1.

In barley, mutation breeding has substantially contributed to improvements in productivity, lodging resistance and quality for malting purposes. The FAO/IAEA records show 314 mutant barley varieties released globally, with 13 from India. The first barley mutant, 'Pallas' (1960), derived from X-ray treatment of 'Bonus', showed superior lodging resistance and yield. 'Diamant', released in Czechoslovakia in 1965, was 15 cm shorter and 12% more productive, covering 2.86 million ha in Europe by 1972. In the UK, 'Golden Promise' (1966) emerged as a key model for transformation studies. Mutants like 'Valticky', which carried the sdw1 dwarfing gene induced by X-rays in 1956, now dominate spring barley cultivation in Western Europe, accounting for more than 90% of the area (Mlčochová et al. 2004). Barley lines deficient in anthocyanins and proanthocyanidins, developed using sodium azide, have contributed to biochemical trait enhancements (Olsen et al. 1993).

Overall, mutation breeding has emerged as a powerful complementary tool to conventional and molecular breeding strategies for cereal improvement. The widespread adoption of mutagenic agents such as gamma radiation, EMS and advanced physical mutagens has led to the release of over 1,500 cereal mutant varieties worldwide, including 87 from India. These advances have played a critical role in developing cultivars that are high yielding, stress-resilient and nutritionally enriched, thereby contributing meaningfully to global food security efforts. Mutant varieties of cereals developed in India is tabulated in Table 1.

3.2 Pulses

Mutation breeding represents an essential strategy for genetically refining pulse crops, securing notable progress in yield potential, disease resistance, earliness and grain quality. Both

Table 1 Mutant varieties of cereals developed in India

Crop	Variety name	Year of release	Description
Rice	Trombay Chhattisgarh Sonagathi Mutant (TCSM)	2022	A mutant rice variety derived through induced mutagenesis techniques at Trombay, specifically targeting the local cultivar Sonagathi
	Trombay Chhattisgarh Vishnubhog Mutant (TCVM)	2022	This variety originated from induced mutation in the popular rice cultivar Vishnubhog, resulting in desirable agronomic enhancements
	CG Jawaphool Trombay	2021	Developed from the traditional Jawaphool variety by exposing seeds to gamma radiation, leading to improved traits
	Vikram-TCR (Trombay Chhattisgarh Rice)	2021	A mutant line obtained from the non- mutant variety Safri 17 through gamma irradiation, aimed at enhancing yield and adaptability
	CRM 2007-1	2006	Developed by subjecting seeds to gamma radiation at 100, 150 and 200 Gy. Exhibited significant yield improvement while maintaining basmati grain quality
	Anashwara	2006	A photoperiod-sensitive, semi-tall rice variety derived as a mutant of PTB 20, ideally suited for cultivation in the Rabi season
	Shivam	2006	Resulted from hybridization involving mutant CR 314-5-10 (an open-floret mutant); recommended for irrigated conditions in eastern Indian states
Emmer wheat	HW1098	2012	A mutant genotype developed by irradiating the tall traditional wheat cultivar NP201 with gamma rays at doses of 200, 250 and 300 Gy at BARC, Mumbai
	COW2	2010	A dicoccum wheat variety evolved from NP200 through gamma irradiation at 100–400 Gy; the selected mutant line HW1095 was released for cultivation in Tamil Nadu

Data are collected from FAO/IAEA-MVD (2022). Available online: https://nucleus.iaea.org/sites/mvd

physical agents (e.g. gamma rays) and chemical mutagens such as ethyl methanesulphonate (EMS) and sodium azide have been harnessed to create advantageous allelic changes across multiple pulses. These efforts have produced cultivars with higher productivity, greater stress tolerance and superior nutritional profiles, thereby bolstering food security and advancing sustainable agriculture. India alone has released 56 mutant pulse varieties to date.

In chickpea, mutation breeding has enabled the release of numerous high-yielding, disease-resistant lines. At the Indian Agricultural Research Institute (IARI), gamma-irradiated mutants such as Pusa-408, Pusa-413 and Pusa-417, combine early maturity with Fusarium wilt resistance and superior yields. Jawaharlal Nehru Krishi Vishwavidyalaya (JNKVV) produced drought-tolerant selections such as JG-62 and JG-74 using 200 Gy gamma rays (Wani et al. 2023). EMS mutagenesis at the Rajasthan Agricultural Research Institute yielded GNG-1958 and GNG-1969, noted for their larger seeds and elevated protein. Despite these successes, India accounts for just 6 of the 27 mutant chickpea varieties registered worldwide, suggesting significant untapped potential.

Likewise, mutation breeding has greatly advanced pigeonpea improvement, especially for productivity and wilt resistance. IARI employed 200–400 Gy gamma rays to develop short-duration, high-yielding cultivars including T-21, AL-15, IPA-204 and Pusa-992. At Mahatma Phule Krishi Vidyapeeth (MPKV), EMS-derived mutants BDN-2 and BDN-3 show earlier flowering, reduced height and higher yields—valuable traits for ongoing breeding programmes.

For mung bean, mutagenesis has generated a suite of resistant, high-yielding varieties. India's gamma-ray lines CO-4, Pant Mung-2 and TAP-7 are widely cultivated, while the Bhabha Atomic Research Centre (BARC) released powdery-mildew-resistant mutants TARM-1, TARM-2 and TARM-18. Other key examples include TMB-37, which matures early and resists yellow-mosaic disease, and TJM-3, tolerant to powdery mildew, yellow-mosaic virus and Rhizoctonia root rot. TM 96-2, bred for rice-fallow systems, synchronizes maturity and withstands powdery mildew. Mutant cultivars occupy roughly 10% of India's mung bean area; in Bangladesh, Binamoog-5 covers about 15,000 ha (Bangladesh Institute of Nuclear Agriculture, Mymensingh).

Gamma-induced mutants CO-4 and CO-5 from Tamil Nadu Agricultural University have upgraded black gram (urd bean) performance through higher yield and Cercospora-leaf-spot resistance. EMS mutants LBG-623 and LBG-685, developed at Acharya N.G. Ranga Agricultural University (ANGRAU), combine earliness with boosted biomass. TAU-1 now dominates ~500,000 ha in Maharashtra, contributing about ₹300 crores annually, while TU-94-2 displays broad adaptation across central and southern India.

In lentil, mutagenesis has delivered lines with better disease resistance and agronomic traits. Punjab Agricultural University's gamma-ray selections PL-4 and PL-7 resist Fusarium wilt and bear bold seeds. Sodium azide mutants IPL-81 and IPL-316 from the Indian Institute of Pulses Research exhibit short duration and high protein. EMS work at the International Center for Agricultural Research in the Dry Areas (ICARDA) produced populations resistant to pod shatter, tolerant to herbicides and able to withstand Orobanche (Kumar et al. 2015). Spontaneous variants of DPL-62 have added diversity in earliness, seed-coat colour and seed size (Punia et al. 2014).

Overall, mutation breeding continues to supply pulse crops with higher yields, stronger resilience and better nutritional quality, thereby reinforcing global food security. Ongoing research that couples classical mutagenesis with modern genomic tools is poised to deliver even greater advances for sustainable pulse production. Mutant varieties of pulses developed in India is tabulated in Table 2.

3.3 Oilseed Crops

India is ranked fourth in terms of production of oilseeds next to the USA, China and Brazil. The most important oilseeds are soybean, sesame, groundnut, rapeseed-mustard, sunflower, castor, safflower and linseed.

3.3.1 Soybean

Soybean botanically named *Glycine max* L. is ranked first in vegetable oil production in the world and is widely grown in the Brazil, the United States, China, Argentina and India. In India, soybean is generally sown as a kharif season crop with an area of 11.50 million hectares and production of 10.50 million tonnes which is about 25% to the total domestic edible oil pool. Mutants with desirable characters like morphological traits, low Trypsin inhibitor content (TI), high oil content percentage, low phytic acid content (lpa), high harvest index, altered protein profile, altered fatty acid content and high seed yield have been identified and characterized and also used as a parent in hybridization programme to develop new varieties and high-yielding recombinant lines which are evaluated under the state or national level field trials. Soybean seed contains 38–40% protein along with 18–23% oil, despite of its excellent nutritional quality, soybean proteins are often associated with some toxic compounds, called the anti-nutritional factors (flatulence factors, goitrogens, antivitamins, protease inhibitors, lectins, oestrogens, lysinoalanine, allergens, saponins, tannins phytate) which are harmful to the animal body. Hence, cultivars with high-yield potential along with minimum anti-nutritional factors will be the foremost breeding goal for export and domestic use. TAMS-38 and TAMS-98-21 are two important soybean cultivars has been released for commercial purpose.

Table 2 Mutant varieties of pulses developed in India

Crop	Variety name	Year of release	Description
Mungbean	TMB-37	2005	A short-duration, high-yielding mung-bean line resistant to yellow-mosaic virus and producing large seeds; released for commercial cultivation in eastern Uttar Pradesh, Bihar, Jharkhand, Assam and West Bengal
	TJM-3	2007	Developed through a cross that included the gamma-irradiated mutant line TARM-1. It performs reliably in both kharif and summer seasons and has been released for cultivation in Madhya Pradesh
	TM-96-2	2007	A hybrid derivative involving mutant line TARM-2, developed through gamma radiation. Specifically recommended for rice-fallow cropping in Andhra Pradesh
	TM-2000-2	2010	The mutant line is highly resistant to powdery mildew and is expressly suited to rice-fallow agro-ecosystems
Black gram	DU-1	2007	Originated from a cross between TAU-1 (a mutant derivative) and No. 169, followed by gamma irradiation (200 Gy) of F1 hybrid seeds. It possesses high grain yield potential and enhanced insect resistance
Soybean	TAMS-98-21	2007	Developed via seed irradiation at 250 Gy, this mutant outperformed the check variety JS-335 with a 21% yield increase (2243 kg/ha vs. 1853 kg/ha) in multilocation trials across Maharashtra
Chickpea	Pusa-547	2006	Derived through exposure to 600 Gy of gamma radiation. Displays notably high yield, excellent cooking quality and broad tolerance to Fusarium wilt, stunt virus and root rot
	BGM-547	2005	Developed via gamma-ray mutagenesis, this cultivar possesses bold, golden-brown grains, delivers a high yield and offers moderate resistance to wilt, root rot, Stunt disease and the pod-borer Helicoverpa armigera
Cowpea	TC-901	2018	TC-901 originated from the exotic, virus-resistant accession EC394763 following exposure to 250 Gy of gamma irradiation. The line delivers roughly a 15% yield boost compared with the national check RC-101 and was cleared for commercial cultivation in 2018
	TRC-77–4 (Kalleshwari)	2007	This mutant selection demonstrates enhanced yield potential over existing varieties
Pigeon pea	TT-401	2007	A high-yielding mutant variety displaying moderate resistance to pod borer and pod fly, offering improved protection against biotic stress

Data are collected from FAO/IAEA-MVD (2022). Available online: https://nucleus.iaea.org/sites/mvd

3.3.2 Groundnut

Mutation breeding has key role in peanut or *Arachis hypogaea* L. to enhance yield potential, tolerance to several abiotic stresses, disease resistance and different quality parameters, have led to release of numerous improved high-yielding and stress-resilient cultivars worldwide. The FAO/IAEA Mutant Variety Database registered 79 mutant varieties of peanut, with 26 developed by India. More or less 80% of mutant varieties have been developed through gamma irradiation, with the optimal mutagenic dose ranging from 100 to 450 Gy for both X-rays and gamma rays, depending on genotype and physiological status. In China, mutation breeding has contributed to approximately 20% of the total groundnut cultivation area, significantly improving national production and farmer livelihoods. In India, Trombay groundnut (TG) mutant cultivars, including TG-26, TAG-24, TG-37A, TKG-19A and TPG-41, have had a major impact on groundnut productivity, particularly in Gujarat, Andhra Pradesh, Karnataka, Maharashtra, Odisha and Rajasthan. Argentina has successfully utilized X-ray-induced mutation breeding to develop Colorado Irradiado, which occupied over 80% (~280,000 ha) of the national groundnut cultivation area in the 1970s, significantly boosting production and export potential. In Sudan, the drought-tolerant mutant Barberton-B-303 demonstrated superior performance in multilocation trials, making it suitable for arid and semi-arid regions.

Mutation breeding has significantly contributed to yield enhancement, as observed in high-yielding mutant varieties such as TAG-24 and TG-26, which exhibit superior pod and kernel yield. Drought tolerance has been achieved in cultivars like Barberton-B-303, while several mutant varieties exhibit enhanced resistance to fungal and bacterial pathogens. Improvement in oil quality has been another major achievement, with the development of varieties containing higher oleic acid content and improved shelf life. Chemical mutagenesis, particularly EMS, has played a key role in genetically improved peanut varieties. The EMS-induced mutant Huayu 40, derived from Huayu 16 after injecting 0.3% EMS into its flower, exhibits a faster growth rate, erect with sequential branching along with darker green foliage. Additionally, EMS mutagenesis has been employed for the development of high-oleic peanut lines and large-seeded breeding lines in the Georgia Browne cultivar.

Table 3 Mutant varieties of groundnut developed in India

Mutant variety	Registration year	Short description
GPBD-5	2010	Spanish bunch variety developed by hybridization between TG-49 and GPBD-4. TG-49, a mutant derivative has showed resistance against late leaf spot along with rust diseases
TDG-39	2009	Mutant variety with large seeds, matured within 120 days along with high-oleic acid content
TG-39	2008	Confectionary large-seeded cultivar suitable for arid as well as semi-arid zones of Rajasthan in rainy (June–October) season. Characters like semi-erectness with optimum desirable height having alternate flowering (ssp. hypogaea var. hypogaea; Virginia bunch) and quality parameters like 12.6% carbohydrate, 50% oil content, 26.5% protein and 4.5% sucrose are ideal for consumption. Its oil contains 23% linoleic acid and 59% oleic acid
TG-51	2008	Ideal for Rabi/summer season (October–May) mainly in the area West Bengal, Orissa and Bihar. Seed contains 49% oil (43% oleic acid and 37% linoleic acid)
TLG-45	2007	Cross between TG-19 and TAG-24 and developed through pedigree method at Bhabha Atomic Research Centre (BARC), Mumbai. TG-19 is a cross between TG-1 (direct X-ray induced mutant of Spanish Improved) and TG-17 (a mutant which is a cross of dark green leaf mutant with TG-1 of Spanish Improved)
TG-38	2006	Developed through treating seeds with gamma rays (300 Gy) having erect growth with sequential branching, dark green leaflets, semi-dwarf height, compact pod setting along with smooth pod surface and exhibits resistance to stem rot and dry root rot
TG-37A	2004	It is a hybridization-based mutant variety of TG-26, i.e. gamma ray (300 Gy)-induced mutant. Main improved traits are compact pod, semi-dwarfness, high yield
TPG-41	2004	It was obtained by hybridization between TG-22 obtained by treating seeds with X-rays (150 Gy) and TG-28A
Mutant 28–2	2003	It is a chemical mutagen induced mutant (0.5% EMS) having improved attributes such as resistance against late leaf spot, large seed, tolerance to Spodoptera litura and thrips

Data are collected from FAO/IAEA-MVD (2022). Available online: https://nucleus.iaea.org/sites/mvd

Advancements in molecular mutations have further improved oil quality traits in groundnut. The *ahFAD2A* gene mutation (G448A) leads to a missense mutation (D150N), resulting in a higher oleic/linoleic acid ratio. Naturally occurring mutant lines PI342666 and PI342664 contain 80% oleic acid due to mutations in *ahFAD2A* (G448A) and *ahFAD2B* (C301G). Furthermore, the miniature inverted-repeat transposable elements (MITE), has been introgressed in the background *ahFAD2B*, has been identified in mutant lines. Mutant varieties of groundnut developed in India is tabulated in Table 3.

3.3.3 Linseed

Linseed (*Linum usitatissimum* Linn.) is an important rabi oilseed crops cultivated over an area of 0.29 million hectares with a production of around 0.14 million tonnes with productivity of 525 kg/ha which is very low as compared to the world average due to lack of high-yielding varieties.

The early maturing mutant TL-145 (7.6 gm) showed significantly higher seed yield per plant as against the parent NL-97 (4.5 gm). TL-145 was evaluated in ICAR trials and it gave yield of 1422 kg/ha as compared to best check Divya (1289 kg/ha) and is promoted to Advanced varietal trials. NL-97 and NL-264 are two linseed genotypes irradiated with gamma rays (600 Gy) for improving yield and different biochemical traits for oil quality. The early maturing mutant line TL-145 was identified with maximum yield (1422 kg/ha) as evaluated in ICAR trials was promoted to Advanced varietal trials (AVT). Low linolenic acid mutants in linseed are the prerequisite for an efficient breeding programme so that a modified and rapid spot test technique was developed at BARC for determining the linolenic acid content. The line TL-99 having low linolenic acid (2%) was evaluated in advanced varietal trial was promoted for one more year testing in ICAR trials.

3.3.4 Mustard

Brassica juncea called as Indian mustard is most important oilseeds crops in India. Induced mutagenesis is a practice used to widen and utilize the desired allele combinations for effective crop improvement to enhance the production and productivity of crop plants and also for the improvement of qualitative and quantitative traits in oleiferous Brassica species. A large spectrum of chlorophyll mutations induced by ethyl methanesulphonate in *B. napus*, chlorophyll deficient mutations in *B. campestris* and chlorophyll-reduced (Cr) mutations treating with fast neutrons and diethyl sulphonate (DES) in *B. napus* were isolated for the development hybrids and superior varieties. Different mutants of chlorophyll variegation, viz., albina, tigrina, maculata, xantha, viridis, variegated, xantho-viridis, virido-xantha, albo- xantha, xantho-alba and virido-alba were obtained with EMS treatment in Brassica juncea (Gupta et al. 2012). However, there is no report whether these mutations are useful to develop high-yielding varieties or not. A mutant, namely, NDF-1, was obtained from DH line 3529 by treating with chemical inducers along with particle bombardment of fast neutrons in *B. napus*. Recently, early maturing dwarf yellow seeded mutants with low erucic acid content variety varuna was isolated by irradiating with gamma rays in *B. juncea.* Tian et al. in the year 2014 was screened out an inbred line Y1127 with low (3.8%) linolenic acid (C18:3) content. Inheritance studies detected two fatty acid desaturase3 (FAD3) gene loci controlling the variation of C18:3 content. Two mutants RLM-198 having 25% higher yield along with higher oil content, moderate resistance against aphids and leaf miners and RLM-214 having 22% more yield along with no shattering characteristic were developed by physical mutagens gamma-rays and X-rays, respectively. M-11, a low linolenic acid mutant was backcrossed with a cultivar, namely, Regent to develop low linolenic acid content along with high-yield potential cultivars Stellar and Apollo.

3.3.5 Sunflower

Sunflower or *Helianthus annuus* L., belongs to family Asteraceae, is the fourth most important oil producing crop among the oilseeds and has originated from North America having genome size of 3000 Mbp. The sunflower variety Pervenets (oleic acid content 75%) was isolated by treating with the chemical mutagen DMS at the All-Union Research Institute of Oil Crops in the former USSR. Genes governing resistance against AHAS-inhibiting (Aceto-hydroxy acid synthase) herbicides in sunflower have been transferred from resistant wild genotypes (ANN-PUR and ANN-KAN) into elite inbreds for the development herbicide resistant cultivars and hybrids (Miller and Al-Khatib, 2000; 2004). EMS concentration of 70 mM (0.87% EMS) was required to isolate mutagenic sunflower line in CAS-3. Two mutants, viz., ML31-12, ML31-12 exhibited high seed yield as compared to the wild sunflower genotype. The oil content in Sunflower closely linked to its seed yield is the main desire of sunflower breeding. Higher oil content was observed in the mutant line ML-31-13 since no drastic mutation has been identified for oil content in sunflower. 200 Gy gamma-ray treatment is more desirable in sunflower for obtaining highly resistant plants against Alternaria leaf blight disease reaction (Sujatha et al. 2019).

3.3.6 Sesame

Sesame (*Sesamum indicum* L.), the 'Queen of oilseeds' is grown worldwide for its nutritional and medicinal value, but low productivity is the major constraints for its cultivation. The mutagens used for irradiating the sesame seed are generally gamma rays, EMS and sodium azide. Sesame seeds can be irradiated with high dose of mutagens as it performs well even after the exposure of high value dose of mutagens, and there are genotypic differences in the sensitivity to mutagenic treatment between different cultivars. The value of gamma irradiation dose for sesame has generally been persisted between 250 and 350 Gy showed most effective against most of the traits to induce mutagenic variation. Sodium azide is a potent mutagen used to develop nine sesame varieties, viz., TMV-3, CO-1, VRI-1, TMV-4, TMV-5, TMV-6, VRI (Sv)-2, TMV (Sv)-7 and Paiyur-1, among these mutants, TMV-5 was isolated as extra early maturing mutant that completely matured within 70 DAS. TVS-1401 was identified as extra early maturing mutant line confirmed with the dissected fatty acid profile and its distinctiveness in molecular level was also confirmed through RAPD markers, viz., $OPC03_{550}$ and $OPA04_{550}$ (Jayaramachandran et al. 2020).

3.3.7 Safflower

Safflower (*Carthamus tinctorius* L.) is a neglected hardy oilseed crop but very much suited for tropical and dry regions like Indian continent. Moreover, the crop can produce good quality oil which is good for taste, cooking and more importantly for our health with some limitations like spininess, low seed yield and low oil content. Two varieties of safflower, viz., AKS-207 and Bhima were treated with some mutagens such as sodium azide, ethyl methanesulphonate and gamma rays. The selection of traits like early bolting, dwarf, high test weight, thin hull, highly branched, large head, high oil content and high-oleic acid for the genetic improvement of safflower, sodium azide has been proved to be most effective for the creation of variation (Rampure et al. 2017).

3.4 Fibre Crops

3.4.1 Cotton

Physical mutagens such as gamma rays, X-rays and chemical mutagens such as alkylating agents (e.g. Ethyl Methanesulphonate, Sodium Azide) have been widely used to isolate desired mutants having the characteristics like heat resistance, early maturity and high yielding can be used to develop targeted cotton varieties. The CRISPR/Cas9 is an effective gene editing technique for targeted mutagenesis with high efficiency in cotton breeding so that the useful traits of wild mutated cotton can be introduced into cultivated cotton through backcrossing and marker-assisted breeding. In cotton research and breeding, notable works have been achieved through EMS-induced mutant libraries of different cotton species such as *G. barbadense*, *G. arboreum*, *G. herbaceum* and *G. hirsutum.* Mutagenesis with gamma-ray irradiation has played a key role for the development of heat resistance and early maturing cotton variety 'Lumian1' has been known for its consistent yield production and the variety was isolated from the X-ray-treated hybrid progenies derived from crossing between 'Zhongmian 2' and '1195' lines. TM-1 (Texas Marker-1) has been derived from linear electron acceleration-based radiation mutagenesis to develop TM-1 as stable cotton inbred line.

3.4.2 Jute

Jute (*Corchorus* spp.), known as 'golden fibre', is the main source of bast fibre with almost 80% of the world fibre production and among these Asia produced 90% of world production with commercial trade cost of 2.3 billion US dollar. Among the species of Corchorus, most popular and economically viable species are *C. capsularis* L. and *C. olitorius* L. cultivated commercially to gain profits. Chemical mutagenesis with EMS (Ethyl Methanesulphonate) of Capsularis jute seed has resulted a novel mutant, namely, Komola showed a shorter flowering time (70 days) than its ancestors (120 days) and the histochemical analysis have showed lower content of lignin along with photo insensitive pattern which can be helpful for future breeding programme for the development of year-round cultivable jute. Physical mutagens like gamma and X-rays result random mutation in the genome may produce abnormalities rather than desired traits. Capsularis jute having lower lignin content along with deficient secondary phloem fibre and Olitorius jute with lower lignin content (7.23%) in comparison with wild genotype (13.7%) showed multiple abnormalities due to mutagen treatment. EMS mutagen treated *C. capsularis* advance line namely CBS-718 has possessed snow-white fibre without fibre bleaching|| for fine yarn production (Hossen et al. 2022).

3.4.3 Sunhemp

Crotalaria juncea L. ($2n = 16$) belongs to family Fabaceae, known as sunnhemp, is a multipurpose crop species due to its dual usefulness such as fibre and forage production can also be used as green manure crop along with biofuel production due to its high heating value. It has a natural metal hyper-accumulator characteristics in the roots and leaves along with promising potential of phytoremediation. Ionizing radiation-induced hormesis and its cytological evaluation of microsporogenesis in sunnhemp and different morphometric traits, viz., seedling height, seed yield, plant height, lateral branches, number of pods and biomass were analysed and showed the range of 100–300 Gy is most useful to induce hormesis (Kumar and Dwivedi 2016).

3.5 Vegetable Crops

3.5.1 Tomato

The development of HYV with enhanced quality and stress is difficult because narrow genetic base present in the cultivated tomatoes. So, mutation breeding offers one of the best alternatives to extend the genetic diversity. To overcome the problem, wild species like *Solanum pimpinellifolium* is generally used for investigation. Some physical and chemical mutagens like gamma rays, EMS (ethyl methanesulphonate) have been utilized for mutation induction in tomato. Globally, extensive genetic resources of tomato mutant lines have been developed utilizing EMS, gamma-ray and fast-neutron mutagenesis. Publicly accessible resources, including TOMATOMA for Micro-Tom, 'Genes that make tomatoes' for M82, and LycoTILL for Red Setter, provide a diverse array of tomato mutants (Chaudhary et al., 2019). The traditional mutagenesis techniques induce random mutations in the genome and produce several non-target mutations, so it is very much problematic to get the desired one. Recent advancements in mutagenesis techniques merged with genetic engineering tools, enable highly specific modifications to target genes. Gene editing by CRISPR/Cas9 and TALENs are the most commonly used approach of targeted mutagenesis due to the availability of genomic sequences of *S. lycopersicum* and its wild relative *S. pimpinellifolium* (Brooks et al. 2014). EMS is the most extensively used chemical mutagen in plants, because of its exceptional efficacy in inducing point mutations and deletions within chromosomal segments. Through EMS mutagenesis, tomato lines have been developed with enhanced resistance to potyvirus which is a highly destructive and widespread viral pathogen. Breeders have utilized the *rin* (ripening inhibitor) mutation to minimize fruit softening, resulting in firmer fruits with delayed ripening. But, this mutation also leads to a decrease in nutritional value, colour and flavour. Cobalt-60 is used to generate tomato mutants with doses of 20, 40 and 60 Gy (Hastari et al. 2014) and resulting in several morphological and genetic changes in the characteristics of tomato. The accretion of lycopene in tomato is attained by high carotenoid synthesis at the time of ripening. Two mutant gene of tomato *hp-1* and *hp-2* have been reported for carotenoid biosynthesis and used in carotene rich tomato hybrid development (Giovannoni 2004). Mutant varieties of tomato developed in India is tabulated in Table 4.

3.5.2 Brinjal

One of the major obstacles to develop improved varieties in brinjal is the narrow genetic diversity of cultivars despite its wide morphological variation. Mutagenesis yielded promising varieties in brinjal, displaying a wide range of agronomic characteristics, including variations in plant height, flower morphology, fruit shape-size, anthocyanin content and insect resistance. These materials offer valuable resources for breeding unique brinjal varieties and elucidating gene function. Brinjal is a good reservoir of phenolic compounds such as chlorogenic acid. This chlorogenic acid was found to be the major phenolic with a range of 50–70% of total phenolics in brinjal. It has wide range of potential health benefits including health promoting activities like anti-diabetic effect, cardio protective, anti-carcinogenic, anti-oxidant and anti-inflammatory. Positive influence of the He–Ne laser was found on cholorogenic acid content of Brinjal. Seeds of the Surya cultivar were treated with EMS to generate genetic

Table 4 Mutant varieties of tomato developed in India

Tomato cultivar	Mutagen	Dose	No. of mutants	Remarks
Moneymaker	EMS	60 mM	–	Isolated *Adh1* null mutant of tomato
M82	EMS	0.5%	2552	Used in functional genomic studies
M82	Fast neutron	15 Gy	865	Used in functional genomic studies
L. esculentum Mill	EMS	1.5%	16	Used in resistance against Orobanche
Micro-tom	Gamma ray	300 Gy	6347	Used in functional genomic studies
Micro-tom	EMS	1%	–	In forward and reverse genetic studies
Red setter	EMS	1%	4500	To develop Red setter TILLING platform
S-12 (Pusa S-12)	X-ray	–	–	Sioux variety is exposed to mutation
Pusalaalmeeruti	Gamma ray	–	–	Obtained from improved meeruti

Choudhary et al. (2017)

variation and the treated seeds produced plants with variations in leaf, flower, fruit and plant height (Subramaniam and Kumar 2023). PKM-1 is an induced mutant (gamma rays) of a local cultivars called PuzhuthiKathiri and developed at TNAU, Periakulam. This variety exhibits exceptional drought tolerance and can withstand prolonged transportation. It yields an average of 34.75 tonnes per hectare, maturing in 150–155 days and weighs between 45 and 65 g.

3.5.3 Chilli

Mutagenesis is considered one of the most effective approaches for inducing morphological and genetic variability in chilli, particularly due to the crop's naturally limited genetic diversity. Its widespread use is attributed to its simplicity, cost-effectiveness, technical feasibility and broad applicability across various plant species. Chilli seeds were treated with the chemical mutagens EMS (ethyl methanesulphonate) and DES (diethyl sulphate). It was observed that higher concentrations of both EMS and DES negatively affected morphological and yield-related traits. Traits such as germination rate, plant height, number of primary and secondary branches, days to first flowering, fruit length and girth, total fruit count per plant, seeds per fruit, seed weight per fruit, 100-seed weight and pericarp thickness exhibited variation under the influence of EMS and DES treatments. Mullainathan and Aruldoss in 2015 treated Chilli (*Capsicum annuum* L). var-K-1 on M2 generation by gamma rays and EMS. Consequently, it was found that treatment with 40 kR of gamma rays and 30 mM EMS led to an increase in chlorophyll, capsaicin, oleoresin, capsanthin and ascorbic acid content in chilli plants. The chilli variety MDU-1 was developed through mutation breeding and released for commercial cultivation in 1977 by the Agricultural College and Research Institute, Madurai. This variety originated from the gamma irradiation of seeds from the parent variety K-1. MDU-1 is characterized by a compact plant architecture, enhanced yield potential and elevated capsaicin content.

3.5.4 Okra

MDU-1 is developed in TNAU through gamma irradiation of seeds of PusaSawani. The plants are compact with close arrangements of nodes, five ridged fruits, light green in colour, medium length and productivity is 10–11t/ha. Punjab 8 (EMS-8) is an induced mutant from PusaSawani by treating seeds with EMS 1% and the variety is resistant to YVMV and also tolerant to fruit borer. Seeds of ParbhaniKranti are treated with EMS solution at different concentrations. In the M2 generation, phenotypic variability exhibited a broader range in both positive and negative directions. Under various mutagenic treatments, most traits showed substantial heritability and genetic advance, indicating effective induction of genetic variation. Three okra genotypes, ParbhaniKranti, HisarUnnat and Satdhari were subjected to treatments with gamma rays (15, 30, 45 and 60 kR) and EMS (0.25, 0.50, 0.75 and 1.00%). It was observed that higher doses of both gamma radiation and EMS had deleterious effects on seed germination, plant survival, seedling height and pollen and ovule fertility. Mutant varieties of vegetables developed in India is tabulated in Table 5.

3.6 Fruit Crops

Mutant varieties of fruits developed in India is tabulated in Table 6.

3.6.1 Mango

Mutations can be successfully done asexually. The techniques need to be achieved in mango to make mutation breeding more purposeful. Naturally happening useful mutants like Rosica (bud mutant) from the Peruvian variety Rosado de Ica has been isolated. Davis Haden is mutant variety of Haden. Bud sticks of Dashehari are irradiated with EMS and NMU and their behaviour is studied. It is observed that dormant buds of Langra to gamma radiations and bud graft

Table 5 Mutant varieties of vegetables developed in India

Sl No.	Crop	Variety	Mutant type
1	Okra	EMS-8	EMS treated mutant of Pusa Sawani
2	French bean	PusaParvati	X-ray mutant of wax pod
3	Tomato	Maruthan (Co-3)	Mutant of Co-1
		PKM-1	Mutant of Annanji
4	Chilli	MDU-1	Gamma ray mutant of K-1
5	Hyacinth bean	Co-10	Gamma ray mutant Of Co-6
6	Bitter gourd	MDU-1	Gamma ray mutant of MC-103
7	Palak	Jobner green	Spontaneous mutant from local collection
8	Pea	MUP-1	Gamma ray mutant of T-163
9	Sweet Potato	Port Blaco	Mutant from Port Rico having white flesh and high starch

Table 6 Mutant varieties of fruits developed in India

Crops	Mutant variety
Mango	Rosica from Rosado de Ica
Papaya	Pusa Nanha from local type
Grape	Marvel seedling from delight
Banana	Highgate from gross michel
	Motta Poovan from Poovan
	Grand Naine from Dwarf Cavendish
Grapefruits	Marsh from Thompson

to 3.0 kR of radiations produce heavier and larger fruits than normal. ArkaPuneet, a mutant hybrid is developed through ethyl methanesulphonate (EMS). The mutants have shorter inter-nodal length, longer leaves and thicker stems. Nekkare is produced by using gamma rays to create mutant populations of the poly-embryonic mango genotype. They have a variety of morphological changes, including reduced plant height.

3.6.2 Papaya

A notable example of mutation breeding in papaya is the development of the 'Sunrise Solo' variety, which exhibits resistance to the papaya ringspot virus (PRSV). This was achieved through induced mutagenesis using gamma radiation. PusaNanha is a popular variety of papaya developed by the IARI, New Delhi, through gamma-ray mutation. It is a compact, dwarf variety of papaya, making it ideal for small gardens, containers and high- density plantations. PusaNanha is known for its high yield, producing around 30–40 fruits per plant. The fruits are medium-sized, weighing around 1–1.5 kg, with a sweet and juicy pulp.

3.6.3 Apple

Mutation breeding can be used to develop apple varieties with improved resistance to diseases such as powdery mildew, scab and fire blight. To develop apple varieties with improved fruit quality characteristics such as flavour, texture and colour mutation techniques are widely used. The 'Starkrimson' apple variety was developed through mutation breeding from the popular Red Delicious variety. The apples were treated with gamma radiation to induce genetic mutations. Starkrimson apples have a deeper red colour, sweeter and more complex flavour profile than Red Delicious apples. Starkrimson apples also have improved resistance to certain diseases, such as powdery mildew. Gala apple variety was developed through mutation breeding and is known for its sweet flavour and crisp texture. Fuji is a mutant variety and is known for its longer shelf life.

3.6.4 Banana

Mutation can increase the genetic diversity of bananas, allowing for the development of new varieties with desirable traits. It also utilized to develop banana varieties with reduced pesticide use, making them more sustainable and environmentally friendly. Gamma radiation is widely utilized to induce mutations in bananas, while chemical mutagens like ethyl methanesulphonate (EMS) also serve as effective agents for mutation induction. FHIA-01 and FHIA-02, mutant variety of bananas are developed through mutation breeding and is known for its high yield, disease resistance and improved fruit quality. Gross Michel variety was developed through mutation breeding and was once a popular export banana, but it has largely been replaced by Cavendish due to its susceptibility to Panama disease. Cavendish is one of the most widely cultivated banana varieties worldwide.

3.6.5 Grapes

Mutation breeding has improved the quality of grape varieties. It has made the fruit more attractive to consumers and suitable for winemaking. Mutagenesis has led to grapes with more intense pigmentation, higher sugar content and better balance of acidity. Fleshless berry is a mutant variety having reduced weight of the pericarp at ripening without affecting fertility or seed size and number.

3.7 Ornamental Crops

Ornamental crops are plants grown primarily for their aesthetic value, including flowers, shrubs, trees and foliage plants that enhance gardens, parks and indoor spaces. These plants contribute significantly to landscaping, floriculture and the global ornamental plant industry, which supports economies through trade, tourism and horticultural businesses. Their visual appeal, fragrance and unique forms make them integral to urban beautification, environmental sustainability and psychological well-being. The ornamental plant business is growing due to urbanization, increased interest in home gardening and the demand for aesthetically pleasing and air-purifying plants in homes and workplaces. The industry is further boosted by floriculture demand, sustainable landscaping trends and the rising appeal of rare plants and innovative business models.

Mutations are the primary drivers of genetic variation, forming the basis of evolution and serving as a valuable tool for improving economically important traits in plants. While these genetic alterations naturally occur at a low frequency, they can be artificially induced using mutagens to accelerate desirable changes. Mutation breeding has emerged as a powerful tool for improving ornamental crops by inducing genetic variations that lead to desirable traits such as novel flower colours, improved shapes, enhanced fragrance, compact growth and increased resistance to diseases and pests. Unlike traditional breeding methods, which rely on natural genetic recombination, mutation breeding accelerates the development of new and unique varieties by exposing plant material to physical (e.g. gamma ray, X-ray, UV ray, alpha particle, thermal neutron) or chemical (e.g. ethyl methane sulphonate, diethyl sulphate, ethylene imine, methyl methanesulphonate) mutagens. This technique has been widely used to create commercially valuable cultivars in species like chrysanthemum, rose, tuberose, carnation, gladiolus, gerbera, bougainvillea and hibiscus.

Mutation breeding offers several advantages over conventional breeding for ornamental plants, primarily due to their unique reproductive characteristics. Vegetatively propagated ornamental plants, like chrysanthemum, lily, rose, typically exhibit low levels of genetic variability since they do not undergo sexual recombination. Mutation breeding introduces new and unique traits, such as novel flower colours, patterns and shapes, that would be difficult to achieve through conventional crossbreeding. Many ornamental plants like lilies, orchids are sterile, polyploid or have complex genetic structures that prevent successful crossbreeding. Mutation breeding bypasses this limitation by directly inducing genetic changes without the need for sexual reproduction. Conventional breeding leads to genetic recombination, which often result in the loss of desirable traits such as fragrance, flower size, shape, colour or growth habit. While, mutation breeding allows for targeted improvements retaining the original desirable characteristics of the parent plant. Mutation breeding is highly effective in modifying plant architecture, such as developing dwarf, compact or bushy growth forms that are ideal for potted plants, urban gardening and landscaping. Conventional breeding for these traits can be slow and unpredictable. Mutation breeding can be effectively combined with tissue culture techniques to produce large numbers of improved plants quickly.

3.7.1 Physical Mutagens in Ornamental Plants

Physical mutagens, such as gamma rays, X-rays, fast neutron, thermal neutron and UV radiation, are widely used to induce genetic variations in ornamental flowering plants. These mutagens create random mutations in plant DNA due to deletion or chromosomal breakage, leading to desirable traits such as novel flower colours, altered plant architecture, improved disease resistance and extended flowering duration. Gamma ray treatment is categorized into acute and chronic irradiation based on dose rate differences. Acute irradiation involves applying an optimal dose in a single exposure, whereas chronic irradiation uses a lower dose that is fractionated and delivered over extended periods, ranging from minutes to months. In general, chronic irradiation tends to result in a lower mutation frequency than delivering the same total dose in a single acute treatment. However, research by Nagatomi (2003) on chrysanthemums revealed that chronic gamma irradiation produced mutants with a broader range of flower colours, whereas acute irradiation resulted in a more limited variation in flower pigmentation. The optimum mutagen dose varies depending on plant species, cultivar and plant part used for mutation induction. Studies have shown that moderate to high doses of gamma rays (1.5–2.5 kR) are effective in generating stable mutants. Research on *Chrysanthemum morifolium* revealed that gamma irradiation (100–250 Gy) caused a reduction in survival, plant height, branching, leaf and flower number and size, while increasing foliage and floral abnormalities along with chromosomal aberrations. Similar mutation studies in *Hibiscus rosa-sinensis* showed that a 10 Gy gamma dose induced a mutant with significantly smaller flowers due to reduced petal length. Mutation breeding in Bougainvillea led to the development of variegated leaf mutants and somatic mutations in bract colour through gamma irradiation of stem cutting. For various ornamental crops, lower gamma ray doses have been recommended to obtain desirable mutants, including 5–7.5 kR for roses, 1.5–2 kR for chrysanthemums, 1–5 kR for gladiolus, 0.5–1 kR for tuberose and 0.5–1 kR for orchids and carnations in vitro (Patil et al. 2019). In tuberose, doses of 1–1.5 kR produced chlorophyll and broad leaf mutants, while doses of 1–2 kR induced branched spike mutants (Kayalvizhi et al. 2017). Additionally, irradiation of four tuberose varieties with

Table 7 Physical mutagen for development of mutant variety in ornamental plants

Sl No.	Crop	Mutagen	Mutant variety	Characteristics
1	Chrysanthemum	Gamma ray	Alankar, Batik, Cosmonaut, Shabnam, Agnisikha, Alankar	Altered flower shape and variety of colour
2	Rose	Gamma ray	Curio, Contempo, Yellow Contempo, Pink Contempo, Tangerine, Twinkle, Abhisarika, Angar, Christiana, Madhosh, PusaSharada, Sukumari	Altered flower colour, enhanced fragrance
3	Tuberose	Gamma ray	Rajatrekha, Swarnarekha	Variegated leaf, large flower
4	Bougainvillea	Gamma ray	Pallavi, Arjuna, Mahara Variegata, Los Banos Variegata	Chlorophyll variegated leaf
5	Portulaca	Gamma ray	Mukta, Karan Pali, Ratnam	Induced flower size and colour mutant

Table 8 Chemical mutagen for development of mutant variety in ornamental plants

Sl No.	Crop	Mutagen	Mutant variety	Characteristics
1	Chrysanthemum	Colchicine	Colchi Bahar	Early blooming and small flower shape
2	Rose	EMS	Christian Dior Mutant	Large flower size and improved colour
3	Marigold	EMS	Arka Saurabh	Compact plant growth, bright flower colour, large shape
4	Tuberose	EMS	Shrinagar	Early flowering, compact plant habit, enhanced fragrance
5	Carnation	EMS	Orange Triumph	Flower colour, size and strong Fragrance
6	Gerbera	EMS	Red Garnet	Large petals and intense colour

gamma rays (0.5–1.5 kR) and X-rays (0.6–1.2 kR) showed that lower doses stimulated sprouting, but higher doses negatively affected survival rate, leaf length, spike number, floret count, flowering duration and vase life (Kainthura and Srivastava 2015). Research also demonstrated that ion beams are generally more efficient at causing mutations than gamma radiation, X-rays and electrons (Shikazono et al. 2002). High frequencies of mutation has been seen for flower colour and shape of rose, chrysanthemum, carnation using ion beam radiation.

Different physical mutagen induced mutant of flower crops are listed in Table 7.

3.7.2 Chemical Mutagens in Ornamental Plants

Chemical mutagens cause point mutations or chromosomal aberrations, leading to beneficial changes in ornamental plants. Some commonly used chemical mutagens include ethyl methanesulphonate (EMS), sodium azide (NaN_3), 1-methyl-1-nitrosourea (MNU), 1-ethyl-1-nitrosourea (ENU), dimethyl sulphate (DMS), ethyleneimine (EI), diethyl sulphate (DES) and colchicine. These are widely used in ornamental plant breeding to induce genetic variations that result in desirable traits such as novel flower colours, altered plant morphology, improved disease resistance and enhanced fragrance. Chemical mutagenesis can be applied to various types of plant materials, including seeds, seedlings, tubers, corms, cuttings, anthers and in vitro cultured cells. However, seeds are the most frequently used material for mutation induction.

EMS treatment has successfully induced flower colour mutations and compact plant growth in chrysanthemum, petunia; early flowering and modified growth in tuberose, fragrance variations in rose. The concentrations of 0.02–5% EMS is mostly useful for creating useful variations in ornamental plants. DES has been used for new petal colour patterns for rose and chrysanthemum. Sodium azide used to obtain early flowering and dwarf mutant of marigold and shorter stem and profuse flowering in petunia, whereas Colchicine treatment successfully induced polyploid mutants with thicker leaves and more robust flowers in orchids and larger flowers and thicker petals in gladiolus.

Some of the important chemical mutagen induced ornamental crops are listed in Table 8.

Mutation breeding has played a crucial role in expanding the diversity and market appeal of ornamental plants. By introducing new flower colours, shapes, compact growth habits and stress resistance, this technique has significantly enhanced the global floriculture industry. The continued use of mutation breeding, combined with modern biotechnological approaches, will ensure the development of even more unique and resilient ornamental plant varieties in the future.

4 Conclusion

Mutation breeding has shown to be an effective tool in crop improvement, providing an effective means to induce genetic variation that may not be achievable through conventional breeding methods. By exposing plants to physical or chemical mutagens, breeders are able to develop new plant varieties with desirable traits like enhanced resistance to diseases, increased yield, superior quality and better adaptation to environmental conditions. This approach has successfully contributed to the development of numerous crop varieties with enhanced traits, benefiting agriculture and food production. The application of mutation breeding in crops like cereals, pulses, oilseeds and horticultural crops highlights its versatility. Despite some challenges, such as the time and cost involved, mutation breeding remains an essential strategy in the ongoing effort to address global food security and environmental sustainability. Moreover, it holds significant potential for accelerating crop innovation and responding to evolving agricultural needs.

References

Ahloowalia BS, Maluszynski M, Nichterlein K (2004) Global impact of mutation-derived varieties. Euphytica 135(2):187–204. https://doi.org/10.1023/B:EUPH.0000014914.85465.4f

Beyaz R (2020) Impact of gamma irradiation pretreatment on the growth of common vetch (Vicia sativa L.) seedlings grown under salt and drought stress. Int J Radiat 96:257–266. https://doi.org/10.1080/09553002.2020.1688885

Brooks C, Nekrasov V, Lippman ZB, Van Eck J (2014) Efficient gene editing in tomato in the first generation using the clustered regularly interspaced short palindromic repeats/CRISPR-associated9 system. Plant Physiol 166(3):1292–1297. https://doi.org/10.1104/pp.114.247577

Bughio HR, Asad MA, Odhano IA, Bughio MS, Khan MA, Mastoi NN (2007) Sustainable rice production through the use of mutation breeding. Pak J Bot 39(7):2457–2461. https://www.pakbs.org/pjbot/PDFs/39(7)/PJB39(7)2457.pdf

Chaudhary J, Alisha A, Bhatt VD, Chandanshive S, Kumar N, Mir ZA, Kumar AP, Yadav SK, Shivaraj SM, Sonah H, Deshmukh RK (2019) Mutation breeding in tomato: advances, applicability and challenges. Plants 8. https://doi.org/10.3390/plants8050128

Choudhary SB, Chowdhury I, Singh RK, Pandey SP, Sharma HK, Anil Kumar A et al (2017) Morphological, histobiochemical and molecular characterisation of low lignin phloem fibre (llpf) mutant of dark jute (Corchorusolitorius L.). Appl Biochem Biotechnol 183:980–992. https://doi.org/10.1007/s12010-017-2477-5

Dube KG, Bajaj AS, Gawande AM (2011) Mutagenic efficiency and effectiveness of gamma rays and EMS in *Cyamopsistetragonoloba* (L.) Var. Sharda. Asiatic J Biotech Res 2:436–440. https://www.cabidigitallibrary.org/doi/full/10.5555/20133380042

Dubey S, Bist R, Misra S (2017) Sodium azide induced mutagenesis in wheat plant. World J Pharm Pharm Sci 6:294–304. https://doi.org/10.20959/wjpps201710-10199

FAO/IAEA-MVD (2022) Food and agriculture organization of the United Nations/International atomic energy agency—Mutant variety database. https://nucleus.iaea.org/sites/mvd

Friebe B, Hatchett JH, Gill BS, Mukai Y, Sebesta EE (1991) Transfer of Hessian fly resistance from rye to wheat via radiation-induced terminal and intercalary chromosomal translocations. Theor Appl Genet 83:33–40. https://doi.org/10.1007/BF00229223

Giovannoni JJ (2004) Genetic regulation of fruit development and ripening. Plant Cell 16(suppl_1):S170–S180. https://doi.org/10.1105/tpc.019158

Gupta S, Tyagi DVS, Rawat RS (2012) Induced Chlorophyll Mutations and Mutagenic Effectiveness in Indian Mustard (*Brassica juncea* L. Czern & Coss). Progr Agric 12(1):128–131. https://www.cabidigitallibrary.org/doi/full/10.5555/20123151297

Hastari RP, Ambarwati E, Murti RH (2014) Evaluation of genetic and environment interaction of six promising mutant tomato lines (*Lycopersicon esculentum*) in low land and high land. Vegetalika 3:63–76. http://download.garuda.kemdikbud.go.id/article.php?article=1334226&val=291&title=Evaluasi%20Interaksi%20Genotipe%20dan%20Lingkungan%20Enam%20Galur%20Mutan%20Harapan%20Tomat%20Solanum%20lycopersicum%20L%20di%20Datar an%20Rendah%20dan%20Dataran%20Tinggi

Haun W, Coffman A, Clasen BM, Demorest ZL, Lowy A, Ray E, Retterath A, Stoddard T, Juillerat A, Cedrone F, Mathis L (2014) Improved soybean oil quality by targeted mutagenesis of the fatty acid desaturase 2 gene family. Plant Biotechnol J 12(7):934–940. https://doi.org/10.1111/pbi.12201

Hossen QMM, Rahman SB, Rahman MN, Sarker MDH, Moniruzzaman M, Tareq MZ et al (2022) Development of early flowering, short life-spanned jute (Corchorus spp.) mutant via ethyl methane sulfonate mutagenesis. J Crop Sci Biotechnol 25(4):489–500. https://doi.org/10.1007/s12892-022-00146-4

Im J, Ryu J, Kim WJ, Kim SH, Kang SY, Ha BK (2018) Growth characteristics and biological responses of Korean Elite Soybean (*Glycine max* L. Merr.) cultivars exposed to Gamma-Rays. Plant Breed Biotech 2:109–118. https://doi.org/10.9787/PBB.2018.6.2.109

Jayaramachandran M, Saravanan S, Motilal A, Prabhu PC, Hepziba SJ, Swain H, Boopathi NM (2020) Genetic improvement of a neglected and underutilised oilseed crop: sesame (*Sesamum indicum* L.) through mutation breeding. The Nucleus 63:293–302. https://doi.org/10.1007/s13237-020-00329-w

Kainthura P, Srivastava R (2015) Induction of genetic variability and isolation of mutants in tuberose (*Polianthes tuberosa* L.). Trop Agric Res 26(4):721–732. https://doi.org/10.4038/tar.v26i4.8135

Kayalvizhi K, Kannan M, Ganga M (2017) Effect of mutagens on vegetative and floral characters in M1V2 generation of tuberose (*Polianthes tuberosa* L.). Bull Enviorn Pharmacol Life Sci 6(1):422–429. https://doi.org/10.20546/ijcmas.2017.604.290

Khawar A, Bhatti IA, Khan QM, Bhatti HN, Sheikh MA (2010) A germination test: an easy approach to know the irradiation history of seeds. Pak J Agri Sci 47(3):279–285. https://api.pakjas.com.pk/downloadPaper/48.pdf

Kumar S, Rajendran K, Kumar J, Hamwieh A, Baum M (2015) Current knowledge in lentil genomics and its application for crop improvement. Front Plant Sci 6:78. https://doi.org/10.3389/fpls.2015.00078

Kumar G, Dwivedi S (2016) Radiation induced hormesis and its cytological evaluation in sunnhemp *Crotalaria juncea* L. Chromosome Botany 11(2):31–36. https://doi.org/10.3199/iscb.11.31

Li S, Zheng YC, Cui HR, Fu HW, Shu QY, Huang JZ (2016) Frequency and type of inheritable mutations induced by γ rays in rice as revealed by whole genome sequencing. J Zhejiang Univ Sci B 17(12):905. https://doi.org/10.1631/jzus.B1600125

Li G, Jain R, Chern M, Pham NT, Martin JA, Wei T et al (2017) The sequences of 1504 mutants in the model rice variety Kitaake facilitate rapid functional genomic studies. Plant Cell 29(6):1218–1231. https://doi.org/10.1105/tpc.17.00154

Lu X, Liu J, Ren W, Yang Q, Chai Z, Chen R et al (2018) Gene-indexed mutations in maize. Mol Plant 11(3):496–504. https://doi.org/10.1016/j.molp.2017.11.013

Marcu D, Damian G, Cosma C, Cristea V (2013) Gamma radiation effects on seed germination, growth and pigment content, and ESR study of induced free radicals in maize (*Zea mays*). J Biol Phys 39:625–634. https://doi.org/10.1007/s10867-013-9322-z

Miller FJ, Al-Khatib K (2000) Development of herbicide resistant germplasm in sunflower. In: Proceedings of the 15th international sunflower conference, Toulouse, France, vol I, 0-37-42. https://www.ars.usda.gov/research/publications/publication/?seqNo115=111466

Miller JF, Al-Khatib K (2004) Registration of two oilseed sunflower genetic stocks, SURES-1 and SURES-2 resistant to tribenuron herbicide. Crop Sci 44(3):1037–1039. https://doi.org/10.2135/cropsci2004.1037

Mlčochová L, Chloupek O, Uptmoor R, Ordon F, Friedt W (2004) Molecular analysis of the barley cv.'Valticky' and its X-ray-derived semidwarf-mutant 'Diamant'. Plant Breeding 123(5):421–427. https://doi.org/10.1111/j.1439-0523.2004.01023.x

Morita R, Kusaba M, Iida S, Yamaguchi H, Nishio T, Nishimura M (2009) Molecular characterization of mutations induced by gamma irradiation in rice. Genes Genet Syst 84(5):361–370. https://doi.org/10.1266/ggs.84.361

Mullainathan L, Aruldoss T (2015) Effect of gamma rays in induced morphological mutants on M2 generation of chilli (*Capsicum annuum* L.) var K1. Int Lett Nat Sci 3. https://doi.org/10.56431/p-lny1ps

Nagatomi S, Yamaguchi H, Degi K, Morishita T, Watanabe H, Tanaka A (2003) Six mutant varieties induced by ion beams in chrysanthemum. https://www.cabidigitallibrary.org/doi/full/10.5555/20043066988

Njau PN, Kinyua MG, Kimurto PK, Okwaro HK, Maluszynski M (2005) Drought tolerant wheat varieties developed through mutation breeding technique. J Agric Sci Technol 7(1):18–29. https://doi.org/10.4314/jagst.v7i1.31717

Olsen O, Wang X, von Wettstein D (1993) Sodium azide mutagenesis: preferential generation of AT–> GC transitions in the barley Ant18 gene. Proc Natl Acad Sci 90(17):8043–8047. https://doi.org/10.1073/pnas.90.17.8043

Patil UH, Masalkar SD, Patil AH (2019) Effect of chemical mutagens on growth and flowering of carnation. J Pharmacogn Phytochem 8(2):1982–1984. https://www.phytojournal.com/archives/2019.v8.i2.7966/effect-of-chemical-mutagens-on-growth-and-flowering-of-carnation

Pillai MA, Subramaniam M, Murgan S (1993) Effectiveness and efficiency of gamma rays and EMS for chlorophyll mutants in upland Rice. Ann Agric Res 14:302–303. https://www.cabidigitallibrary.org/doi/full/10.5555/19941610129

Punia SS, Ram B, Dheer M, Jain NK, Koli NR, Khedar OP (2014) Hyper-variable spontaneous genetic variation for earliness, seed characters and other yield-contributing traits in lentil (*Lens culinaris* Med.). Curr Sci 75–83. https://mail.i-scholar.in/index.php/CURS/article/view/86187/76226

Rampure NH, Choudhary AD, Jambhulkar SJ, Badere RS (2017) Isolation of desirable mutants in safflower for crop improvement. Indian J Genet Plant Breed 77(01):134–144. https://doi.org/10.5958/0975-6906.2017.00018.9

Shan Q, Zhang Y, Chen K, Zhang K, Gao C (2015) Creation of fragrant rice by targeted knockout of the OsBADH2 gene using TALEN technology. Plant Biotechnol J 13:791–800. https://doi.org/10.1111/pbi.12312

Shikazono N, Tanaka A, Kitayama S, Watanabe H, Tano S (2002) LET dependence of lethality in *Arabidopsis thaliana* irradiated by heavy ions. Radiat Environ Biophys 41:159–162. https://doi.org/10.1007/s00411-002-0157-4

Subramaniam R, Kumar VS (2023) Ethyl methane sulphonate (EMS)-mediated mutagenesis induces genetic and morphological variations in eggplant (*Solanum melongena* L.). Int J Plant Biol 14(3):714–728. https://doi.org/10.3390/ijpb14030053

Sujatha K, Nadaf HL, Srinivasa N (2019) Induced mutagenesis in creating genetic variability for seed yield and Alternaria leaf blight resistance in sunflower (*Helianthus annus*). Ann Agric Res 40(1):36–44. https://epubs.icar.org.in/index.php/AAR/article/view/127478

Tian E, Zeng F, MacKay K, Roslinsky V, Cheng B (2014) Detection and molecular characterization of two FAD3 genes controlling linolenic acid content and development of allele-specific markers in yellow mustard (*Sinapis alba*). Plos One 9(5):e97430. https://doi.org/10.1371/journal.pone.0097430

Tran QH, Bui NH, Kappel C, Dau NTN, Nguyen LT, Tran TT et al (2020) Mapping-by-sequencing via MutMap identifies a mutation in ZmCLE7 underlying fasciation in a newly developed EMS mutant population in an elite tropical maize inbred. Genes 11(3):281. https://doi.org/10.3390/genes11030281

Wani MR, Raina A, Yaqoob S, Laskar RA, Khan S, Bhat TA (2023) Induced mutagenesis: a successful breeding strategy for crop improvement. In: Biotechnologies and genetics in plant mutation breeding. Apple Academic Press, pp 69–87. https://doi.org/10.1201/9781003305064-4?urlappend=%3Futm_source%3Dresearchgate.net%26utm_medium%3Darticle

Warghat AR, Rampure NH, Wagh P (2011) Effect of sodium azide and gamma rays treatments on percentage germination, survival, morphological variation and chlorophyll mutation in musk okra (*Abelmoschus moschatus* L.). Int J Pharm Pharm Sci 3(5):483–486. https://consensus.app/papers/effect-of-sodium-azide-and-gamma-rays-treatments-on-warghat-rampure/d1fc111045915e99b93f4316d5a43cb6/

Wolverton C, Kisss JZ (2009) An update on plant space biology. Gravit Space Biol 22(2):13–21. http://gravitationalandspacebiology.org/index.php/journal/article/view/464

Enhancing Crop Resilience to Combined Abiotic and Biotic Stress Through Induced Mutation

Nusrat Perveen, Hamidullah Itoo, and Hidayatullah Mir

Abstract

Crop plants are continuously threatened by various biotic and abiotic stresses, resulting in significant reduction in quality, yield, and productivity. Abiotic factors such as drought, salinity, extremities of temperatures, and heavy metal toxicity hamper growth and development, often predisposing plants to biotic pathogens. Climate change has further intensified these challenges, leading to more frequent and severe stress events that traditional breeding approaches struggle to address. Although conventional breeding methods like hybridization and marker-assisted selection (MAS) have contributed to the development of stress-resilient varieties, they are often constrained by long breeding cycles, limited genetic diversity, and compromise between stress tolerance and agronomic traits. Induced mutation breeding has emerged as an effective alternative for generating novel genetic variations to enhance stress resilience. This approach has successfully led to the development of drought-tolerant bananas, salt-resistant grapevines, and disease-resistant papayas. Recent advancements, including CRISPR/Cas9 and TALEN-based genome editing, have further revolutionized mutation breeding, allowing precise genetic modifications to enhance crop resistance against multiple stress inducing factors. Additionally, omics technologies and high-throughput screening have accelerated the identification of beneficial mutations for stress tolerance. This chapter highlights the importance of induced mutation technologies in breeding climate-resilient crops and discusses recent innovations that can help in sustainable crop production in the face of environmental challenges.

Keywords

Mutation breeding · Gamma radiation · Biotic stress · Abiotic stress

1 Introduction

Crop plants are continuously exposed to an array of environmental challenges that significantly limit their growth, yield, and quality. These challenges arise from both biotic and abiotic stress factors, often occurring simultaneously and exacerbating their effects. Biotic stress, caused by living organisms such as fungi, bacteria, viruses, nematodes, and insect pests, directly affects plant health and productivity. The combined impact of abiotic and biotic stress poses a significant threat to global crop production. It is estimated that abiotic stresses alone contribute to 50–70% of global yield losses in staple crops such as wheat, rice, and maize. Drought stress has been reported to cause an annual yield reduction of up to 40% in major cereals (Lesk et al. 2016), while more than 20% of irrigated agricultural land worldwide is affected by soil and water salinity (Perveen et al. 2024). Similarly, biotic stress factors such as plant pathogenic fungi, bacteria, and viruses cause crop losses exceeding 15–20% annually. The occurrence of multiple stresses simultaneously exacerbates the situation as plants normally exhibit trade-offs between stress tolerance and yield, making it difficult to achieve stable yields in stress-prone environments (Atkinson and Urwin 2012). In fruit crops, fungal pathogens like *Colletotrichum* (anthracnose) in mango and *Botrytis cinerea* (gray mold)

N. Perveen
Central Institute of Temperate Horticulture (ICAR-CITH), Srinagar, J&K, India

H. Itoo
Ambri Apple Research Centre, Sher-e-Kashmir University of Agricultural Sciences and Technology of Kashmir, Srinagar, J&K, India

H. Mir (✉)
Regional Horticulture Research Sub-Station, Sher-e-Kashmir University of Agricultural Sciences and Technology of Jammu, Bhaderwah, J&K, India
e-mail: hidayatmay14@yahoo.co.in

J. Devi (ed.), *Mutation Breeding in Crop Improvement*, Sustainable Landscape Planning and Natural Resources Management,
https://doi.org/10.1007/978-3-032-17911-1_5

in strawberries lead to substantial post-harvest losses, while bacterial infections such as *Xanthomonas campestris* (bacterial spot) in citrus and *Erwinia amylovora* (fire blight) in apples severely affect yield and quality of produce (Fischer et al. 2014). Other biotic stresses like banana bunchy top virus (BBTV) in bananas and citrus tristeza virus (CTV) in citrus pose significant threats to global fruit production. Meanwhile, insect pests such as the Mediterranean fruit fly and the mango fruit borer contribute to substantial pre- and post-harvest losses.

Further, abiotic stresses such as drought, salinity, extreme temperatures, heavy metal toxicity, and oxidative stress further increase these challenges. Drought stress has been reported to significantly reduce fruit set and size in grapes, apples, and citrus, while salinity stress leads to poor fruit development in bananas, guavas, and papayas by disrupting water uptake and ion balance (Lesk et al. 2016). Similarly, sunburn in apples and berries caused by heat stress or chilling injuries in citrus, banana, and avocado caused by cold stress results in reduced commercial value of the fruits. Heavy metal contamination in polluted regions like cadmium accumulation in strawberries and arsenic stress in apples negatively impact fruit crops. The interplay of biotic and abiotic stresses often exacerbates the severity of their effects, for example, heat stress increases the susceptibility of strawberries to *Botrytis cinerea* and drought stress weaken the plant defenses against insect infestations (Atkinson and Urwin 2012).

The impact of these stress factors on global crop productivity is profound. It is estimated that abiotic stress alone contributes to 50–70% of worldwide yield losses, with drought reducing fruit production by 30–50% in crops such as grapes, mangoes, and citrus (Lesk et al. 2016). Soil salinity negatively affects over 833 million hectares of agricultural land, further restricting fruit cultivation in arid and semi-arid regions (FAO 2021). Biotic stressors are equally detrimental, as fungal pathogens cause annual global crop losses exceeding 15–20%, amounting to economic damages of over $200 billion, while insect pests contribute to 15–25% of total yield losses. In fruit crops, these stresses not only reduce productivity but also deteriorate fruit quality, shorten post-harvest shelf life, and increase production costs. For example, bacterial spot in peaches can lead to yield losses of up to 50%, while powdery mildew in grapes significantly lowers wine quality and market value.

To mitigate these challenges, plant breeding has focused on developing stress-resilient varieties through conventional approaches such as hybridization, selection breeding, and marker-assisted selection (MAS). However, these methods have several limitations. The development of stress-tolerant varieties through traditional breeding can take over a decade, making it inadequate for addressing rapidly emerging threats such as new viral diseases in fruit crops. Additionally, many clonally propagated fruit species, including bananas and citrus, exhibit limited genetic diversity, restricting the potential for breeding stress-resistant varieties. Furthermore, breeding for stress resistance often leads to trade-offs with other desirable traits, such as reduced fruit quality or lower yield potential. Moreover, conventional breeding struggles to address newly evolving pathogens, such as *Fusarium oxysporum* tropical race 4 (TR4), which is causing severe damage to global banana production. Given these limitations, alternative breeding strategies are essential to accelerate the development of climate-resilient varieties.

Induced mutation has emerged as a powerful breeding tool for enhancing stress resilience in horticultural crops. By exposing plant materials to physical mutagens such as gamma rays, X-rays, and ion beams, or chemical mutagens such as ethyl methanesulfonate (EMS) and sodium azide, novel genetic variations can be generated to introduce beneficial traits. This approach has led to the successful development of stress-tolerant fruit crop varieties, including drought-resistant bananas, salt-tolerant grapevines, and disease-resistant papayas. Cold-tolerant citrus mutants have also been developed, enabling extended seasonal production in regions prone to chilling stress. Mutation breeding has contributed to the release of over 3,300 officially recognized mutant varieties worldwide, demonstrating its efficacy in improving stress resilience. Recent advancements in mutation breeding technologies have further enhanced their precision and efficiency. CRISPR/Cas9 genome editing has been successfully employed to modify stress-related genes in fruit crops, such as banana for resistance to Fusarium wilt and grapevine for improved drought tolerance (Chen et al. 2019). Similarly, TALEN (Transcription Activator-Like Effector Nuclease) technology has been applied to develop virus-resistant papaya and salt-tolerant apple rootstocks. The use of ion-beam and space-induced mutagenesis has led to the development of fruit crops with unique traits, such as enhanced fragrance in strawberries and improved shelf life in grapes. High-throughput screening techniques and omics-based approaches including genomics, transcriptomics, and proteomics are further accelerating the identification of beneficial mutations for stress tolerance. Additionally, the TILLING (Targeting Induced Local Lesions in Genomes) approach has proven valuable in efficiently identifying and utilizing stress-resilient mutations in fruit crops such as apple, citrus, and grapevine. With these technological advancements, induced mutation breeding, combined with modern genome-editing tools is revolutionizing the development of climate-resilient crop plants. Given the increasing frequency and severity of abiotic and biotic stresses, leveraging these innovative breeding approaches is essential for ensuring sustainable crop production and global food security in the face of climate change.

2 Mutation: Definition and Historical Account

Mutagenesis is the process by which genetic material undergoes alteration, leading to mutations. These changes can occur naturally due to environmental factors such as radiation or chemicals or can be induced artificially in a controlled setting. Mutagenesis plays a fundamental role in evolution, genetic diversity, and various applications in medicine, agriculture, and biotechnology. The concept of mutagenesis emerged in the early twentieth century with the rediscovery of Gregor Mendel's genetic principles in 1900. In 1901, Hugo de Vries introduced the "mutation theory," suggesting that genetic changes drive evolution. This idea laid the groundwork for future studies on how mutations occur and their effects on organisms. The breakthrough in understanding induced mutations came in 1927 when Hermann J. Muller demonstrated that exposure to X-rays could cause genetic mutations in fruit flies (*Drosophila melanogaster*), proving that external factors could alter DNA. Around the same time, Lewis Stadler confirmed similar radiation-induced mutations in plants, expanding the scope of mutagenesis research. Further advancements occurred in the mid-20th century with the discovery of chemical mutagens. During World War II, scientists found that mustard gas could induce genetic mutations, leading to a new era of research into chemical agents that could alter DNA. By the 1950s, bacterial and viral genetic studies, pioneered by researchers like Salvador Luria and Max Delbrück, provided deeper insights into spontaneous and induced mutations. The discovery of the DNA double-helix structure by Watson and Crick in 1953 further solidified the molecular understanding of how genetic material changes. The 1970s marked the development of site-directed mutagenesis, a technique that allowed researchers to create specific mutations for experimental and therapeutic purposes. This was followed by the introduction of Polymerase Chain Reaction (PCR) technology in the 1980s, which revolutionized DNA manipulation and enabled more precise studies on mutagenesis. These advancements paved the way for modern gene-editing techniques, leading to more controlled and targeted genetic modifications.

Table 1 Different types of mutations

S.No.	Type of mutation	Remarks
01	Point mutations	Single nucleotide base is altered, inserted or deleted in the DNA sequence
02	Chromosomal mutations	Results in changes in chromosome structure or chromosome number. Eg. Aneuploidy
03	Inversions	When a single chromosome undergoes breakage and rearrangement within itself
04	Insertions	The addition of one or more nucleotides into a DNA sequence
05	Translocation	Caused by rearrangement of parts between non-homologous chromosomes
06	Somatic mutations	Mutations in somatic tissues; not heritable
07	Germline mutations	Mutations in germ cells; Heritable
08	Spontaneous mutations	Mutations occurring in the natural populations without human intervention
09	Induced mutations	Mutations induced by treatments with certain physical, chemical and biological agents known as mutagens
10	Macro-mutations	Mutations producing a large phenotypic effect; oligogenic in nature; can be easily selected in the M_2 generation
11	Micro-mutations	Mutations producing small phenotypic effect that cannot be recognized on individual plant basis, are polygenic in nature and selection for them is delayed till M_3 or later generation

3 Applications of Induced Mutations

Induced mutations are a valuable tool in crop improvement to develop new plant varieties with improved traits such as higher yield, disease resistance, improved nutritional quality, and better adaptability to environmental conditions. Treatment of plant (seeds or vegetative parts) to physical mutagens like radiation (X-rays, gamma rays, and ultraviolet light) or chemical mutagens such as ethyl methanesulfonate (EMS) and sodium azide, scientists can introduce genetic variations that may not occur naturally within a short time frame. These mutations are then screened and selected for desirable traits, significantly accelerating the breeding process compared to traditional methods. One of the most significant applications of induced mutations is the development of varieties resistant to biotic or abiotic stress and altered plant architecture like dwarfness (Perveen et al. 2022). Another major advantage of induced mutations is the improvement of nutritional quality in food crops and enhanced oil content in oilseed crops, improves starch quality in tubers, and alters secondary metabolite production in medicinal plants, broadening their commercial value and utility. Mutations can alter morphological, biochemical, and anatomical features of the plants to better suit the requirement of plant breeder (Perveen et al. 2025). Different types of mutations have been provided in Table 1 and Figs. 1, 2 and 3.

4 Mutagens

The agents that induce mutations are known as mutagens. Mutagens may be different kinds of radiation (physical mutagens), certain chemicals (chemical mutagens), or biological

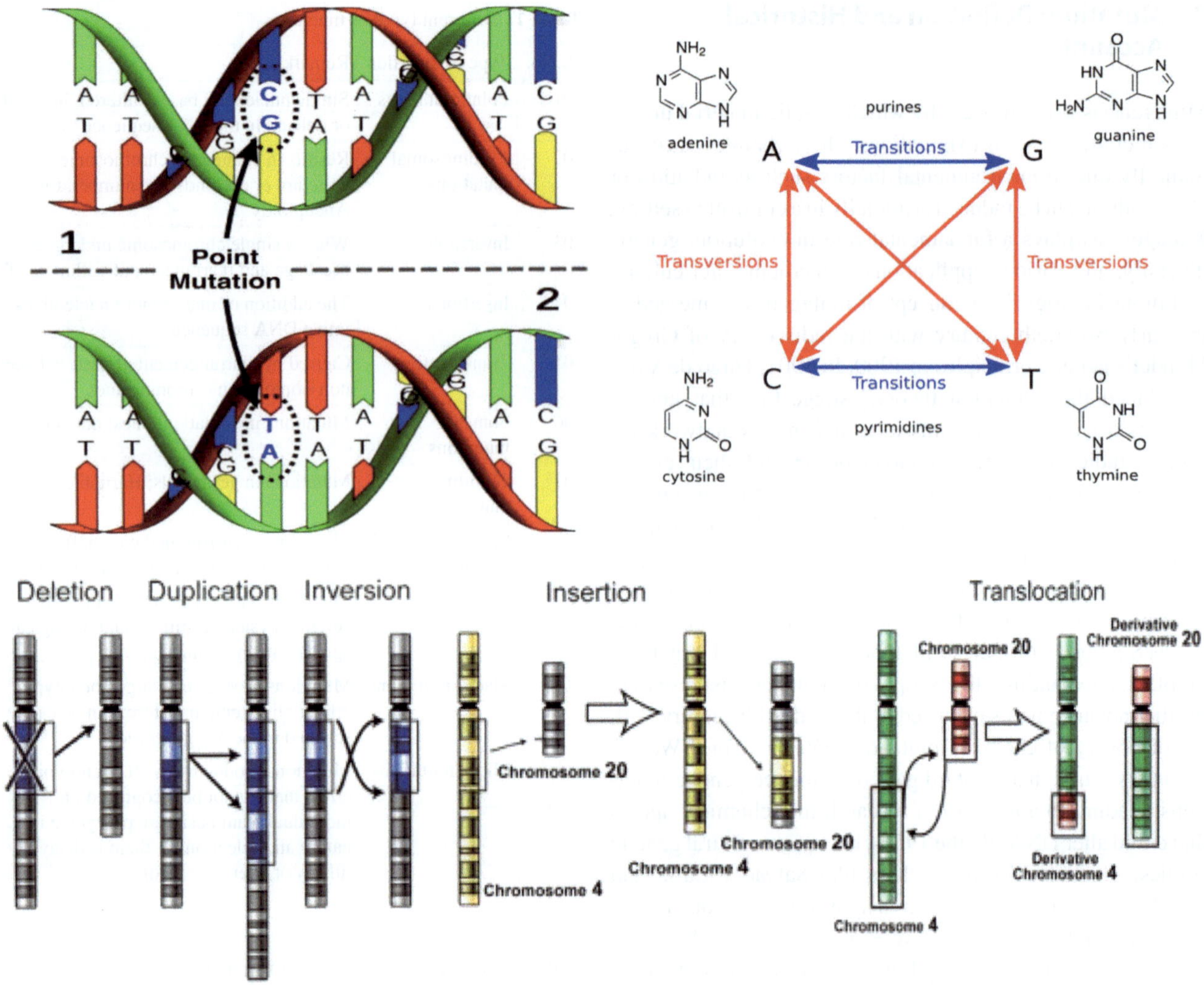

Fig. 1 Different types of mutations

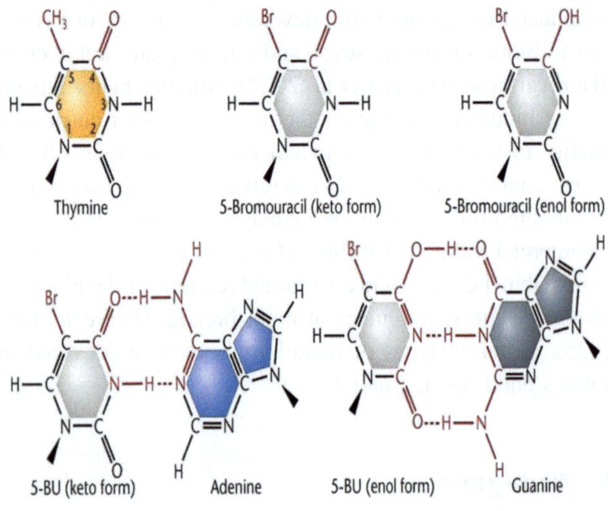

Fig. 2 Base analogues

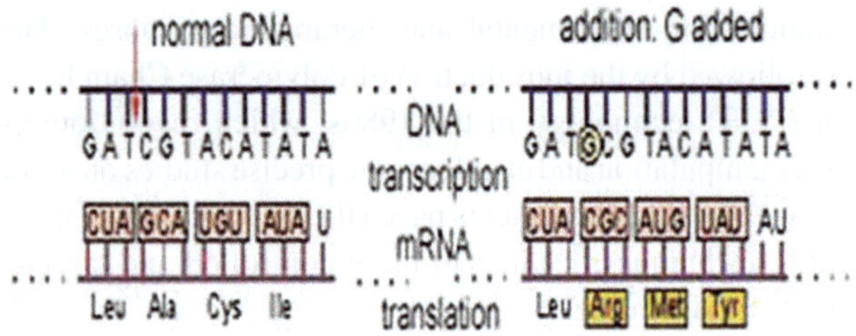

Fig. 3 Frame shift mutations

agents like viruses (insertion of virus DNA into the genome; disruption of genetic function) or bacteria (e.g., *Helicobacter pylori* produces oxidative species causing DNA damage and reduction in efficiency of DNA repair systems). Different categories of mutagens have been provided in Table 2.

Table 2 Different groups of mutagens and their examples

Types	Examples
A. Physical	
1. Ionizing radiations a. Particulate b. Non-particulate (Electromagnetic)	α-rays (DI), β-rays (SI), Fast neutrons* (DI), Thermal neutrons (DI) X-rays* (SI), γ-rays* (SI)
2. Non-ionizing radiations	Ultraviolet (UV) radiation
B. Chemical mutagens	
1. Alkylating agents	Sulfur mustard, Nitrogen mustard, Imines (Ethylene imine or EI*), Ethyl methane sulphonate (EMS*), Methyl methane sulphonate (MMS)
2. Acridine dyes	Acriflavin, Acridine Yellow, Ethidium Bromide
3. Base analogues	5-bromouracil, 5-chlorouracil
4. Others	Nitrous acid, hydroxyl amine, sodium azide*
C. Biological	Transposons, Virus, Bacteria, insertion elements

4.1 Physical Mutagens

Physical mutagens are those mutagens that induce genetic alterations through physical means, primarily radiation and extreme environmental conditions. These mutagens include ionizing radiation (e.g., X-rays, gamma rays, alpha and beta particles, and thermal neutrons), non-ionizing radiation (e.g., ultraviolet light), and extreme physical stress factors (e.g., heat and mechanical shear stress).

4.1.1 Ionizing Radiation

Ionizing radiation are high energy radiations that can directly break molecular bonds, causing single-strand and double-strand DNA breaks, which can lead to chromosomal aberrations, mutations, or cell death. Different types of ionizing radiations include the following.

4.1.2 X-Rays and Gamma Rays

X-rays and gamma rays are forms of electromagnetic radiation with high energy and penetrating power. They can ionize biological molecules, leading to base modifications (e.g., oxidation of guanine to 8-oxo-guanine), chromosomal rearrangements resulting from DNA strand breaks, formation of reactive oxygen species (ROS) that induce indirect DNA damage (Hall and Giaccia 2006). Gamma rays have stronger penetration ability than X-rays and can affect deeper tissues, increasing the risk of mutations and cancer.

4.1.3 Alpha and Beta Particles

Alpha particles (helium nuclei) and beta particles (high-energy electrons) are emitted during radioactive decay. These particles can only cause localized damage, as alpha particles have low penetration power but high ionization potential.

4.1.4 Thermal Neutrons

Thermal neutrons are low-energy neutrons that interact with atomic nuclei, especially boron, nitrogen, and hydrogen, triggering nuclear reactions that release ionizing particles. These reactions can lead to chromosomal aberrations, especially in rapidly dividing cells and DNA strand breaks, primarily through neutron capture reactions that release secondary radiation.

4.2 Non-ionizing Radiation

Unlike ionizing radiation, non-ionizing radiation lacks sufficient energy to remove electrons from atoms but can still cause molecular changes that lead to mutations. Different categories of non-ionizing radiations include:

4.2.1 Ultraviolet (UV) Radiation

The lower energy concentration of UV radiation makes them less capable of penetrating deep into plant tissues than gamma radiation. The lack of a protective barrier makes it commonly employed for the treatment of callus, cell suspension cultures, and protoplasts. Different UV radiations have different effects, for example, UV-A (315–400nm) indirectly contributes to mutagenesis by generating ROS, UV-B (280–315 nm) causes direct DNA damage, leading to pyrimidine dimmers while UV-C (100–280 nm) is highly mutagenic but mostly absorbed by the atmosphere. UV radiation primarily induces cyclobutane pyrimidine dimers (CPDs) and 6-4 photoproducts, leading to errors in DNA replication and an increased risk of skin cancer.

4.2.2 Microwave and Infrared Radiation

Microwaves and infrared radiation do not directly alter DNA but can cause local tissue heating, which may lead to indirect genetic damage and disruption of cellular processes, such as protein denaturation and membrane instability.

4.3 Chemical Mutagens

Chemical mutagens are substances capable of interacting with DNA to induce genetic mutations, leading to alterations in gene function and chromosomal instability. These agents can modify nucleotide bases, interfere with DNA replication, or cause structural changes in genetic material.

The impact of chemical mutagens depends on factors such as their concentration, duration of exposure, and the efficiency of cellular repair mechanisms. They are broadly categorized into base analogs, alkylating agents, deaminating agents, intercalating agents, and oxidative agents, each inducing mutations through distinct mechanisms.

4.3.1 Alkylating Agents

Alkylating agents transfer alkyl groups to DNA bases, altering their pairing properties and leading to mispairing during replication. Two most important alkylating agents include ethyl methanesulfonate (EMS) and methyl methanesulfonate (MMS), which add alkyl groups to guanine, causing it to mispair with thymine instead of cytosine. Mustard gas and nitrosamines are other potent alkylating agents that induce cross-linking of DNA strands, disrupting normal replication and transcription processes. EMS is most commonly used mutagen in induced mutagenesis specifically for in-vitro mutagenesis.

4.3.2 Deaminating Agents

Deaminating agents remove amino groups from DNA bases, altering their hydrogen-bonding properties. Nitrous acid (HNO_2) deaminates adenine to hypoxanthine, which pairs with cytosine instead of thymine, leading to base substitution mutations. Similarly, cytosine is converted into uracil, which pairs with adenine instead of guanine. These changes can lead to missense or non-sense mutations, potentially affecting critical genes involved in cell cycle regulation and tumor suppression.

4.3.3 Intercalating Agents

Intercalating agents are planar molecules that insert themselves between base pairs in the DNA double helix, causing distortions that lead to insertions or deletions during replication. Ethidium bromide, acridine dyes, and proflavine are well-known intercalating agents. These substances disrupt DNA integrity, resulting in frameshift mutations that can completely alter the reading frame of a gene, leading to dysfunctional proteins. Such mutations are particularly dangerous because they can create premature stop codons, truncating protein synthesis and leading to loss of function.

5 Molecular Basis of Gene Mutations

Various characters of organisms are produced due to the proteins produced by their genes. The properties of a protein depend on the type and the sequence of amino acids present in that protein molecule, which itself is determined by the base sequence of the concerned gene. Thus, mutation in a character would generally be produced due to the change in the amino acid sequence of the protein which in turn would result from a change in the base sequence of the gene encoding the protein. The change in base sequence of a gene may occur due to the following: a. base substitution, b. base addition or deletion, c. transposition. These form the molecular basis of gene mutation. Base analogue incorporation occurs when chemically similar molecules to normal DNA bases are incorporated into the DNA during replication (Fig. 2). These analogues, such as 5-bromouracil (5-BU) and 2-aminopurine, resemble thymine and adenine, respectively, but have different base-pairing properties. When incorporated, they can lead to mispairing during subsequent replication cycles, resulting in transition mutations where purines are replaced with purines or pyrimidines with pyrimidines. Base alteration involves the chemical modification of nucleotide bases, leading to mispairing during replication. Deamination is a common form of base alteration in which cytosine is converted into uracil, leading to an adenine being incorporated instead of guanine in the next replication cycle. Oxidative damage caused by reactive oxygen species (ROS) can modify guanine to 8-oxoguanine, which mispairs with adenine instead of cytosine, causing transversion mutations.

Alkylation, another type of base alteration, occurs when alkyl groups are added to bases, leading to abnormal base-pairing and strand instability. Base insertion and deletion mutations result from the addition or removal of one or more nucleotides in the DNA sequence. These mutations can lead to frameshift mutations if the number of inserted or deleted bases is not a multiple of three, altering the reading frame of the genetic code and potentially producing a nonfunctional or truncated protein (Fig. 3). Transpositions involve the movement of transposable elements or "jumping genes" within the genome. These mobile genetic elements, such as DNA transposons and retrotransposons, can insert themselves into new genomic locations, disrupting gene function or regulatory sequences (Fig. 4). In some cases, transpositions can create new mutations by inserting into coding or promoter regions, altering gene expression, or even causing chromosomal rearrangements. Distortion of DNA structure can result from various factors, including chemical modifications, intercalating agents, and DNA damage. Intercalating agents like ethidium bromide and acridine dyes insert themselves between base pairs, causing DNA helix distortion and leading to insertions or deletions during replication.

6 Radio Sensitivity and LD_{50} Dose

The term radio sensitivity is a relative measure of the quantum of recognizable effects on the irradiated material, and it is essential to observe the growth responses after irradiation to determine optimal dosages. Factors such as genome size, ploidy level, water content, and cell cycle stage influence a plant's response to radiation. Generally, diploid plants are more radiosensitive than polyploid plants due to their lower genetic redundancy, while actively dividing cells, such as meristematic tissues, are more susceptible to radiation-

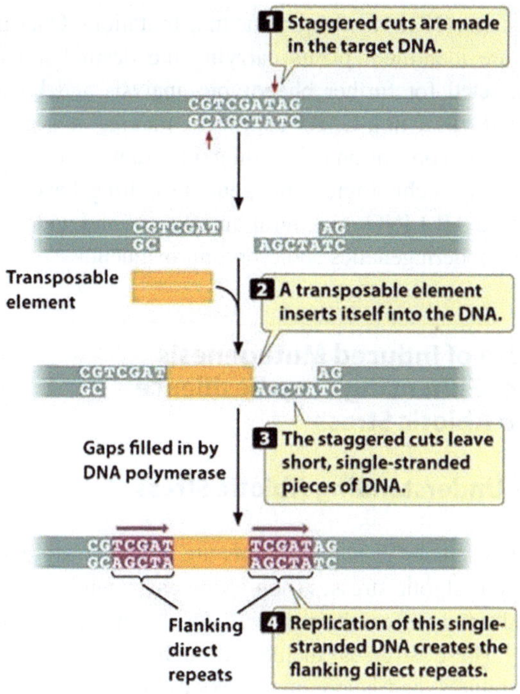

Fig. 4 Mechanism of action: transposons

induced mutations. The lethal dose for 50% of individuals (LD_{50}) is a key parameter in mutation breeding, representing the radiation dose at which 50% of the treated material fails to survive. Determining the LD_{50} dose for a particular plant species is essential to optimize mutagenic treatments, ensuring a balance between mutation induction and plant viability. Very high dose of mutagen cause excessive damage, leading to reduced germination, growth abnormalities, or sterility, whereas too low a dose may not induce significant genetic variation. The LD_{50} dose is often used as a reference to determine the optimal mutagenic dose (ranging from LD_{50} to sublethal doses), which maximizes beneficial mutations while maintaining plant regeneration capacity. LD_{50} doses for some of the fruit crops are presented in Table 3.

7 Biotechnological Interventions in Induced Mutagenesis

7.1 Insertional Mutagenesis

Insertional mutagenesis is a process in which a foreign DNA sequence, such as a transposon, viral genome, or artificially introduced genetic element, integrates into the plant genome disrupting normal gene function. This type of mutation can lead to gene inactivation, altered gene expression, or chromosomal instability, depending on the insertion site. Transposable elements, or "jumping genes," are a common source of insertional mutagenesis. These mobile DNA sequences can move within the genome, inserting into coding or regulatory regions of genes and altering their function. Depending on where they integrate, transposons may disrupt essential genes, create new regulatory interactions, or cause chromosomal rearrangements. Insertional mutagenesis is most commonly used tool for functional genomics. Transposon-based systems, viral vectors, or CRISPR-based approaches are used to introduce insertions into specific genes in order to study gene function and identify genes associated with particular traits. Insertional mutagenesis is also a method used in forward genetic screens, where random insertions help identify genes responsible for specific phenotypes.

Another method of insertional mutagenesis is T-DNA mediated which involves integration of transfer DNA (T-DNA) from *Agrobacterium tumefaciens* into a plant genome, leading to gene disruption or altered gene expression. This method is widely used in plant molecular biology and functional genomics to study gene function by creating loss-of-function or gain-of-function mutations. *Agrobacterium tumefaciens* is a naturally occurring soil bacterium that infects plants and transfers a segment of its Ti (tumor-inducing) plasmid, known as T-DNA, into the plant genome. In genetic engineering, the tumor-causing genes within the T- DNA region is replaced with selectable marker genes or reporter genes, allowing controlled insertion of foreign DNA into plant chromosomes. When *Agrobacterium* infects plant

Table 3 LD_{50} doses for different fruit crops for different plant tissues

Crop	Target tissue	Mutagen	Sensitivity dose (LD_{50})	References
Shamouti sweet orange	Seeds Decapitated young seedlings Calli (in vitro)	Gamma rays	80–100 Gy 20–40 Gy 240 Gy	Spiegel-Roy and Kochba (1973)
Rough lemon	Calli (in vitro)	Gamma rays	20 Gy	Kumar et al. (2010)
Grapes cv. "Muscat"	Cuttings	Gamma rays	15–25 Gy	Surakshithaa et al. (2017)
Mango cv. Langra cv. Mallika	Dormant buds	Gamma rays	30 Gy 39 Gy	Sharma et al. (1983)
Mango cv. Bappakkai, Nekkare, Kurukkan	Seeds	Gamma rays	22.39 Gy, 19.95 Gy and 19.95 Gy respectively	Perveen et al. (2023a)

cells, the modified T-DNA is randomly integrated into the genome, potentially disrupting coding sequences or regulatory regions, leading to insertional mutagenesis. T-DNA insertional mutagenesis is a valuable tool for studying gene function because it creates stable, heritable mutations. The inserted T-DNA often contains reporter genes such as GUS (β-glucuronidase) or GFP (green fluorescent protein), which help track gene expression. Additionally, T-DNA tags the mutation site, making it possible to identify disrupted genes through PCR-based techniques like T-DNA border sequencing.

In vitro mutagenesis is used to introduce specific or random mutations into genome and is useful for the study of gene function, protein structure, and genetic regulation by creating targeted modifications in DNA. Unlike natural or induced mutagenesis, in vitro mutagenesis is performed under controlled conditions. Site-directed mutagenesis is a common approach of in vitro mutagenesis which enables the introduction of precise nucleotide changes at specific locations within a gene. Chemical mutagens like EMS are most preferably used mutagens for in vitro mutagenesis. Amenability of target plants to in vitro culture is the pre-requisite for using this technique.

8 TILLING: Moving Forward in Reverse!

TILLING (Targeting Induced Local Lesions in Genomes) is a reverse genetics technique used to identify and analyze induced mutations in specific genes of interest. It combines chemical or physical mutagenesis with high-throughput screening methods to detect point mutations in target genes without the need for transgenic modifications. TILLING is widely used for functional genomics and crop improvement. In this technique seeds or cells are treated with chemical mutagens like ethyl methanesulfonate (EMS) or physical mutagens such as gamma radiation. EMS is commonly used because it induces point mutations, primarily G/C to A/T transitions, which can lead to missense or nonsense mutations in genes. After mutagenesis, the treated seeds are grown into M_1 plants, which are then self-pollinated to generate the M_2 population, where mutations become homozygous and stable for screening. Further the DNA is extracted from M_2 plants and pooled for efficient mutation detection. The target gene is amplified using gene-specific primers through polymerase chain reaction (PCR). The amplified DNA is then subjected to heteroduplex formation, where mutant and wild-type DNA strands anneal to form mismatched regions. These mismatches are detected using enzymatic cleavage methods, such as CEL I endonuclease digestion, which specifically cuts at the mutation site, or through high-throughput sequencing techniques.

The resulting DNA fragments are analyzed by gel electrophoresis, capillary electrophoresis, or next-generation sequencing (NGS) to identify and map mutations. Once mutations are identified, plants carrying the desired mutations are selected for further phenotypic analysis and breeding. TILLING is a non-GMO approach, making it acceptable for use in conventional breeding programs. Advances in sequencing technologies and genome editing have further enhanced TILLING, making it an efficient and widely used tool in modern genetics and crop improvement.

9 Use of Induced Mutagenesis for Enhancing Crop Resilience to Abiotic Stress

9.1 Understanding Abiotic Stress

Being sedentary in nature, plants are continuously exposed to various abiotic stresses (non-living environmental factors) that adversely affect their growth, development, and productivity. These stresses include drought, salinity, extreme temperatures, and heavy metal toxicity. Understanding how plants perceive and respond to these challenges is crucial for developing strategies to enhance crop resilience and ensure food security. Abiotic stress includes drought stress, salinity stress, temperature stress, and heavy metal stress. Drought stress refers to a condition of acute water shortage in the rhizosphere which results in reduced cell turgor, stomatal closure, and impaired photosynthesis, ultimately affecting plant growth and yield. Salinity refers to the accumulation of soluble salts, primarily sodium chloride (NaCl), calcium sulfate ($CaSO_4$), magnesium chloride ($MgCl_2$), and sodium sulfate (Na_2SO_4), in the soil to levels that negatively impact plant growth, soil health, and agricultural productivity. High salinity reduces the ability of plant roots to absorb water due to osmotic stress, leading to dehydration, disrupts ion homeostasis leading to ion toxicity, and nutrient imbalances (Zhang et al. 2025). Another abiotic stress factor include temperature stress which can be heat stress where elevated temperatures can denature proteins, disrupt membrane integrity, and impair photosynthetic efficiency (Zhang et al. 2022) or cold stress where low temperatures can cause membrane rigidity, reduced enzymatic activities, and oxidative stress. Besides these accumulation of heavy metals like cadmium and lead at levels higher than recommended interferes with essential physiological processes and induces oxidative damage.

Under stress conditions, plants have evolved intricate mechanisms to sense and respond to abiotic stresses. Plants utilize membrane-bound receptors and sensors to detect environmental changes. For instance, calcium (Ca^{2+}) channels play a pivotal role in perceiving osmotic changes during drought and salinity stress (Zhang et al. 2025). Once the plants perceive stress, secondary messengers like Ca^{2+} ions, reactive oxygen species (ROS), and phyto-hormones mainly abscisic acid transmit signals, activating specific protein kinases and phosphatases. Further, at the molecular level stress-responsive transcription factors regulate the expression of genes involved in protective functions, such as osmoprotectant synthesis, antioxidant defense, and molecular chaperones (Zhang et al. 2022). Additionally, abiotic stress factors can induce epigenetic changes like DNA methylation and histone modifications that can alter gene expressions. Plants employ various strategies to mitigate the adverse effects of abiotic stresses including accumulation of compatible solutes like proline and glycine betaine helps maintain cell turgor and enzyme function under osmotic stress (Zhang et al. 2025), enhanced production of antioxidants, such as superoxide dismutase and catalase, scavenges ROS, protecting cells from oxidative damage. Under temperature stress, plants synthesize specific proteins like heat shock proteins and late embryogenesis abundant proteins that help in protein folding and protection during stress conditions. Also, alterations in membrane lipid composition help maintain fluidity and functionality under temperature extremes (Zhang et al., 2022).

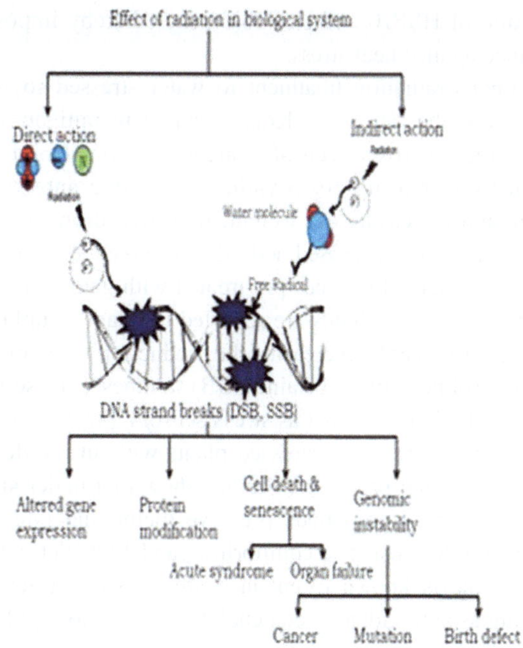

Fig. 5 Mechanism of action: gamma rays

9.2 Use of Induced Mutagenesis for Improving Crop Resilience to Abiotic Stress

Abiotic stresses significantly impact crop productivity, posing challenges for sustainable agriculture and food security. Climate change exacerbates these stresses, making it important to develop crops with enhanced resilience. Recent advances in omics studies and genetics have generated a lot of information about molecular mechanisms underlying plant stress responses, offering potential targets for genetic improvement (Zhang et al. 2025). Traditional breeding methods are often time-consuming and limited by genetic variation within existing germplasm. Gamma irradiation induced mutagenesis has been used for generating novel genetic variability and improving crop resilience to abiotic stresses (Bharat et al. 2024). Mechanism of action of gamma rays induced mutation is illustrated in Fig. 5. Hormesis effect is observed in gamma irradiation that refers to an effect where lower doses of any agent result in positive effect while higher doses exhibit toxicity. Low to moderate doses (e.g., 50–300 Gy) are often beneficial for improving stress tolerance, while higher doses may cause detrimental effects such as reduced germination, growth inhibition, and sterility. In rice *(Oryza sativa)*, 92% of all mutants created with physical agents were created only with gamma rays. According to Jan et al. (2011), gamma rays are comparatively more successful at removing tiny deletions (1–16 bp) from the *Oryza sativa* genome than they are at removing big fragments and deletions (9.4–129.7 kb).

Numerous studies have demonstrated that gamma rays can produce genetic variety in crops, hence enhancing their resistance to abiotic stressors. *Oryza sativa* mutant line ST-87/ST-301 has been reported to exhibit tolerance to salinity (Song et al. 2012). Heavy metals like arsenic (As), cadmium, mercury (Hg), lead (Pb), chromium (Cr), aluminum (Al), and beryllium (Be) have been proven to cause metabolic problems and morphological abnormalities that ultimately result in low productivity. *Oryza sativa* mutant line arsenic-tolerant type 1 (*aatt1*) has been shown to exhibit increased tolerance to arsenic stress (Hwang et al. 2017). Pre-treatment of seeds or seedlings with lower doses of gamma irradiation (5–100 Gy) have been found to increase tolerance against heavy metal stress in several crops like barley (Wang et al., 2017), *Arabidopsis thaliana* (Qi et al. 2015), and pea. Lower doses of gamma irradiation have been reported to trigger auxin responsive genes, RNA, protein synthesis, and elevated production of ROS (Qi et al. 2015; Wang et al. 2017). A considerable up-regulation of the expression of ABA biosynthesis genes and related signal transduction processes has been reported in plants in response to stress conditions (Qi et al. 2015). Pre-treatment of seeds with lower doses of gamma rays has been found to increase the production of enzymatic antioxidants like superoxide dismutase, peroxidase, and catalase along with proline resulting in improved tolerance to heavy metals by alleviating oxidative stress and preventing heavy metal-induced destruction of chloroplasts (Wang et al. 2017). Treatment of heat stressed *Arabidopsis thaliana* with lower doses of gamma rays resulted in an increased level of enzymatic and non-enzymatic antioxidants along with up-regulation of HSP70 and HSP18.2 genes thereby imposing resistance against heat stress.

Gamma irradiation treatment to water stressed-soybean seedlings at the rate of (5 Krad) resulted in mitigation of some of the negative effects of drought stress on plant development by increasing photosynthesis, boosting antioxidant system, and increasing dry weight. Improved drought tolerance mediated by increased activities of enzymatic antioxidants was observed in seeds pre-treated with gamma rays at the rate of 1–5 Gy. Plants are reported to adapt to unfavorable circumstances by increasing the production of osmolytes such as proline, glycinebetaine (GB), amines, and soluble sugars. Osmolytes also act as stress sensors, preserving cell volume and hydration in stressed plants without interfering with their normal processes. Studies show that under stress condition, these compounds preserve membrane integrity, protein stability, osmotic equilibrium, and ROS detoxification. Proline is known to enhance abiotic stress tolerance by acting as antioxidant, metal chelator, and osmo-regulator, giving plants resistance to stress. Increased proline contents have been reported in gamma irradiated mango, *Arabidopsis thaliana*and soybean leading to increased tolerance to salinity and drought respectively. Choline monooxygenase (CMO) and betaine aldehyde dehydrogenase (BADH) are the key enzymes catalyzing the rate limiting steps in the biosynthesis of glycinebetaine, another important osmoprotectant that enhances plant's tolerance to abiotic stress. Gamma irradiation lead to changes in expression of CMO and BADH gene resulting in enhanced production of glycinebetaine in gamma irradiated salt-stressed wheat. In another study, gamma irradiated population of polyembryonic mango variety bappakkai recorded higher K^+, highest fold increase in proline content (+7.27 fold), highest percent increase in chlorogenic acid (+510%), protocatechuic acid (+750%), and ferulic acid along with lower Na^+/K^+ ratio and lower reduction in the levels of caffeic and sinapic acid at higher level of salt stress suggesting that mutations were better at tolerating salt stress as compared to the other two genotypes (Perveen et al. 2024). Salinity results in modified expression of genes related to stress further affecting the roles of proteins involved in signal transduction, antioxidative defence, osmolyte synthesis. Gamma-ray treatment (50 Gy) has been reported to increase protein synthesis under salinity stress. RNA sequencing-based transcriptome analysis observed differential expression of several genes in response to salt stress in both the genotypes. Quantitative real-time PCR (qRT–PCR) based validation of transcriptome data showed contrasting expression pattern of Auxin Response Factor 2 (ARF2), Late Embryogenesis Abundant (LEA), and CDPK genes were observed between Turpentine and Mylepelian (Perveen et al. 2023b). Chemical mutagen EMS has also been used frequently for generating useful mutations in plants to improve abiotic stress tolerance. EMS induced mutants have been reported to possess good level of salt tolerance and mutants N22-L-1013, N22-L-806, and N22-L 1010 were identified as typical Na excluders (Shankar et al. 2021).

Putative mutants generated through EMS treatment (0.1%) of sugarcane calli were subjected to drought stress induced by 28% PEG-6000. Two mutant lines of sugarcane cultivar ROC22 (i.e., MR22-15 and MR22-20) were found to possess better physiological adaptation under drought stress and possess great potential for drought resistance breeding. EMS derived mutants of spring wheat and rice (Nasser et al. 2024) were found to be more tolerant to drought stress than control plants. Using CRISPR system in combination of Cas9 and a truncated sgRNA (tru-sgRNA), *ost2* mutants were generated with greater than 32% extraordinary efficiency exhibiting a substantially higher rate of stomatal closure with reduced rate of transpirational water loss thus increasing drought tolerance. Negative regulators of abiotic stress also known as S genes could be potential targets for genome editing. Knocking out of such genes can result in increased stress tolerance by disrupting the pathways involved. CRISPR/Cas9-mediated knock out of S gene in tomato *SlAGAMOUS-LIKE 6* (*SlAGL6*) resulted in heat tolerance and improved fruit set under heat stress (Klap et al. 2017). CRISPR/dCas9 fused with *histone acetyltransferase* was introduced to enhance the expression of *AREB1*-activated plants exhibited increased drought tolerance, higher chlorophyll content and a fast stomatal closure under a water deficit. *SlLBD40* belonging to subfamily II *LBD* family in tomato is involved in jasmonic acid signaling and is known to be a negative regulator of drought tolerance. CRISPR/Cas9 mediated knockout of *SlLBD40* has been shown to increase water holding capacity and enhance drought tolerance in tomato (Biswas et al. 2021).

10 Use of Induced Mutagenesis for Enhancing Crop Resilience to Biotic Stress

10.1 Understanding Biotic Stress

Plants constantly interact with a diverse range of living organisms, some of which pose significant threats to their growth and survival. Biotic stress, caused by pathogens, insect pests, nematodes, and parasitic plants, disrupts normal physiological processes, leading to reduced yield and crop quality. These stresses trigger complex defense responses in plants, including physical barriers, chemical signaling, and immune responses. However, the increasing impact of climate change has intensified biotic stress, altering pest and pathogen dynamics and increasing crop vulnerability. Among the most devastating biotic stress factors are bacterial, fungal, and viral pathogens. Bacterial diseases, such as bacterial blight in rice caused by *Xanthomonas oryzae*, interfere with the vascular system, resulting in wilting and leaf necrosis. Similarly, fungal pathogens like *Magnaporthe oryzae*, responsible for rice blast disease, and *Fusarium oxysporum*, which causes vascular wilt, pose serious threats to global food security by damaging plant tissues and obstructing nutrient transport. Insect herbivory represents another major biotic stress, with pests such as aphids, whiteflies, and caterpillars consuming plant tissues and acting as vectors for secondary infections. Root-feeding insects, including wireworms and root weevils, damage underground structures, limiting water and nutrient uptake. In response, plants activate defense mechanisms such as the production of toxic secondary metabolites, trichomes, and signaling molecules like jasmonic acid to deter herbivores (Li et al. 2023). However, these natural defenses are often insufficient against large-scale infestations, necessitating integrated pest management strategies. Nematodes also contribute significantly to biotic stress, particularly root-knot nematodes (*Meloidogyne spp.*), which form galls on roots, disrupting normal function and predisposing plants to secondary infections. They severely impact crops such as citrus, grapes, bananas, strawberries, apples, and peaches, often resulting in poor fruit quality and reduced harvests. In citrus orchards, *Tylenchulus semipenetrans*, commonly known as the citrus nematode, is a major concern. This nematode causes slow decline in trees, leading to reduced fruit size and lower yields over time. Similarly, *Radopholussimilis*, the burrowing nematode, poses a severe threat to tropical fruit crops like bananas and plantains by causing root necrosis and weakening the plant's stability, often leading to toppling disease. The lesion nematodes (*Pratylenchus* spp.) are another group of damaging nematodes that create necrotic lesions on roots, predisposing crops such as apples, pears, peaches, and strawberries to secondary infections by soilborne pathogens. Beyond direct root damage, some nematodes act as vectors for plant viruses, exacerbating their impact. *Xiphinema* spp., commonly known as dagger nematodes, are notorious for transmitting viral diseases such as grape fanleaf virus, which significantly reduces vine productivity and fruit quality. Likewise, reniform nematodes (*Rotylenchulusreniformis*) weaken root systems in crops like bananas, citrus, and pineapples, leading to poor water uptake and diminished fruit development.

To counteract biotic stress, plants have developed several defense mechanisms, ranging from structural barriers such as waxy cuticles and lignified cell walls to chemical defenses involving antimicrobial compounds and phytohormones. The activation of systemic acquired resistance (SAR) and induced systemic resistance (ISR) further enhances plant immunity, allowing them to respond more effectively to repeated pathogen attacks. Hypersensitive response (HR) is one of the common defense mechanisms in plants which results in rapid and localized cell death at the infection site (Wu et al. 2008). However, climate change is complicating these interactions by altering pest and pathogen distribution, influencing plant immune responses, and accelerating the evolution of virulent pathogen strains. Rising temperatures and increased atmospheric CO_2 levels have been shown to affect insect feeding behavior, pathogen infection rates, and plant resistance pathways, leading to more unpredictable outbreaks of biotic stress factors (Leisner et al. 2023). Given the severity of biotic stress and its implications for global agriculture, research efforts are increasingly focused on improving plant resistance through breeding for disease-resistant traits, and exploring plant-microbe interactions to develop sustainable disease management strategies. Techniques such as induced mutagenesis, CRISPR-based genome editing hold promise for enhancing plant immunity by introducing targeted mutations in susceptibility genes or enhancing defense-related pathways (Li et al. 2022).

10.2 Use of Induced Mutagenesis for Improving Crop Resilience to Biotic Stress

Induced mutations using physical and chemical mutagens have been used to develop disease and pest resistant crop varieties. Kaur et al (2016) reported use of gamma irradiation for improvement of sugarcane hybrid Co J 64 (Co 976 × Co 617) for resistance against red rot. Lower doses of gamma ray (10 and 20 Gy) were found to generate mutants moderately resistant against red rot suggesting the effectiveness of lower doses of irradiation in generating disease resistant mutants (Kaur et al. 2016). In another study, gamma irradiation was used to generate rice mutant (mutant R917) resistant against fungal disease blast caused by *Magnaporthe oryzae*. Improved potato cultivars Kufri Jyoti and Kufri Chandramukhi were irradiated with gamma rays (40 Gy) to induce heat tolerance and resistance against late blight disease (Gosal et al. 2001). Fusarium wilt resistant banana mutants have been generated using gamma irradiation mediated mutations. Mutagens like ultraviolet and ozone rays have been reported to increase synthesis of pathogenesis related proteins and salicylic acid in tobacco plants. Resulting increased resistance against viral (TMV) infections. After a pathogen invasion, a plant cannot quickly activate all of its defence mechanisms. Along with the buildup of SA and PR-1 to boost TMV resistance in plants, radiation treatment can stimulate a variety of enzyme synthesis pathways that aid in the synthesis of phytoalexins to fight the pathogens. The soil-borne fungus *Fusarium oxysporum* f. sp. cubense (Foc) was the cause of Fusarium wilt (Panama disease), which decimated the banana export sector that included Gros Michel. The strain that affects Gros Michel, Foc race 1, was found to be ineffective against Cavendish banana. But a new strain known as Foc tropical race 4 (TR4) started to seriously harm Cavendish bananas in Australia's Northern Territories and five Asian nations, and it is currently spreading throughout the world. Jankowicz-Cieslak et al. (2021) reported generation of banana mutants tolerant to Foc TR4 by using gamma irradiation treatment at the rate of 30–40 Gy. CRISPR was used to target *eIF4E* gene to generate cucumber mutants immune to cucumber vein yellow virus (Chandrasekaran et al. 2016). *PthA4* is the main transcription activator-like (TAL) effector of *Xcc* which specifically binds to the effector binding element (EBE_{PthA4}) in the *CsLOB1* promoter and activates the expression of *CsLOB1* which leads to citrus canker development (Hu et al. 2014). TALEN-mediated knockout of three *TaMLO* homeologues conferred strong resistance against powdery mildew (PM) in wheat but caused growth and yield penalties, restricting its use in plant breeding. Later on CRISPR/Cas9 mediated genome editing produced a knockout mutant with deletion in TaMLO-B1, showing strong resistance against PM while avoiding yield penalties (Zhang et al. 2017). Also, CRISPR/Cas9 knockout of all *TaEDR1*homeologues provided moderate resistance to powdery mildew besides enhancing stripe rust resistance in wheat without affecting growth and yield parameters making it suitable for breeding (Liu et al. 2024). In banana CRISPR was used to inactivate an endogenous banana streak virus in the variety Gonja.

11 Conclusion

Enhancing crop resilience to abiotic and biotic stresses is essential for sustainable agriculture. Induced mutation breeding is a powerful tool that introduces genetic variability, enabling the development of stress-tolerant crop varieties. Unlike conventional breeding, it generates novel traits without altering desirable characteristics, making it effective for improving resistance to multiple stresses. Abiotic factors like drought, salinity, and heat, combined with biotic pressures from pathogens and pests, significantly impact crop yields. Induced mutations through physical (gamma rays, X-rays) and chemical (EMS) methods help develop resilient crops by enhancing their physiological and molecular responses. Integrating advanced techniques such as CRISPR and TALEN further strengthens biotic stress resistance by precisely editing genes associated with disease tolerance, offering a faster and more targeted breeding approach. High-throughput sequencing, marker-assisted selection, and phenomics accelerate the identification and utilization of beneficial mutations. Combining mutation breeding with modern gene-editing tools ensures a comprehensive strategy for crop improvement. As climate change intensifies agricultural challenges, induced mutation, along with CRISPR and TALEN-mediated breeding, provides sustainable solutions for developing resilient crops. Continued research and collaboration are vital to harness these innovations for global food security and agricultural sustainability.

References

Atkinson NJ, Urwin PE (2012) The interaction of plant biotic and abiotic stresses: from genes to the field. J Exper Bot 63(10):3523–3543. https://doi.org/10.1093/jxb/ers100

Bharat RA, Prathmesh SP, Sarsu F, Suprasanna P (2024) Induced mutagenesis using gamma rays: biological features and applications in crop improvement. OBM Gen 8(2):1–27. https://doi.org/10.21926/obm.genet.2402233

Biswas D, Saha SC, Dey A (2021) CRISPR-Cas genome-editing tool in plant abiotic stress-tolerance. Plant Gene 26:100286. https://doi.org/10.1016/j.plgene.2021.100286

Chandrasekaran J, Brumin M, Wolf D, Leibman D, Klap C, Pearlsman M, Sherman A, Arazi T Gal-On A (2016) Development of broad virus resistance in non-transgenic cucumber using CRISPR/Cas9 technology. Mol Plant Pathol 17:1140–1153. https://doi.org/10.1111/mpp.12375

Chen K, Wang Y, Zhang R, Zhang H, Gao C (2019) CRISPR/Cas genome editing and precision plant breeding in agriculture. Ann Rev Pl Bio 70(1):667–697. https://doi.org/10.1146/annurev-arplant-050718-100049

FAO (2021) The state of food and agriculture: climate change and agriculture. Food and Agriculture Organization of the United Nations

Fischer RA, Byerlee D, Edmeades G (2014) Crop yields and global food security. ACIAR: Canberra ACT 8–11

Gosal SS, Das A, Gopal J, Minocha JL, Chopra HR, Dhaliwal HS (2001) In vitro induction of variability through radiation for late blight resistance and heat tolerance in potato

Hall EJ, Giaccia AJ (2006) Radiobiology for the radiologist. Int J Radiat Oncol Biol Phys 66(627):10–1016

Hu Y, Zhang JL, Jia HG, Sosso D, Li T, Frommer WB, Yang B et al (2014) Lateral organ boundaries 1 is a disease susceptibility gene for citrus bacterial canker disease. PNAS 111:E521–E529. https://doi.org/10.1073/pnas.1313271111

Hwang SG, Chapagain S, Lee JW, Han AR, Jang CS (2017) Genome-wide transcriptome profiling of genes associated with arsenate toxicity in an arsenic-tolerant rice mutant. Plant Physiol Biochem 120:40–51. https://doi.org/10.1016/j.plaphy.2017.09.019

Jan S, Parween T, Siddiqi TO, Mahmooduzzafar (2011). Effect of gamma radiation on morphological, biochemical, and physiological aspects of plants and plant products. Environ Rev 20:17–39. https://doi.org/10.1139/a11-021

Jankowicz-Cieslak J, Goessnitzer F, Datta S, Viljoen A, Ingelbrecht I, Till BJ (2021) Induced mutations for generating bananas resistant to Fusarium wilt tropical race 4. In Mutation breeding, genetic diversity and crop adaptation to climate change (pp. 366-378). Wallingford UK: CABI

Kaur M, Thind KS, Sanghera GS, Kumar R, Kashyap L (2016) Gamma rays induced variability for economic traits, quality and red rot resistance in sugarcane (*Saccarum* spp.). Int J Environ Sci Technol 5:355–365

Klap C, Yeshayahou E, Bolger AM, Arazi T, Gupta SK, Shabtai S, Usadel B, Salts Y, Barg R (2017) Tomato facultative parthenocarpy results from SlAGAMOUS-LIKE 6 loss of function. Pl Biotech J 15:634–647. https://doi.org/10.1111/pbi.12662

Kumar K, Gill MIS, Choudhary OP (2010) In vitro mutagenesis and somaclonal variation assisted salt tolerance in 'Rough lemon' (*Citrus jambhiri* Lush.). Eur J Horticult Sci 75(6):233–238

Leisner CP, Potnis N, Sanz-Saez A (2023) Crosstalk and trade-offs: plant responses to climate change-associated abiotic and biotic stresses.

Plant Cell Environ 46(10):2946–2963. https://doi.org/10.1111/pce.14532

Lesk C, Rowhani P, Ramankutty N (2016) Influence of extreme weather disasters on global crop production. Nature 529(7584):84–87. https://doi.org/10.1038/nature16467

Li Y, Wu X, Zhang Y, Zhang Q (2022) CRISPR/Cas genome editing improves abiotic and biotic stress tolerance of crops. Front Gen Ed 4:987817. https://doi.org/10.3389/fqeed.2022.987817

Li T, Feng M, Chi Y, Shi X, Sun Z, Wu Z, Li A, Shi W (2023) Defensive resistance of cowpea *Vigna unguiculata* control megalurothrips usitatus mediated by jasmonic acid or insect damage. Plants 12(4):942. https://doi.org/10.3390/plants12040942

Liu S, Zhang F, Su J, Fang A, Tian B, Yu Y, Bi C, Ma D, Xiao S, Yang Y (2024) CRISPR-targeted mutagenesis of mitogen-activated protein kinase phosphatase 1 improves both immunity and yield in wheat. Pl Biotech J 22(7):1929–1941. https://doi.org/10.1111/pbi.14312

Nasser H, Youssef MS, El-Kholy AS, Haroun SA (2024) EMS-mutagenesis, in vitro selection for drought (PEG) tolerance and molecular characterization of mutants in rice (*Oryza sativa* L.) employing qRT-PCR and ISSR markers. Egypt J Bot 64(1):87–105. https://doi.org/10.21608/ejbo.2023.225773.2436

Perveen N, Dinesh MR, Sankaran M, Shivashankara KS,Venugopalan R (2022) Characterization and evaluation of putative mutant populations of polyembryonic mango genotype Nekkare for dwarfing rootstock traits. J Hortic Sci 17(2):261–271. https://doi.org/10.24154/jhs.v17i2.1456

Perveen N, Dinesh MR, Sankaran M, Venugopalan R (2023a) Mutagenic-sensitivity and variability in the putative mutants of polyembryonic mango genotypes. Indian J Hortic 80(1):3–9. https://doi.org/10.58993/ijh/2023.80.1.1

Perveen N, Dinesh MR, Sankaran M, Ravishankar KV, Krishna jee HG, Hanur VS, Alamri S, Kesawat MS, Irfan M (2023b) Comparative transcriptome analysis provides novel insights into molecular response of salt-tolerant and sensitive polyembryonic mango genotypes to salinity stress at seedling stage. Front Plant Sci 14:1152485. https://doi.org/10.3389/fpls.2023.1152485

Perveen N, Dinesh MR, Sankaran M, Lakkireddy V, Shivashankara KS, Venugopal R, Mir H (2024) Phenolics signatures in response to salinity stress provide novel insights into physiological basis of salt tolerance in mango (*Mangifera indica* L.). J Pl Growth Regul 43(12):4866–4885. https://doi.org/10.1007/s00344-024-11440-6

Perveen N, Dinesh MR, Sankaran M. Bindu KH, Shivashankara KS, Ravishankar KV, Kanade NM, Kalal P, Mir H (2025) Radiation induced mutations alter morpho-biochemical, anatomical and molecular responses in polyembryonic mango genotype Bappakkai. Plant Physiol Rep 30:284–295. https://doi.org/10.1007/s40502-025-00860-5

Qi W, Zhang L, Wang L, Xu H, Jin Q, Jiao Z (2015) Pretreatment with low-dose gamma irradiation enhances tolerance to the stress of cadmium and lead in *Arabidopsis thaliana* seedlings. Ecotoxicol Environ Saf 115:243–249. https://doi.org/10.1016/j.ecoenv.2015.02.026

Shankar A, Choudhary O, Singh K (2021). Effect of EMS induced mutation in rice cultivar nagina 22 on salinity tolerance. bioRxiv. https://doi.org/10.1101/2021.08.03.455004

Sharma DK, Majumdar PK, Singh RN (1983) Induction of mutation in mango (*Mangifera indica* L.). J Nuclear Agric Biol 12:14–17

Song JY, Kim DS, Lee MC, Lee KJ, Kim JB, Kim SH (2012) Physiological characterization of gamma-ray induced salt tolerant rice mutants. Aust J Crop Sci 6:421–429. https://doi.org/10.3316/INFORMIT.360742565763762

Spiegel-Roy P, Kochba J (1973). Mutation breeding in citrus. In: FAO/IAEA Induced mutations in vegetative-propagated plants. IAEA, pp 91–105. https://doi.org/10.5555/19751630606

Surakshithaa NC, Soorianathasundarama K, Meenakshi Ganesan N (2017) Determination of mutagenic sensitivity of hardwood cuttings of grapes 'Red Globe' and 'Muscat' (*Vitis vinifera* L.) to gamma rays. Sci Horticult 226:152–156. https://doi.org/10.1016/j.scienta.2017.08.040

Wang X, Ma R, Cui D, Cao Q, Shan Z, Jiao Z (2017) Physio-biochemical and molecular mechanism underlying the enhanced heavy metal tolerance in highland barley seedlings pre-treated with low-dose gamma irradiation. Sci Rep 7(1):14233. https://doi.org/10.1038/s41598-017-14601-8

Wu C, Bordeos A, Madamba MRS, Baraoidan M, Ramos M, Wang G, Leach JE, Leung H (2008) Rice lesion mimic mutants with enhanced resistance to diseases. Mol Genet Genom 279:605–619. https://doi.org/10.1007/s00438-008-0337-2

Zhang Y, Bai Y, Wu G, Zou S, Chen Y, Gao C et al. (2017) Simultaneous modification of three homoeologs of TaEDR1 by genome editing enhances powdery mildew resistance in wheat. Plant J 91:714–724. https://doi.org/10.1111/tpj.13599

Zhang H, Zhu J, Gong Z, Zhu JK (2022) Abiotic stress responses in plants. Nat Rev Genet 23(2):104–119. https://doi.org/10.1038/s41576-021-00413-0

Zhang H, Lang Z, Zhu JK, Wang P (2025) Tackling abiotic stress in plants: recent insights and trends. Stress Biol 5:8. https://doi.org/10.1007/s44154-025-00216-x

Mutation Breeding for Crop Adaptation to Climate Change

C. Akshaya Prakash, Nair G. Sarath, Delse P. Sebastian, and Satheesh George

Abstract

Population growth, combined with climate change and natural catastrophes, has seriously affected the food security worldwide. As per the reports of the Food and Agricultural Organization of the United Nations, one in nine people of the world suffers from undernourishment. Ensuring food security in the current scenario can be achieved to some extent by using appropriate crop improvement methods. Mutation breeding is an appropriate strategy to enhance the germplasm collection of crops because the technique can be exploited to develop and create new varieties. Moreover, the time required to develop new varieties is comparatively less compared to conventional hybridization techniques. Physical, chemical and biological mutagens have been used to induce mutations in plants. A number of studies have reported that mutation breeding is a potential breeding technique that has the capability to improve the quantitative and qualitative characteristics of crops significantly in a shorter span of time compared to conventional breeding techniques. This technique can be exploited for improving the resilience of crops towards biotic and abiotic stresses induced by climate change. Further efforts to advance mutation breeding using appropriate methods will enable humanity to accomplish the primary objective of plant breeding, thus promoting global food security.

C. Akshaya Prakash · D. P. Sebastian (✉) · S. George
Department of Botany, St. Joseph's College (Autonomous) Devagiri (Affiliated to University of Calicut), Kozhikode, Kerala, India
e-mail: delsbotany@gmail.com

N. G. Sarath
Department of Botany, Mar Athanasius College (Autonomous), Kothamangalam, Kerala, India

Keywords

Abiotic and biotic stresses · Crop improvement · Food security · Mutagens · Population growth

1 Introduction

Presently, the world is experiencing a tremendous growth in human population exerting severe stress on various resources. According to the medium-variant estimate, the annual population growth is 1.1% and is expected to reach 9.7 billion by the year 2050 (UN 2018).

One of the greatest challenges of world today is to ensure food availability to every single person while focusing on environmental and social sustainability. Extreme changes in climate are the major driving force of global hunger (FAO 2018). Climate change directly affects crop productivity, reducing 3.1–7.4% of crop yields for each degree-Celsius rise in global mean temperature (Zhao et al. 2017).

Population growth, combined with climate change and natural catastrophes, has seriously affected the food security worldwide. Globally, the percentage of undernourished individuals has risen to 10.6% in the year 2015 and to 11% in the year 2016 (UN 2018). As per the reports of the Food and Agricultural Organization of the United Nations, one in nine people of the world suffers from undernourishment (FAO 2018). Recent projections using 2014 baseline announced that, in order to meet the growing crop demand, the crop production requires to be increased by 25–70% by 2050 (Hunter et al. 2017). Moreover, FAO (2016) declared that if the current situation of climate change prevails or aggravates, the yields of major cereal crops like maize, wheat and rice would reduce by 25–45%, 5–50% and 20–30% respectively, by 2100. Hence, it is high time that appropriate strategies are taken for facing climate change induced issues and ensuring food security for all living beings.

J. Devi (ed.), *Mutation Breeding in Crop Improvement*, Sustainable Landscape Planning and Natural Resources Management,
https://doi.org/10.1007/978-3-032-17911-1_6

Ensuring food security in the current scenario can be achieved to some extent by using appropriate crop improvement methods. Crop improvement deals with genetic modifications of plants for meeting the food requirements of expanding population (Tahir ul Qamar et al. 2020). Since time immemorial, farmers have been practicing various conventional crop improvement techniques for enhancing the quality and quantity of crops. Here they created the best-fit cultivars by combining particular characteristics from closely related plants, such as increased yields, disease resistance, tasty fruits and quicker growth. Common conventional crop improvement techniques practised included pure line selection, hybrid breeding, population breeding and pedigree breeding.

Pure line selection involves the cross breeding of two closely lines exhibiting desired features. The progenies are selected from each selected plant for examination of desired characteristics and further screened, as a result of which crops with desired traits are developed. Hybrid breeding involves intraspecific or interspecific breeding of two different lines resulting in hybrid progenies with augmented occurrence of desired features (Reif et al. 2005). Population breeding leads to increasing the frequency of advantageous alleles, thereby enhancing the performance of crops (Breseghello and Coelho 2013). Pedigree breeding uses self-pollinating species to produce several cultivars. Parents having desirable features are crossed, followed by a number of selection events to get different inbred lines. The inbred lines can have desirable characteristics of both the parents (Breseghello and Coelho 2013).

Even though conventional breeding techniques have various advantages like being economical and easier, some limitations call for the need for modern techniques for crop improvement. Some of the limitations include inbreeding depression, probability of appearance of recessive deleterious genes, inability to enhance food quality and time-consuming nature (Tahir ul Qamar et al. 2020). Scientists have found out methods that can overcome the limitations of conventional breeding techniques. These non-conventional or modern plant breeding techniques include mutation breeding, in vitro culture and molecular breeding. Compared to the conventional methods, these techniques have been reported to be faster and effective (Sun et al. 2024). The present study gives a comprehensive account on mutation breeding and its role in enabling crops to adapt climate change.

Mutation refers to a sudden heritable alteration in the DNA of a living organism that is not the result of genetic recombination or segregation (van Harten 1998). The intentional application of mutations in plant breeding is known as mutation breeding (Pathirana 2011). Mutation breeding offers the benefit of correcting a defect in an already superior cultivar without sacrificing its agronomic and qualitative traits when compared to hybridization and selection (Pathirana 2011). Mutation breeding has been found to have applications in improving the characteristics of seedless crops, root and tuber crops, as well as other vegetatively propagated plants (Pathirana 2011).

2 History of Mutation Breeding

According to history, the first account of crop plant mutations was recorded in the book Lula in China around 300 BC. But roughly 2317 years ago, the first naturally occurring mutant plants among cereals were discovered in China (Van Harten 1998). In 1791, an English farmer Seth Write noticed an oddly short-legged male sheep in his livestock at his farm in Dover, Massachusetts, USA. This was the first documented instance of an actual case of dominant germinal mutation in domestic animals and its application in cross-breeding. The short-legged 'Ancon' sheep breed was developed from this mutant animal (Van Harten 1998). In the late nineteenth century, while conducting experiments to rediscover Mendels's rules of inheritance, Hugo de Vries discovered a variant in evening primroses (*Oenothera lamarckiana*) that deviated from the Mendelian pattern of inheritance (3:1), but was heritable (Kharkwal 2023). Hugo de Vries initially recognized mutation as a method for generating variety in 1901. He saw mutations as heritable alterations caused by mechanisms apart from segregation and recombination (Kharkwal 2023).

Hugo de Vries who is credited with discovering mutations, defined them as abrupt changes in organisms that were inherited and had a major impact on phenotypic characteristics. He termed the word 'mutation' and offered a comprehensive theory on the occurrence of abrupt, startling changes (leaps) in preexisting characteristics that resulted in the emergence of new species and diversity (Kharkwal 2023). de Vries discovered several abnormal varieties of evening primroses in his research, which he called 'mutants'. After de Vries' seminal 'Mutationstheorie' (1901–1903) was published, the idea of mutation as a primary source of evolution became widely accepted. de Vries also proposed that X-rays and gamma rays may be used to artificially induce mutations. The history of mutation techniques in the field of plant breeding may be considered to have begun with de Vries's work due to his 'Mutation Theory of Evolution' and his early theories on possible application and utility of mutations for plant breeding. However, it was Muller (1927) and Stadler (1928) who first recognized the relevance of mutations in breeding through their X-ray studies using fruit flies and maize and barley, respectively, to produce genetic alterations.

Lewis John Stadler's (1928, 1930) discoveries of the mutagenic effect of X-rays in maize, barley and wheat initiated the application of the mutation approach to create unique genetic variety in plants by radiations. Despite publishing findings on induced mutations in barley as well, Stadler was rather pessimistic about them and claimed that they would not be useful for plant breeding (Stadler 1928). However, by 1930 Stadler could find hundreds of X-ray mutations in plants. His experimental designs were critical and exquisite, with the most remarkable dimensions (Kharkwal 2023). These findings sparked in-depth research on induced mutations, demonstrated the usefulness of radiation as a plant breeding technique, and nearly immediately led to the recovery of certain commercially valuable wheat mutants (Delaunay 1931). Development of various mutant wheat, oats and rice varieties occurred following this. The production of the dwarf wheat and rice varieties are prime examples of mutation breeding accomplished by effectively utilizing the natural mutant genes Rht-1 and Rht2—Norin in the case of wheat and sd1—dee-gee woo-gen in rice. These genes affected a wide range of traits that give them their superior agronomic responses, and are known as Green Revolution Genes (Kharkwal 2023).

3 Types of Mutations

Mutations are broadly classified as point mutations, structural mutations and mutations resulting in changes in the number of chromosomes (Pathirana 2011) (Fig 1). Point mutations are those mutations that occur when the chemical structure of a gene is altered at the molecular level. These mutations occur within the gene of DNA sequences (Pathirana 2011). They are also known as gene mutations. Structural mutations are those mutations that bring about changes in the structure of chromosomes. They occur due to the breaking and rearrangement of chromosomes (Pathirana 2011). Such mutations include deletions, duplications, inversions and translocations. They are also known as chromosomal mutations. Various studies have reported that deletions, duplications, inversions and translocations have played important roles in the evolution of crop plants (Pathirana 2011).

Crop evolution and plant breeding also heavily depend on genome-level mutations that alter the number of chromosomes (Vavilov 1992). Genome level mutations include allopolyploidy, autopolyploidy and aneuploidy. Allopolyploidy occurs when the genomes of two or more species combine. It happens as a result of interspecific or intergeneric hybridization and chromosome doubling (Pathirana 2011). Autopolyploidy happens when the chromosome number multiplies within the species resulting in more than two homologous chromosomes. Compounds that prevent spindle fibre development during mitosis can be used to induce autopolyploidy, retaining the split chromosomes within the same cell. The two most commonly used substances for artificial polyploidization are colchicine and oryzalin (Pathirana et al. 2011; Van Harten 1998). Addition or deletion of chromosomes results in aneuploidy (Sharma 1990). Crop improvement programmes may also rely upon aneuploidy to obtain desirable features.

Extranuclear or plasmone mutations are the mutations that occur in the genomes of chloroplast and mitochondria. It has been demonstrated that these genomic mutations are useful in agriculture. Numerous publications have reported that nitroso chemicals frequently result in plasmone mutations. Since these mutations frequently arise as chimaeras, they can be advantageous when breeding ornamental crops with different leaf or fruit characteristics (Pathirana 2011).

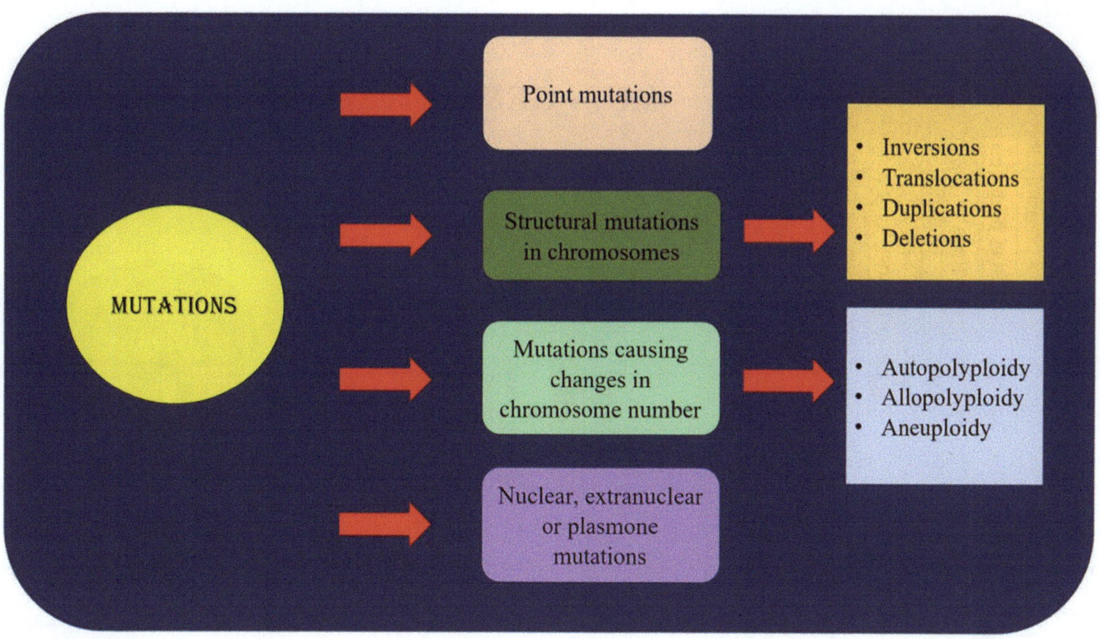

Fig. 1 Types of mutations

4 Mutagens

Mutagens are the substances that can cause mutations. Mutagens can be physical, chemical and biological (Fig. 2). Among these, physical and chemical mutagens are commonly used for inducing mutagenesis in plants. Seeds, whole plants, seedlings or *in vitro* cultured cells can be taken for inducing mutations (Oladosu et al. 2016). Physical mutagens are commonly used to induce chromosomal rearrangements while chemical mutagens are used for inducing point mutations (Kharkwal and Shu 2009). Physical mutagens have some advantages over chemical mutagens like zero waste generation and non-necessity to clean the samples off the mutagen.

Physical mutagens that induce mutagenesis in plants include ionizing and non-ionizing radiations like cosmic rays, alpha particles, beta particles, neutrons, ion beams, laser beams, UV rays, X-rays and gamma rays (Udage 2021). Among these, gamma rays and X-rays are the most widely used physical mutagens for inducing mutations. Gamma rays have large penetrating capability and are utilized for inducing mutations in whole plants as well as soft materials like pollen grains (Oladosu et al. 2016). Because gamma rays have a shorter wavelength than protons and X-rays, they carry greater energy and enter tissues more deeply thus being highly effective. The degree of accuracy and adequate repeatability of physical mutagenesis over chemical mutagenesis are its main advantages, especially for gamma rays, which have a consistent penetrating capacity in the tissue (Yamaguchi et al. 2003). DNA double strand breaks are primarily responsible for the mutagenic effects of physical mutagens. The mutants, thus produced, are employed as a supplementary tool in plant breeding since they have a greater potential to improve plant architecture, which will improve crop quality.

Chemical mutagens include the use of chemicals that can induce mutations in plants. Common chemical mutagens used for mutagenesis include alkylating agents, sodium azide, base analogues and acridine dyes (Udage 2021). Among these, alkylating agents have been reported as the most successful chemical mutagens. Epoxides, ethyleneimines, ethyleneimides, alkyl methanesulphonates, alkylnitrosoureas, alkylnitrosoamines, alkylnitrosoamides, alkyl halides, alkyl sulphates, alkyl phosphates, chloroethyl sulphides, chloroethylamines, diazoalkanes, and sulphur mustards and nitrogen mustards are some examples of alkylating agents (Wani et al. 2014). Alkylating agents alkylate purines, pyrimidines and phosphate groups and thereby induce mutations. Base analogues induce mutation by substituting original bases of nucleic acids. Acridine dyes induce mutations by stretching DNA strands which permits the addition or deletion of nucleotide bases (Kumar and Gorham 1975). An obvious benefit of point mutations induced by chemical mutagens is their ability to produce both gain-of-function and loss-of-function phenotypes (Bradshaw et al. 1997). Moreover, chemical mutagens do not require the use of sophisticated instruments for their application. However, utmost care has to be taken while handling chemical mutagens as they are extremely carcinogenic (Oladosu et al. 2016).

Biological mutagens include DNA transposons, retrotransposons and Insertion Sequence elements (Udage 2021). Insertions, deletions as well as other chromosomal changes are induced by transposons (Krishnan and Damaraju 2019). Using cDNA, which is produced by reverse transcription of the RNA, retrotransposons integrate into new locations of

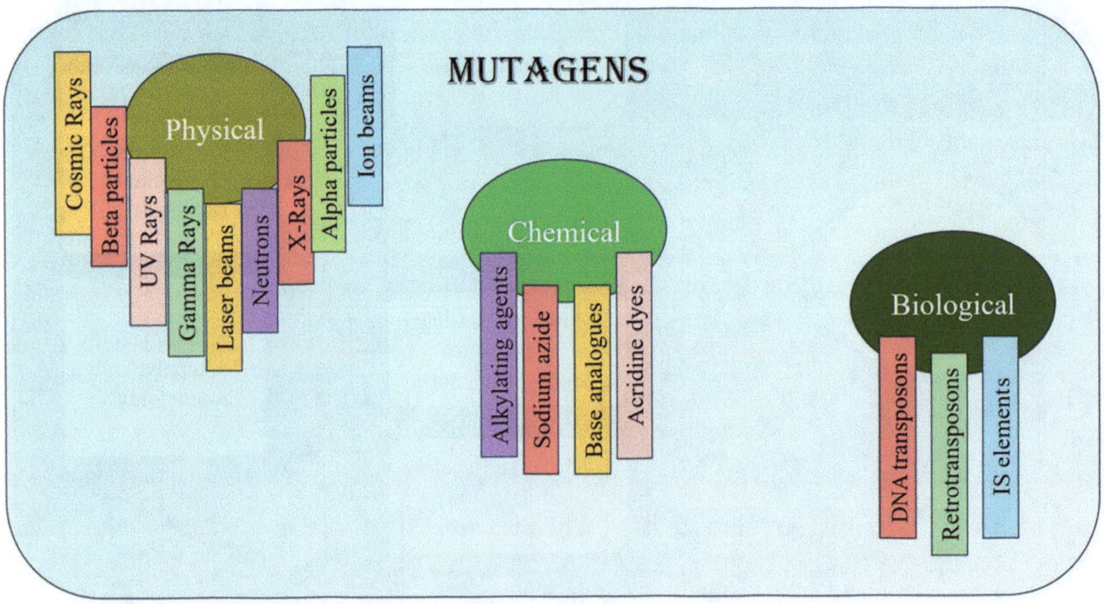

Fig. 2 Different types of mutagens

the host genome after mobilizing via an RNA intermediate, thus bringing about mutations (Qi and Sandmeyer 2012). Insertion Sequence elements are small DNA segments, which upon insertion into a genome, can induce mutations (Williams 2016). Hence, biological mutagens can also be successfully inducing desirable mutations in plants of interest.

5 Application of Mutation Breeding in Crop Improvement Under Climate Change

Climate is an important factor that has a direct role on crop productivity and yield. Therefore, climatic fluctuations can affect the performance of crops adversely. The abiotic stresses induced by climatic changes in plants include heat, cold, salinity and drought stresses. These stresses can severely affect crop yield which in turn affects global food security. It is anticipated that climate change would increase the frequency, duration and severity of abnormally low and high temperatures causing cold and heat stress in plants. Plants suffer heat stress when the temperature rises above its ideal range which can result in reduced yield (Bita and Gerats 2013).

Plants suffer from drought stress when the water available in soil is less and more water is lost through transpiration and evaporation, as a result of climatic conditions. According to Anjum et al. (2011), drought is the most damaging environmental stressor that significantly changes plant production, growth rate, biomass and yield. Salinity stress refers to the detrimental effects that excess mineral salts, such as Na+ and Cl−, have on plants (Parihar et al. 2015). Increasing incidence of salinity stress is anticipated in the future due to various irrigation techniques and changing global environment. Since all these stresses have detrimental effects on crop yield, it has become a necessity to develop crops that are resilient to these stresses.

Various studies have reported the role of mutation breeding in the development of mutants capable of tolerating abiotic stresses. Mutation breeding has been used to more than 232 distinct plant species and crops, including some important crops including wheat, rice, citrus, rapeseed, sunflower, cotton and bananas. According to the Food and Agriculture Organization of the United Nations (FAO)/IAEA (2015), 3222 mutant varieties with enhanced characteristics have been formally released. Panigrahy et al. (2011) reported that treating *Oryza sativa* plants with ethylmethane sulphonate could make the crop heat tolerant. Likewise, a study by Joshi et al. (2016) revealed that gamma irradiation could make *O. sativa* plants salt tolerant. Similar reports are provided in Table 1.

Similar to the case of abiotic stresses, climatic changes can increase the susceptibility of plants to biotic stresses as well. The growth, multiplication, dissemination and severity/infestation of numerous plant pathogens and insect pests are all directly impacted by climatic changes. Climate change has increased the probability of occurrence of spot blotch disease in wheat, Indian cassava mosaic virus, African cassava mosaic virus, Sri Lankan cassava mosaic virus, groundnut bud necrosis virus and *Phytophthora* blight (Chattopadhyay et al. 2019; Kumar et al. 2012). It is anticipated that rising temperatures and precipitation will increase the number of crop plant mites and sucking insect pests (Chattopadhyay et al. 2012).

Through mutation breeding, crops that are tolerant to various insects and pests have been developed. High yield

Table 1 Mutation breeding for crop improvement under abiotic stresses induced by climate change

Sl no	Crop	Mutagen	Stress	Reference
1	*Oryza sativa* L. (Shua-92)	15 Gy fast neutrons	High yield and fine grain quality, salt tolerance	Bughio et al. (2007)
2	*Oryza sativa* L. (IR64)	Gamma rays	Salt tolerance	Joshi et al. (2016)
3	*Oryza sativa* L. (Arsenic-tolerant type 1 (ATT1))	Gamma rays	Arsenic tolerance	Hwang et al. (2017)
4	*Oryza sativa* L (Nagina 22 mutants N22-H- *dgl56*, N22-H-*dgl101*, N22-H- *dgl162* and N22- H-*dgl219*	Ethylmethane sulphonate	Heat tolerance	Panigrahy et al. (2011)
5	*Oryza sativa* L. (Dark green leaf mutant N22-H- dgl219 (NH219)	Ethylmethane sulphonate	Heat stress tolerance	Poli et al. (2013)
6	*Oryza sativa* L. (IR64 mutants 167-1-1, 167-1- 3 and 17-34)	Chemicaland irradiation mutagens	Salinity tolerance	Nakhoda et al. (2012)
7	*Oryza sativa* L. (shs1)	Sodium azide	Salinity tolerance	Lin et al. (2016)
8	*Triticum aestivum* L	Ethyl methane sulfonate	Salt tolerance	Lethin et al. (2020)
9	*Triticum aestivum* L	Gamma irradiation	Salinity tolerance	Karimzadeh et al. (2023)
10	*Triticum aestivum* L. (Sakha 8 and its six M5 derived mutants)	Gamma radiation	Salinity tolerance	El-Mouhamady and Ibrahim (2020)

Table 2 Mutation breeding for crop improvement under biotic stresses induced by climate change

Sl no	Crop	Mutant	Mutagen	Benefit	Reference
1	*Oryza sativa* L	Mutants from Salama and Supa variety	Gamma-rays	High yield potential and resistant to rice yellow mottle virus (RYMV)	Luzi-Kihupi et al. (2009)
2	*Oryza sativa* L	Mutants from Sarshar variety	150 Gy gamma rays	High yielding variety with tolerance to insect pests	Bughio et al. (2007)
3	*Oryza sativa* L	Zhefu802	Gamma-rays	Resistance to rice blast	Kharkwal and Shu (2009)
4	*Oryza sativa* L	Atomita3	Gamma-rays	Resistance to bacterial leaf blight and bacterial leaf stripe	IAEA (2011)
5	Durum wheat	Sredetz	Gamma-rays	Resistanceto Black stem rust	IAEA (2011)
6	Mungbean	MUM2	Gamma-rays	Resistance against Yellow mosaic virus	Ahloowalia et al. (2004)
7	Chickpea	CM98	Gamma-rays	Resistance to Ascochyta blight and Fusarium wilt	Haq et al. (1999)
8	Sugarcane	Co 6608 mutant	Gamma-rays	Resistance against red rot	IAEA (2011)
9	Rape seed	9 mutant lines of Huashuang 3 cultivar	Gamma-rays	Resistance against Stem rot (*Sclerotinia sclerotiorum*)	Liu et al. (2003)
10	Wheat	Eldo Nganol and Eldo Mavunol	Gamma rays	Tolerance against stem rust (*Puccinia graminis* f. sp. *tritici*)	(FAO/IAEA 2014)

potential and resistance to rice yellow mottle virus have been reported in the *O. sativa* mutants upon gamma irradiation (Luzi-Kihupi et al. 2009). A study conducted by Ahloowalia et al. (2004) reported resistance to yellow mosaic virus in mungbean plants exposed to gamma rays. Similar reports on various crops have been provided in Table 2.

6 Conclusion

Mutation breeding is a potential breeding technique that has the capability to improve the quantitative and qualitative characteristics of crops significantly in a shorter span of time compared to conventional breeding techniques. This technique can be exploited for improving the resilience of crops towards biotic and abiotic stresses induced by climate change. In spite of the advantages, mutation breeding technique has some limitations. These include the requirement for large sample size, low frequency of desirable changes in plants, undesirable changes that occur occasionally in the plants, and the need to deal with a number of mutant generations. Further efforts to advance mutation breeding using appropriate methods will enable humanity to accomplish the primary objective of plant breeding, thus promoting global food security.

References

Ahloowalia BS, Maluszynski M, Nichterlein K (2004) Global impact of mutation-derived varieties. Euphytica 135(2):187–204. https://doi.org/10.1023/B:EUPH.0000014914.85465.4f

Anjum SA, Xie XY, Wang LC, Saleem MF, Man C, Lei W (2011) Morphological, physiological and biochemical responses of plants to drought stress. Afr J Agric Res 6(9):2026–2032. https://doi.org/10.5897/AJAR10.027

Bita C, Gerats T (2013) Plant tolerance to high temperature in a changing environment: scientific fundamentals and production of heat stress-tolerant crops. Front Plant Sci 4:273. https://doi.org/10.3389/fpls.2013.00273

Bradshaw LD, Padgette SR, Kimball SL et al (1997) Perspectives on glyphosate resistance. Weed Technol 11:189–198. https://doi.org/10.1017/S0890037X00041567

Breseghello F, Coelho AS (2013) Traditional and modern plant breeding methods with examples in rice (Oryza sativa L.). J Agric Food Chem 61:8277–8286. https://doi.org/10.1021/jf305531j

Brock RD (1977) Prospects and perspectives in mutation breeding. In: Muhammed A, Aksel R, von Borstel RC (eds) Genetic diversity in plants. Plenum Press, New York, pp 117–132. https://doi.org/10.1007/978-1-4684-2886-5_12

Bughio HR, Asad MA, Odhano IA, Bughio MS, Khan MA, Mastoi NN (2007) Sustainable rice production through the use of mutation breeding. Pak J Bot 39(7):2457–2461. https://www.pakbs.org/pjbot/PDFs/39(7)/PJB39(7)2457.pdf

Chattopadhyay C, Bhattacharya BK, Kumar V, Kumar A, Meena PD (2012) Impact of climate change on pests and diseases of oilseeds Brassica—the scenario unfolding in India. J Oilseed Brassica 2:48–55. https://doi.org/10.5555/20123132597

Chattopadhyay C, Birah A, Jalali BL (2019) Climate change: impact on biotic stresses afflicting crop plants. Lecture Notes in Computer Science, pp 133–146. https://doi.org/10.1007/978-3-319-99768-1_8

Delaunay LN (1931) Resultate eines dreijahrigen Röntgen-Versuchs mit Weizen. Der Züchter 3:129–137. https://doi.org/10.1007/BF01812596

El-Mouhamady ABA, Ibrahim HF (2020) Elicitation of salt stress-tolerant mutants in bread wheat (Triticum aestivum L.) by using gamma radiation. Bull Natl Res Cent 44:108. https://doi.org/10.1186/s42269-020-00357-1

FAO (2018) The state of food security and nutrition in the world 2018. Building climate resilience for food security and nutrition. FAO, Rome

Food and Agriculture Organization of the United Nations (FAO) (2016) Climate change, agriculture and food security. The state of food and agriculture. ISBN: 978-92-5-109374-0

Forster BP, Till BJ, Ghanim AMA, Huynh HOA, Burstmayr H, Caligari PDS (2015) Accelerated plant breeding. CABI Rev 2014:1–16. https://doi.org/10.1079/PAVSNNR20149043

Haq MA, Sadiq M, Hassan M (1999) CM 98 (CM31–1/85)–a very high yielding disease resistant mutant variety of chickpea. International Chickpea and Pigeonpea Newsletter 6:9–10

Hunter MC, Smith RG, Schipanski ME, Atwood LW, Mortensen DA (2017) Agriculture in 2050: recalibrating targets for sustainable intensification. Bioscience 67:386–391. https://doi.org/10.1093/biosci/bix010

Hwang SG, Chapagain S, Lee JW, Han AR, Jang CS (2017) Genome-wide transcriptome profiling of genes associated with arsenate toxicity in an arsenic-tolerant rice mutant. Plant Physiol Biochem 120:40–51. https://doi.org/10.1016/j.plaphy.2017.09.019

IAEA (2011) IAEA mutant database. International atomic energy agency, Vienna. http://mvgs.iaea.org/Search.aspx

IAEA (2015) IAEA mutant database. Vienna: International atomic energy agency. http://mvd.iaea.org/

Jankowicz-Cieslak J, Till BJ (2016) Chemical mutagenesis of seed and vegetatively propagated plants using EMS. Curr Protoc Plant Biol 1(4):617–635. https://doi.org/10.1002/cppb.20040

Joshi R, Prashat R, Sharma PC, Singla-Pareek SL, Pareek A (2016) Physiological characterization of gamma-ray induced mutant population of rice to facilitate biomass and yield improvement under salinity stress. Indian J Plant Physiol 21:545–555. https://doi.org/10.1007/s40502-016-0264-x

Karimzadeh H, Borzouei A, Naserian B et al (2023) Investigating the response mechanisms of bread wheat mutants to salt stress. Sci Rep 13:18605. https://doi.org/10.1038/s41598-023-45009-2

Kharkwal MC (2023) History of plant mutation breeding and global impact of mutant varieties. In: Mutation breeding for sustainable food production and climate resilience. Springer Nature Singapore, Singapore, pp 25–55. https://doi.org/10.1007/978-981-16-9720-3_2

Kharkwal MC, Shu QY (2009) The role of induced mutations in world food security. In: Shu QY (ed) Induced plant mutations in the genomics era. Food and agriculture organization of the United Nations, Rome, pp 33–38. https://www.fao.org/4/i0956e/i0956e02.pdf

Krishnan P, Damaraju S (2019) piRNAs in the pathophysiology of disease and potential clinical applications. In: Mallick BB (ed) Advances in non-coding RNA. Academic Press, pp 335–356. https://doi.org/10.1016/B978-0-12-815669-8.00013-0

Kumar A, Chattopadhyay C, Singh KN, Vennila S, Rao VUM (2012) Trend analysis of climate variables for pigeonpea growing area. In: Abstract, international conference on statistics and computation, pp 145–148

Kumar HD, Gorham PR (1975) Effects of acridine dyes and other substances on growth, lysis and toxicity of Anabaena flos-aquae NRC-44-1. Biochem Physiol Pflanz 167(6):473–487. https://doi.org/10.1016/s0015-3796(17)31305-7

Lethin J, Shakil SSM, Hassan S et al (2020) Development and characterization of an EMS- mutagenized wheat population and identification of salt-tolerant wheat lines. BMC Plant Biol 20:18. https://doi.org/10.1186/s12870-019-2137-8

Lightner J, Caspar T (1998) Seed mutagenesis of Arabidopsis. In: Arabidopsis protocols, pp 91–102. https://doi.org/10.1385/0-89603-391-0:91

Lin KC, Jwo WS, Chandrika NNP, Wu TM, Lai MH, Wang CS, Hong CY (2016) A rice mutant defective in antioxidant-defense system and sodium homeostasis possesses increased sensitivity to salt stress. Biol Plant 60(1):86–94. https://doi.org/10.1007/s10535-015-0561-7

Liu L, Shi S, Wu J, Chen Y, Zhou Y (2003) Mutation induction and in vitro screening for stem rot resistant/tolerant materials with oxalic acid in rape (Brassica napus L.). Chin J Oil Crop Sci 25(1):5–8, 13

Luzi-Kihupi A, Zakayo JA, Tusekelege H, Mkuya M, Kibanda NJM, Khatib KJ, Maerere A (2009) Mutation breeding for rice improvement in Tanzania. Induced Plant Mutations in the Genomics Era, pp 385–387

Muller HJ (1927) Artificial transmutation of gene. Science 66:84–87. https://doi.org/10.1126/science.66.1699.84

Nakhoda B, Leung H, Mendioro MS, Mohammadi-nejad G, Ismail AM (2012) Isolation, characterization, and field evaluation of rice (Oryza sativa L., Var. IR64) mutants with altered responses to salt stress. Field Crop Res 127:191–202. https://doi.org/10.1016/j.fcr.2011.11.004

Oladosu Y, Mohd Y Rafii, Abdullah N, Hussin G, Ramli A, Rahim HA, Miah G, Usman M (2016) Principle and application of plant mutagenesis in crop improvement: a review. Biotechnol Biotechnol Equip 30(1):1–16. https://doi.org/10.1080/13102818.2015.1087333

Panigrahy M, Neelamraju S, Rao DN, Ramanan R (2011) Heat tolerance in rice mutants is associated with reduced accumulation of reactive oxygen species. Biol Plant 55:721–724. https://doi.org/10.1007/s10535-011-0175-7

Parihar P, Singh S, Singh R, Singh VP, Prasad SM (2015) Effect of salinity stress on plants and its tolerance strategies: a review. Environ Sci Pollut Res 22(6):4056–4075. https://doi.org/10.1007/s11356-014-3739-1

Pathirana R (2011) Plant mutation breeding in agriculture. CAB Rev: Perspect Agric VetY Sci Nutr Nat Resour 6:032. https://doi.org/10.1079/PAVSNNR20116032

Pathirana R, Frew T, Hedderley D, Timmerman-Vaughan G, Morgan E (2011) Haploid and doubled haploid plants from developing male and female gametes of *Gentiana triflora*. Plant Cell Rep 30(6):1055–1065. https://doi.org/10.1007/s00299-011-1012-3

Poli Y, Basava R, Panigrahy M, Vinukonda V, Dokula N, Voleti S et al (2013) Characterization of a Nagina22 rice mutant for heat tolerance and mapping of yield traits. Rice 6:36. https://doi.org/10.1186/1939-8433-6-36

Qi X, Sandmeyer S (2012) Nonhomologous recombination: retrotransposons. In: Lennarz WJ, Lane MD (eds) Encyclopedia of biological chemistry, 2nd edn. Academic Press, pp 283–291. https://doi.org/10.1016/B978-0-12-378630-2.00508-9

Reif JC, Hailauer AR, Melchinger AE (2005) Heterosis and heterotic patterns in maize. FAO, Rome

Sharma CBSR (1990) Chemically induced aneuploidy in higher plants. Mutagenesis 5(2):105–126. https://doi.org/10.1093/mutage/5.2.105

Sigurbjornsson B (1983) Induced mutations. In: Wood DR (ed) Crop breeding, American society of agronomy and crop science society of America, Madison, Wisconsin, pp 153–176. https://doi.org/10.2135/1983.cropbreeding.c8

Stadler LJ (1928) Genetic effect of X-rays in maize. Proc Natl Acad Sci USA 14:69–75. https://doi.org/10.1073/pnas.14.1.69

Stadler LJ (1930) Some genetic effects of X-rays in plants. J Hered 21:3–19. https://doi.org/10.1093/oxfordjournals.jhered.a103249

Stubbe H (1934) Einige Kleinmutationen von Anthirrinum majus L. Zuechter 6:299–303. https://doi.org/10.1007/BF01813299

Sun L, Lai M, Ghouri F, Nawaz MA, Ali F, Baloch FS, Nadeem MA, Aasim M, Shahid MQ (2024) Modern plant breeding techniques in crop improvement and genetic diversity: from molecular markers and gene editing to artificial intelligence—a critical review. Plants 13(19):2676. https://doi.org/10.3390/plants13192676

Suprasanna P, Mirajkar SJ, Bhagwat SG (2015) Induced mutations and crop improvement. In: Plant biology and biotechnology: Volume I: Plant diversity, organization, function and improvement. Springer, pp 593–617. https://doi.org/10.1007/978-81-322-2286-6_23

Tahir ul Qamar M, Faryad A, Bari A, Zahid B, Zhu X, Chen LL (2020) Effectiveness of conventional crop improvement strategies vs. omics.

In: Environment, climate, plant and vegetation growth, pp 253–284. https://doi.org/10.1007/978-3-030-49732-3_11
Udage, A. C. (2021). Introduction to plant mutation breeding: different approaches and mutagenic agents. J Agric Sci—Sri Lanka 16(03). https://doi.org/10.4038/jas.v16i03.9472
UN (2018) The sustainable development goals report. United Nations, New York. https://doi.org/10.11260/kenkokyoiku.19.77
van Harten AM (1998) Mutation breeding: theory and practical applications. Cambridge University Press, Cambridge, UK. urn:lcp:mutationbreeding0000hart:lcpdf:eba2af42-e0f2-4434-961e-0d46dfd656f8
Vavilov NI (1992) Origin and geography of cultivated plants. Cambridge University Press, Cambridge
Wani MR, Kozgar MI, Tomlekova N, et al. (2014) Mutation breeding: a novel technique for genetic improvement of pulse crops particularly Chickpea (*Cicer arietinum* L.). In: Parvaiz A, Wani MR, Azooz MM, Lam-son PT, editors. Improvement of crops in the era of climatic changes. Springer, New York, pp 217–248. https://doi.org/10.1007/978-1-4614-8824-8_9
Williams AB (2016) Genome instability in bacteria: causes and consequences. In: Kovalchuk I, Kovalchuk O (eds) Genome Stability. Academic Press, pp 69–85. https://doi.org/10.1016/B978-0-12-803309-8.00005-7
Yamaguchi H, Nagatomi S, Morishita T et al (2003) Mutation induced with ion-beam irradiation in rose. Nucl Instrum Methods Phys Res 206:561–564. https://doi.org/10.1016/S0168-583X(03)00825-5
Zhao C, Liu B, Piao S, Wang X, Lobell DB, Huang Y, Huang M, Yao Y, Bassu S, Ciais P, Durand J-L, Elliott J, Ewert F, Janssens IA, Li T, Lin E, Liu Q, Martre P, Müller C, Asseng S (2017) Temperature increase reduces global yields of major crops in four independent estimates. Proc Natl Acad Sci USA 114:9326–9331. https://doi.org/10.1073/pnas.1701762114

Mutation Breeding for Yield and Quality Improvement: A Case Study for Wheat and Barley

Londhe Ranjit, Kamble Sadhana, Shedbale Namrata, Dhole Vinod, Bapat Vishwas, and Gaikwad Nikhil

Abstract

With the rising global demand for food, animal feed, and industrial raw materials and to fulfil this demand, there is an urgent requirement for crop improvement. However, challenges such as shrinking arable land, water scarcity, and the effects of climate change limit the effectiveness of conventional farming approaches. To overcome these obstacles and ensure food security, it is essential to adopt innovative strategies that enhance crop yield, quality, and resilience to different stresses. Mutation breeding has become an effective method in crop improvement programs to increase yield and improve grain quality. The chapter provides an in-depth exploration of mutation breeding with a special focus on cereal crops, particularly wheat (*Triticum aestivum* L.) and barley (*Hordeum vulgare* L.). Application of physical and chemical mutagens has enabled the development of novel genetic variations that are otherwise rare in natural populations. Induced mutations have been very helpful in improving important plant traits like plant height, grain size, stress resistance, and nutritional value. A comprehensive overview of the role of mutation breeding in increasing yield and quality traits in wheat and barley is presented. In addition, traditional mutation breeding has significantly increased the accuracy and speed of identifying beneficial mutants. Using wheat and barley as a representative cereal crop, the chapter highlights the practical significance and future potential of mutation breeding in ensuring food security and advancing sustainable agriculture.

Keywords

Cereals · Crop · Mutation breeding · Mutagens · Quality · Yield

L. Ranjit · K. Sadhana · S. Namrata · G. Nikhil (✉)
Department of Botany, Shivaji University, Kolhapur, India
e-mail: nbgaikwadsu@gmail.com

D. Vinod
Nuclear Agriculture and Biotechnology Division, Bhabha Atomic Research Centre, Mumbai, India

B. Vishwas
Department of Biotechnology, Shivaji University, Kolhapur, India

1 Introduction

Food security is a big concern in the twenty-first century because the world is getting more crowded, and by 2050, the population might hit 9.1 billion (FAO 2009). Agricultural yield must increase by around 70% to achieve the future requirement (FAO 2009). However, reaching this goal is difficult because weather and environmental conditions are becoming more unpredictable (Charles et al. 2014). Now, mutation breeding is one of the most commonly used methods to create improved crop varieties. Induced mutation helps to create genetic differences in plants to develop useful traits which are not usually occuring in nature Horn and Shimelis (2013). The genetic variation is created by applying chemical and physical mutagens. For many years, scientists have employed chemical mutagens to enhance crop yields. In 1939, Thom and Steinberger were the first to use nitrous acid to induce mutations in the fungus *Aspergillus*. There are hundreds of known chemical mutagens, including alkylating agents, azides, base analogs, acridine dyes, and nitroso compounds (Kharkwal 2012). The use of physical mutagens began before 1920, when scientists found that radium could cause mutations in fruit flies (Kharkwal et al. 2004). Common physical mutagens used in plants include fast and slow neutrons, gamma rays, X-rays, UV rays, and beta particles. Cereal crops are essential for global food security because they provide most of the calories and nutrients for billions of people. Many people around the world depend on cereals for food. Cereals, come from the grass family,

J. Devi (ed.), *Mutation Breeding in Crop Improvement*, Sustainable Landscape Planning and Natural Resources Management,
https://doi.org/10.1007/978-3-032-17911-1_7

called Poaceae, are grown for edible grains. These grains are rich in energy, protein, vitamins, and key minerals. Through induced mutation, more than 851 mutant varieties of rice have been released, followed by barley with 311 and bread wheat with 265 (FAO/IAEA 2022). Wheat (*Triticum aestivum L.*) is the main cereal crop around the world. It is cultivated on approximately 222 million metric hectares, and production is around 775.8 million tons annually. Wheat is not only a source of energy that comes from carbohydrates but also provides important nutrients like protein, vitamins, and fiber. Among all cereal crops, barley (*Hordeum vulgare* L.) has the longest history of cultivation, known for its adaptability to different types of climate. After wheat, rice, and corn, barley is the most valuable cereal crop and usually grows in temperate zones (Namuk et al. 2023). Barley is commonly used as food, fodder, and for malting purposes which is rich in fiber, vitamins, and minerals, it plays an important role in human nutrition and livestock feeding. Today, about 157 million tons of barley are grown on 51 million hectares of land worldwide (FAO/STAT 2022). More than 60% of barley is used as fodder, about 40% is used to make malt, 5% is kept for seed, and 3% is used as food (Ullrich 2011). This chapter aims to explore how mutation breeding can be used to increase the yield and improve the quality of cereal grains. The study will also help future research and crop breeding efforts that aim to develop cereal varieties that can handle environmental changes and provide food for the growing world population.

2 Historical Background of Released Mutant Varieties Across the Globe

2.1 Wheat

Wheat (*Triticum aestivum* L.) is a common cereal food crop and is a key part of the daily meals of people (FAO/STAT 2021). Wheat is extensively grown in temperate regions due to its ability to adapt to a changing environment. India is the second-largest producer of wheat, following China, and it is commonly grown in states like Punjab, Haryana, Uttar Pradesh, Madhya Pradesh, etc. Wheat covers almost 31.36 million hectares of farmland in India (FAO/STAT 2021). Its high nutritional value makes it a vital source of carbohydrates, proteins, vitamins, and minerals. Wheat is important for food and plays a big role in the economic strength of several countries. Its global importance continues to grow due to its increasing demand and role in fighting against hunger and malnutrition.

2.1.1 Wheat Mutant Varieties Developed Worldwide

Through mutation breeding, improved wheat varieties were developed with significant enhancements in yield, quality,

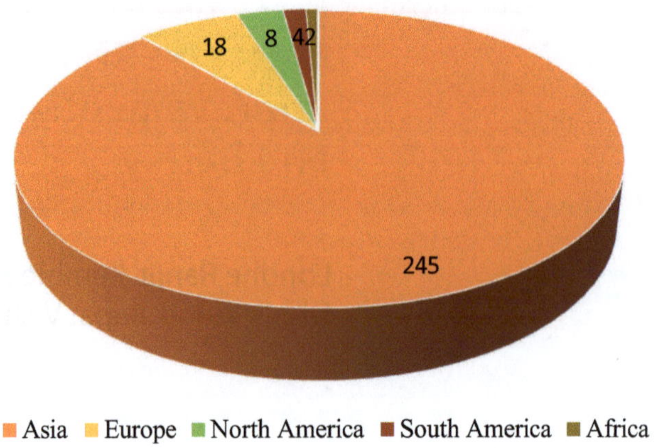

Fig. 1 Wheat mutant varieties developed worldwide

and stress resistance. Through mutation breeding, 277 wheat varieties have been successfully released. Induced mutation breeding has been widely adopted across different continents, with Asia leading the way in developing improved cereal crop varieties. Asia has developed 245 mutant varieties, making it the leading continent in the field of mutation breeding. Countries like China, India, Pakistan, and Mongolia has significantly contributed to this achievement. Europe has developed 18 mutant varieties, showing its efforts to improve crop for better yield and quality. North America has developed eight mutant varieties, mainly from the United States, which have contributed to cereal crop improvement. South America has introduced four mutant varieties, with Brazil and Mexico as major contributors. Africa has developed two mutant crop varieties, showing increasing interest in mutation breeding to improve food security. This distribution highlights the dominance of Asia in mutation breeding, while other continents are also making contributions to improve wheat crop through induced mutation (Fig. 1).

2.1.2 Wheat Mutant Varieties Released by Different Countries

Mutation breeding has been important for increasing the growth of cereal crops, enhancing yield, quality, and stress tolerance. Globally, many countries have successfully developed and released mutant varieties through various mutagenic techniques such as gamma irradiation, EMS (ethyl methanesulfonate), and other physical or chemical methods. Among the 277 mutant wheat varieties developed worldwide, China has made the most significant contribution by developing 175 varieties. It is more than 63% of the total mutant varieties of wheat. China has put a lot of effort and resources into mutation breeding to improve agriculture through intense research and investment and has developed improved crop varieties with better yield, quality, and tolerance to diseases. The Russian Federation follows with 40 mutant varieties of wheat. Other countries that contributed to mutant variety development

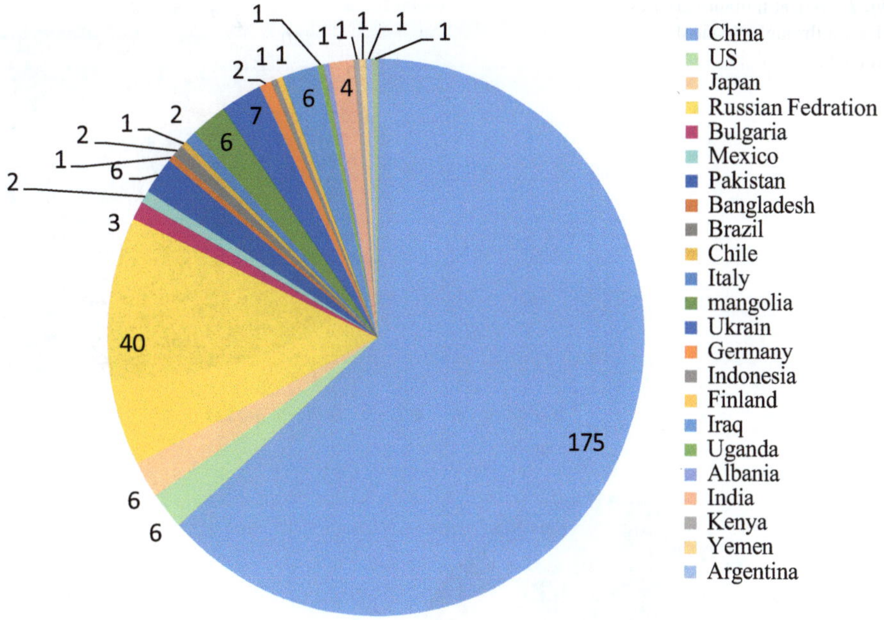

Fig. 2 Wheat mutant varieties released from different countries through physical mutagens

include the United States, Japan, Pakistan, and Mongolia, each of which released six mutant varieties. Ukraine has developed seven mutant varieties, and India has released four mutant varieties, contributing to regional agricultural sustainability. Mexico, Brazil, Germany, and Bulgaria have each developed three mutant varieties, focusing on improving cereal crops through induced mutations. These efforts show the global importance of mutation breeding in enhancing food security, making crops stronger, and adapting to climate change. Mutation breeding remains a valuable tool for developing superior cereal crop varieties worldwide (Fig. 2). These varieties were released using both physical and chemical mutagens.

2.1.3 Varieties Released Through Physical Mutagens

Physical mutagens, primarily radiation-based methods, have been commonly used to cause mutations in wheat. There are 189 varieties of wheat developed using different types of physical mutagens (gamma rays, laser radiation, fast neutrons, space mutagenesis, X-rays, ion beam, thermal neutrons, etc.). Gamma rays is the most extensively used physical mutagen due to its ability to cause stable genetic alterations. It is used to improve agronomic traits such as drought tolerance, disease resistance, and grain quality. Laser-induced mutations have gained attention due to their ability to create stable genetic modifications, leading to enhanced physiological and yield-related traits. Fast neutron irradiation leads to deletions in the genome, which can help to develop stress-tolerant varieties. Space-induced mutations occur under microgravity and cosmic radiation, leading to novel genetic changes. Some wheat varieties have been successfully developed using this technique. X-ray radiation has been used to create genetic variability, though its application is less frequent compared to gamma rays. Ion beam technology has emerged as a precise method for inducing targeted mutations, while thermal neutrons (thN) contribute to genetic variation (Fig. 3).

2.1.4 Varieties Released Through Chemical Mutagens

Chemical mutagens are potent tools for inducing point mutations, leading to specific genetic modifications in wheat. A total of 37 varieties were released using various chemical mutagens.

Chemical Mutagen-Induced Varieties

A total of 37 mutant wheat varieties were developed using chemical mutagens, including n-ethyl-n-nitrosourea (NEU) (19 varieties), n-methyl-n-nitrosourea (NMU) (6 varieties), sodium azide ($NaN3$) (4 varieties), ethyl methanesulfonate (EMS) (3 varieties), n-diethyl-n-nitrosourea (NDEU) (3 varieties), and ethyleneimine (EI), and dimethyl sulfate (DMS) (1 variety each) (Fig. 4).

2.2 Barley

Barley (*Hordeum vulgare* L.) has been grown for a very long time and can adapt to different climate conditions. Barley is mainly grown in temperate climate and ranks fourth among the world's top cereal crops, after wheat, rice, and maize (Namuk et al. 2023). Barley is commonly used for food, fodder, and making malt. Barley is rich in fiber, vitamins, and minerals, and it helps support healthy diets for both people

Fig. 3 Wheat mutant varieties released through physical mutagens

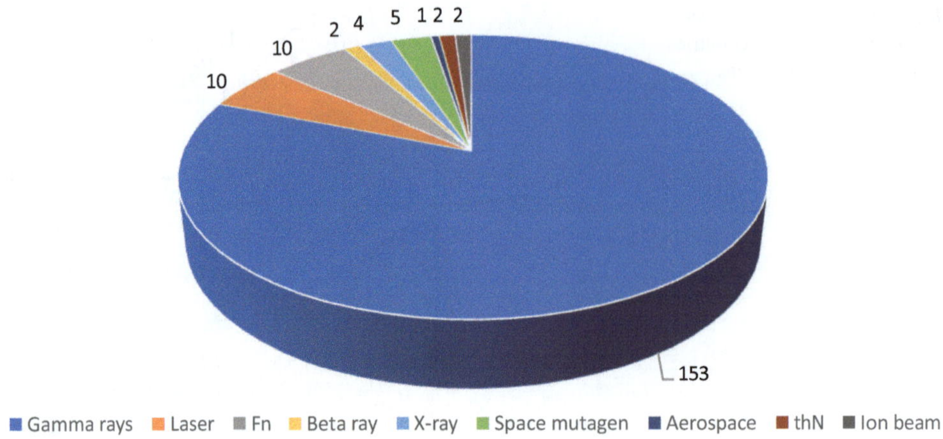

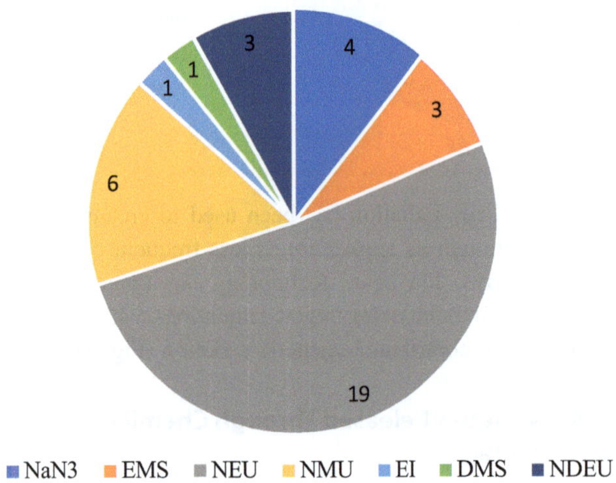

Fig. 4 Wheat mutant varieties released through chemical mutagens

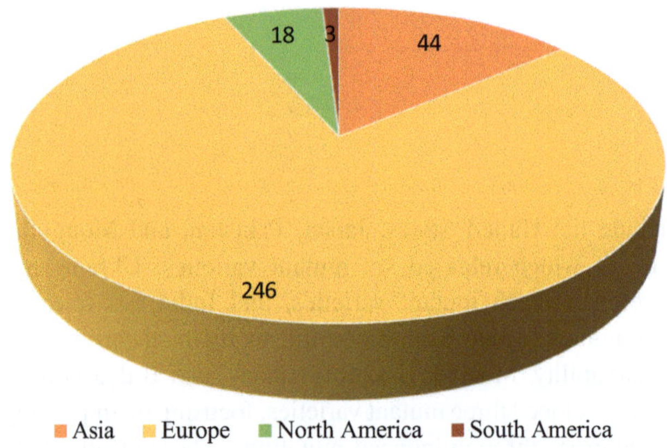

Fig. 5 Barley mutant varieties developed worldwide

and animals. Around 157 million metric tons of barley are grown worldwide on 51 million hectares of land (FAO/STAT 2022). Most of the barley (60%) is used to feed animals, for making malt 40%, for seed 5%, and 3% for food (Ullrich 2011).

2.2.1 Barley Mutant Varieties Released Worldwide

There are 311 mutant barley varieties have been released across the world, through induced mutation breeding. Among the continents Europe leads with 246 mutant varieties, making it the highest contributor to barley crop improvement through induced mutations. Asia ranks second after Europe with 44 mutant varieties, showing significant efforts in developing improved barley crops. North America has developed 18 mutant varieties, contributing to advancements in barley breeding. South America has released three mutant varieties, playing a smaller but notable role in barley improvement (Fig. 5).

2.2.2 Barley Mutant Varieties Released by Different Countries

Several countries have developed mutant varieties of barley through mutation breeding. Germany leads with 67 varieties, showing its dominance in barley improvement. The Russian Federation follows with 34 varieties, while the United Kingdom (UK) has developed 32 varieties, that contribute to the mutation breeding program. Other major contributors include Denmark (21), Sweden (20), Czech Republic (16), The United States, and India each released 13 varieties. Other notable contributors include Japan (11), Slovakia (10), Austria (8), France (8), China (7), Iraq (7), and Ukraine (6) varieties. Several countries, including Canada, Bulgaria, Estonia, Finland, and Turkey, have contributed between 2 and 5 varieties each. Some nations, such as Belgium, Syrian Arab Republic, Greece, Chile, Jordan, Korea, and Norway, have each developed 1 mutant variety. This distribution highlights Germany as the top contributor, with Europe playing a dominant role in barley mutation breeding. Other continents, including Asia and North America, have also made valuable contributions (Fig. 6).

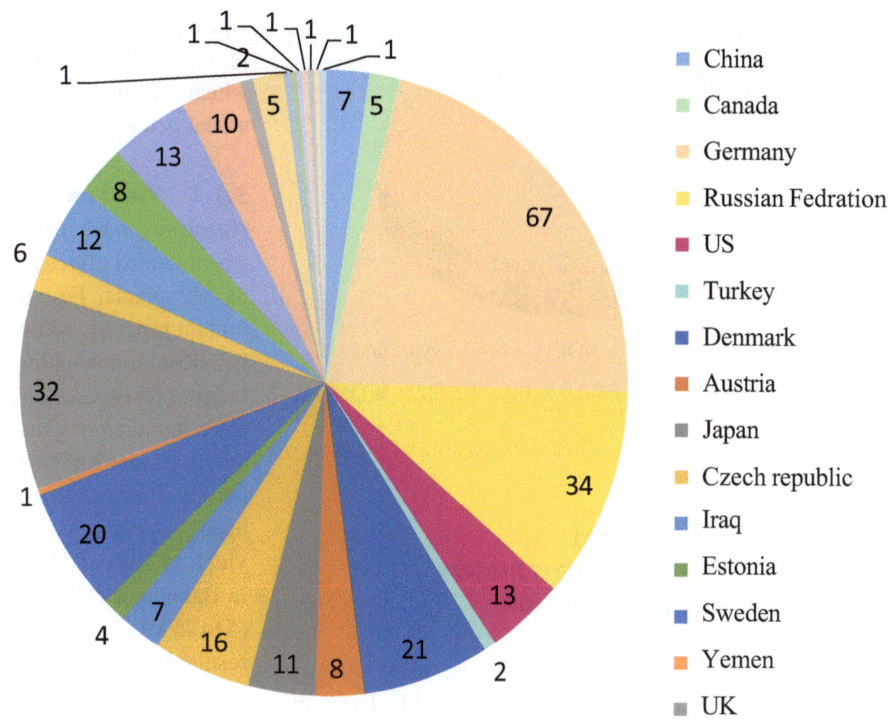

Fig. 6 Barley mutant varieties released from different countries

2.2.3 Barley Mutant Varieties Created Through Physical Mutagens

There are 148 mutant barley varieties released through physical mutagens, which induce genetic variations to improve yield, quality, and stress tolerance. The contribution of different physical mutagens is as follows; X-rays were the most widely used, leading to the development of 79 varieties, making it the dominant physical mutagen for barley improvement. Gamma rays contributed to the development of 48 varieties, highlighting their effectiveness in mutation breeding. Neutron radiation was responsible for 13 varieties, showing its role in inducing beneficial mutations. Thermal neutrons (thN) led to the creation of seven varieties, and fast neutrons (FN) contributed one variety, indicating limited but targeted use (Fig. 7).

It indicates that gamma rays and X-rays are the most extensively applied physical agents, together accounting for over 85% of the total mutant varieties. The use of neutron-based mutagens has been relatively lower but still valuable in creating useful genetic variations in barley.

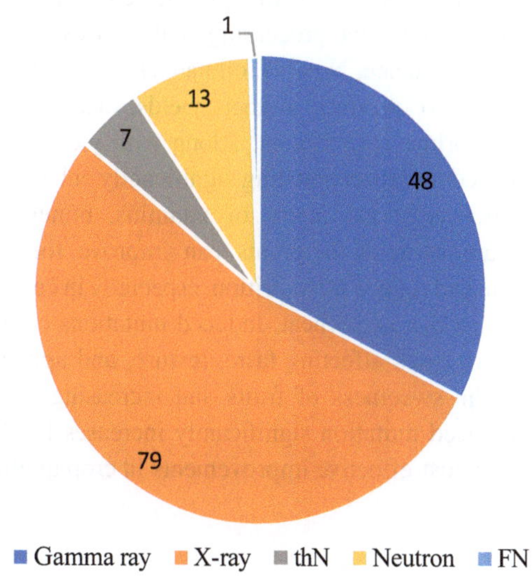

Fig. 7 Barley mutant varieties released through physical mutagens

2.2.4 Barley Mutant Varieties Developed Through Chemical Mutagens

Through the use of chemical mutagens, 41 improved barley varieties have been developed. NEU (nitrosoethylurea) contributed the highest number, leading to 13 varieties. EMS (ethyl methanesulfonate), and EI (ethyleneimine) each resulted in seven varieties, showing their effectiveness in mutation breeding. NDEU (nitrosodeoxyuridine) also contributed seven varieties, playing a significant role in genetic modifications. Other chemical mutagens, including brodeoxyuridine, sodium azide (NaN3), BEO, SA (sulfuric acid), ethylene oxide, NTMU, and DMS (dimethyl sulfate) each contributed 1 variety. It indicates that NEU, EMS, EI, and NDEU are the most commonly used chemical mutagens, collectively accounting for over 80% of the chemically induced mutant varieties in barley (Fig. 8).

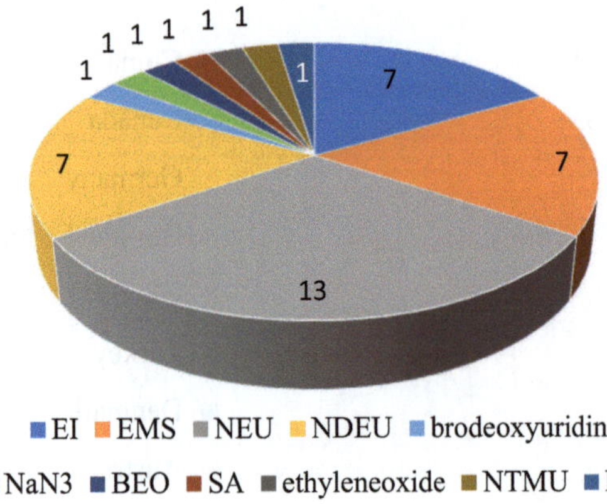

Fig. 8 Barley mutant varieties released through chemical mutagens

3 Improvement in Yield and Quality-Related Parameters Through Induced Mutation

Induced mutation breeding is one of the most effective methods for enhancing crop quality and yield. The technology accelerates genetic progress as well as offers answers to several challenges faced by the agriculture sector, including improved nutritional and processing quality, stress tolerance, and disease resistance. Notably, climate change is impacting the capacity to produce higher yields, and nutritional quality is a prime requirement. Along with yield enhancement, induced mutation breeding significantly enhances crop quality. Increasing the levels of vitamins, minerals, and essential amino acids in plants can improve their nutritional value and reduce malnutrition, especially in cereal food crops such as rice and wheat. Induced mutations could also enhance properties affecting taste, texture, and appearance, including the sweetness of fruits and increasing shelf life period. Induced mutation significantly increases food security through cost-effective improvements in crop quality and yield.

3.1 Wheat

In wheat improvement programs, mutation breeding plays a crucial role in enhancing key agronomic traits, including plant height, number of grains per spike, grain weight, and overall yield per plant. It also plays a significant role in enhancing quality traits like protein content, gluten strength, starch composition, etc., which are essential for both food and industrial uses. Using modern techniques along with mutation breeding, which helps breeders to develop wheat varieties with higher yields and better quality. This method not only improves old breeding ways but also makes it faster to get improved wheat varieties. The biggest challenge in wheat production is to increase grain yield to meet global food requirements. However, it is challenging to develop improved wheat varieties due to the limited genetic diversity in existing wheat genotypes (Dwivedi et al. 2017).

3.1.1 Yield and Yield-Related Traits

According to Khvostova et al. (1965), ethyleneimine at 0.01% was the most effective mutagen, causing mutations in 50.8% of M2 plants. Fast neutrons (740 rad) resulted in 36.5% mutant progeny, while 10 kR gamma rays caused only 6.2%. Ethyleneimine treatment has a diverse variety of mutants, like lodging resistance, disease resistance, and high yield, which were found among the mutants. Kumar (1977) developed the dwarf mutant K68 using 15 kR gamma rays, which exhibited similar tillers and grains per spike compared to the control. It had shorter spikes with heavier grains and produced 56% more yield. Madiha and Umer (2002) carried out induced mutation in Bakhtawar-92 treated with different doses of gamma rays (5 to 25 kR). The best results were reported from 20 kR, which resulted in plant height (82.78 cm), higher grains (59.80), and maximum yield (3926 kg/h) compared with the control. Singh and Balyan (2009) applied Gamma rays to develop mutants in the wheat variety Kharchia 65, reporting improvements in traits such as shorter plants, square heads, awnless ears, amber seeds, and bold grains. These mutants were reported at doses of 10 to 40 kR of gamma rays. Among the mutants, three lines (398, 446–7, and 621) yielded better than the original variety.

3.1.2 Resistance to Abiotic and Biotic Stress

Njau et al. (2005) developed a drought-tolerant wheat variety from the Kenyan variety "Pasa" with gamma rays. Six mutant lines were selected and tested in conjunction with checks "Duma" and "Pasa" in dry areas of Kenya over a two-year period. Mutants KM14 and KM21 reported higher yields compared to parent and checks in most sites. KM14 showed the best overall performance and was later released as the variety Njoro BW1 for use in dry regions. Drought-tolerant wheat mutants were developed with the help of gamma rays. Leaf and stem rust resistance mutants were identified from gamma-ray treatment. A loose smut-resistant mutant of wheat was identified using sodium azide (FAO/IAEA 2022). A semi-dwarf and drought-tolerant wheat mutant (HUW-SDf1) was developed by Boluwatife et al. (2021) using EMS (ethyl methanesulfonate).

3.1.3 Nutritional Quality

Mutants with high-grain-protein content were recorded using gamma rays (FAO/IAEA 1967). Kumar (1977) developed variety K68, which has higher protein content (14.15%) and medium gluten quality. Corpuz et al. (1983) developed four genotypes with high protein content in hard red winter wheat through EMS treatment, where many mutants were dwarf and had lodging resistance. Low-phytic acid mutants of wheat were identified by Guttieri et al. (2004) using EMS treatment. Mishra et al. (2016) developed 101 mutant lines of wheat using EMS, which exhibited differences in amylose and resistant starch content. These lines help to improve starch quality for studying genes related to this trait. Amylose plays a key role in determining the starch properties and grain quality of wheat. The mutation of the *Wx-D1* gene directly affects the formation of amylose, leading to noticeable changes in starch properties. This discovery offers valuable insights for enhancing wheat starch quality and developing more diverse starch characteristics (Chakraborty et al. 2023). Genetic variability was induced through gamma rays, resulting in alterations to traits such as 1000-seed weight, wet gluten content, gluten index, and protein content (Chen et al. 2024).

3.2 Barley

Induced mutation breeding has played a crucial role in developing new barley varieties with improved agronomic traits. Many mutant lines have been created with improved yield-related characteristics such as increased plant height, more tillers, longer panicles, and larger grain size. Moreover, mutation breeding has improved grain quality traits, such as increasing protein content and enhancing malting properties. It has also helped to develop barley varieties that can withstand stressful conditions, ensuring stable yields even in challenging environments.

3.2.1 Yield and Yield-Related Traits

Gustafsson et al. (1977) developed a mutant variety with stiff straw and high yield through X-ray treatment. Various workers have reported the number of early maturing mutants in barley. An early maturing and low-temperature-tolerant mutant of barley was identified by Shevtsov and Dundr (2011) through NEU n-nitroso-n-methylurea treatment. Early maturity and lodging resistance mutants were reported by Zakhrabekova et al. (2012) using X-ray treatments.

There were substantial, significant differences between the mutant lines in traits like days to flowering, plant height, number of spikes plant^{-1}, awn length, peduncle length, spikelet number spike^{-1}, grain number spike^{-1}, grain weight spike^{-1}, 1000-grain weight, and grain yield plant^{-1}. Mutant lines (six) showed significantly higher yields per plant compared to the parent variety Kuber and the standard varieties Obzor and Emon (Boryana and Nikolay, 2020). Using gamma radiation, the seeds from the barley variety Yalin and barley line YAA7050-14 produced mutants with useful variations. These variations included plant height, spike length, grain spike^{-1}, grain weight spike^{-1}, and 1000 seed weight, making them suitable for selection in the appropriate direction (Namuk et al. 2023). These studies demonstrate the effectiveness of various mutagens, including X-rays, gamma rays, ethyl methanesulfonate, neutrons, and sodium azide, in enhancing crop traits such as plant height, yield, protein content, and lodging resistance.

3.2.2 Nutritional Quality

According to Muyuan et al. (2003), barley is known for sensitive to aluminum (Al) toxicity, so developing Al-tolerant varieties is important for successful cultivation in acidic soils. Aluminum-tolerant barley lines were developed using mutagens such as EMS (ethyl methanesulfonate), sodium azide (NaN3), and gamma rays. Alpha-amylase activity is a key quality trait in barley, as it plays a crucial role in breaking down starches into sugars during the malting and brewing process. Bovina et al. (2011) reported 29 mutants in the sodium azide-treated barley population. Among these, five important genes, **Bmy1**, **GBSSI**, **LDA1**, **SSI**, and **SSII** were found to be closely associated with starch metabolism.

3.2.3 Resistance to Abiotic and Biotic Stress

Nematode-resistant barley mutants were developed with the help of fast neutrons (FAO/IAEA, 1981). Powdery mildew resistance mutants were reported with the help of X-rays by Jorgenson (1992). Whereas barley mutants with resistance to powdery mildew and various rusts (brown rust, black rust, and stem rust) were found by Ohnoutkova (2019) with the help of sodium azide treatment.

4 Conclusion

To meet the growing need for food, fodder, and other essential resources, we must enhance cultivated crops, especially with growing challenges like less farmland, water shortages, and environment change. Induced mutation breeding is a valuable technique that introduces new and beneficial traits in plants through the use of physical or chemical treatments. Mutation breeding is considered a faster and lower-cost alternative to traditional breeding, which can take many years of crossing and selection. Mutation breeding can generate useful variations in just a few generations, helping researchers to develop improved crop varieties more efficiently. Induced mutation has been successfully utilized in essential cereal crops, like wheat and barley, to enhance yield, grain quality (including protein, minerals, starch quality, and other nutritional properties), and resistance to various stresses. Induced mutation breeding, combined with modern tools, can help to accelerate crop improvement. Overall, mutation breeding is a valuable approach for increasing food production, ensuring food security, and supporting sustainable farming in the future.

References

Boluwatife M, Hussein S, Mark L, Isack M (2021) Development of Wheat (*Triticum aestivum* L.) populations for drought tolerance and improved biomass allocation through Ethyl methanesulfonate mutagenesis. Front Agron 3:1–16. https://doi.org/10.3389/fagro.2021.655820

Boryana D, Nikolay D (2020) Grain yield and related traits of sodium azide induced barley mutant lines. J Cent Eur Agric 21(1):83–91. https://doi.org/10.5513/JCEA01/21.1.2419

Bovina R, Talame V, Silvio S, Sanguineti MC, Trost P (2011) Starch metabolism mutants in barley: TILLING approach. Plant Gen Res: Char and Uti 9(2):170–173. https://doi.org/10.1016/j.jbiotec.2010.09.773

Chakraborty S, Mohapatra S, Hooi A, Ali N, Satdive R (2023) Determination of median lethal (LD50) and growth reduction (GR50) dose of gamma irradiation for induced mutation in Wheat. Braz Arch Biol Technol 66:1–9. https://doi.org/10.1590/1678-4324-2023220294

Charles H, Godfray H, Garnett T (2014) Food security and sustainable intensification. Philos trans 369(1639):20120273.https://doi.org/10.1098/rstb.2012.0273

Chen X, Shao Y, Jiang Y, Seung D, Guzman C, Xu Q, Zhang Y, Chen Q, Tang H, Qi P, Deng M, Ma J, Chen G, Wang J, Wei Y, Zheng Y, Jiang Q (2024) Reducing amylose content in Wheat (*Triticum aestivum* L.) using a novel Wx-D1 null allel generated by chemical mutagenesis. J Sci Food Agric 7:1–10. https://doi.org/10.1002/jsfa.14003

Corpuz LM, Heyne EG, Paulsen GM (1983) Increasing grain protein content of hard red winter wheat (*Triticum aestivum* L.) by mutation breeding. Theor Appl Genet 65(1):41–61. https://doi.org/10.1007/BF00276260

Dwivedi S, Scheben A, Edwards D, Spillane C, Ortiz R (2017) Assessing and exploiting functional diversity in germplasm pools to enhance abiotic stress adaptation and yield in cereal and legumes. Front Plant Sci 8:1461–1471. https://doi.org/10.3389/fpls.2017.01461

FAO (2009) How to feed the World in 2050.https://doi.org/10.1111/j.1728-4457.2009.00312.x

FAO/IAEA (1967) Mutant Variety Database. https://mvd.iaea.org

FAO/IAEA (1981) Mutant Variety Database. https://mvd.iaea.org/

FAO/IAEA (2022) Mutant Variety Database. https://mvd.iaea.org/! Search

Food and Agriculture Organization (FAO) (2021) FAOSTAT statistical database of the United Nations food and agriculture organization (FAO) statistical division. Rome. http://www.fao.org/faostat/en/data. accessed:10.02.2025

Food and Agriculture Organization (FAO) (2022) FAOSTAT statistical database of the United Nations food and agriculture organization (FAO) statistical division. Rome. http://www.fao.org/faostat/en/data. accessed:10.02.2025

Gustafsson Å, Ekman G, Dormling I (1977) Effects of the Pallas gene in barley: Phene analysis, overdominance, variability. Hereditas 86(2):251–266. https://doi.org/10.1111/j.1601-5223.1977.tb01235

Guttieri M, Bowen D, Dorsch JA, Raboy V, Souza E (2004) Identification and characterization of a low phytic acid wheat. Crop Sci 44(2):423–430. https://doi.org/10.2135/cropsci2004.4180

Horn L, Shimelis H (2013) Radio-sensitivity of selected cowpea (Vigna unguiculata) genotypes to varying gamma irradiation doses. Sci Res Essays. 8(40):1991–1997. https://doi.org/10.5897/SRE2013.5682

Jorgenson IH (1992) Discovery, characterization and exploitation of Mlo powdery mildew resistance in barley. Euphytica 63:141–152. https://doi.org/10.1007/BF00023919

Kharkwal M, Pandey R, Pawar S (2004) Mutation breeding for crop improvement. A book published by Springer, pp 601–645. https://doi.org/10.1007/978-94-007-1040-526

Kharkwal M (2012) A brief history of plant mutagenesis. In: Plant mutation breeding and biotechnology. A book published by CABI, pp 21–30.https://doi.org/10.1079/9781780640853.0021

Khvostova VV, Mozhaeva VS, Aigaes NS, Valeva SA (1965) Mutants induced by ionizing radiations and ethyleneimine in winter wheat. Mutat Res 2(4):339–344. https://doi.org/10.1016/0027-5107(65)90069-2

Kumar D (1977) Promising semi dwarf mutant in wheat variety K68. Env and Exp Bot 17, Pergamon Press. Printed in Great Britain, pp 79–85. https://doi.org/10.1016/0098-8472(77)90024-7

Madiha J, Umer Q (2002) Study of Genetic variation in yield components of wheat cultivar bukhtwar-92 as induced by gamma radiation. Asi Jrnl Plant Sci 1:579–580. https://doi.org/10.3923/ajps.2002.579.580

Mishra A, Singh A, Sharma M, Kumar P, Roy J (2016) Development of EMS-induced mutation population for amylose and resistant starch variation in bread wheat (*Triticum aestivum*) and identification of candidate genes responsible for amylose variation. Plant Bio 16(1):217–222. https://doi.org/10.1186/s12870-016-0896-z

Muyuan Y, Jianwei P, Lilin W, Qing G, Chunyuan H (2003) Mutation-induced enhancement of Al tolerance in barley cell lines. Plant Sci 164:17–23. https://doi.org/10.1016/S0168-9452(02)00317-5

Namuk E, Guray A, Saime U, Sinan A (2023) Determination of optimum gamma ray irradiation doses for hulless Barley (*Hordeum vulgare* var. *nudum* L. Hook. f.) Genotypes. J Agric Sci 33(2):219–230. https://doi.org/10.29133/yyutbd.1248710

Njau PN, Kinyua MG, Kimurto PK, Okwaro HK, Maluszynski M (2005) Drought tolerant wheat varieties developed through mutation breeding technique. J Agric Sci Tech 7:18–29. https://doi.org/10.4314/jagst.v7i1.31717

Ohnoutkova L (2019) Mutation breeding in barley: Historical overview. Barley: Methods and Protocols, pp 7–19. https://doi.org/10.1007/978-1-4939-8944-72

Shevtsov SP, Dundr M (2011) Nucleation of nuclear bodies by RNA. Nat Cell Biol 13(2):167–73.https://doi.org/10.1038/ncb2157

Singh NK, Balyan HS (2009) Induced mutations in bread wheat (*Triticum aestivum* L.) CV. 'Kharchia 65' for reduced plant height and improved grain quality traits. Adv in Bio Res 3(5–6):215–221. https://www.idosi.org/abr/3(5-6)/14.pdf

Ullrich S (2011) Significance adaptation, production, and trade of barley. A book published by Semantic scholar, pp 3–13. https://doi.org/10.1002/9780470958636.ch1

Zakhrabekova S, Gough SP, Braumann I, Müller AH, Lundqvist J, Ahmann K (2012) Induced mutations in circadian clock regulator Mat-a facilitated short-season adaptation and range extension in cultivated barley. Proc Natl Acad Sci 109(11):4326–4331. https://doi.org/10.1073/pnas.1113009109

Mutation Breeding for Yield and Quality Traits in Pulses

Adhimoolam Karthikeyan, Manoharan Akilan, Jeeva Kasiviswanathan Lekshmi, Divya Sankaran, Ariharasutharsan Gunasekaran, Manickam Dhasarathan, and Natesan Senthil

Abstract

Mutation breeding has become a valuable and holistic strategy for enhancing yield and quality traits in pulse crops, which are crucial for food security and sustainable agriculture. Pulses, being rich sources of protein, micronutrients, and dietary fiber, are integral to human nutrition, particularly in developing countries. However, genetic improvement in pulses has traditionally lagged behind that in cereals due to limited genetic diversity. Induced mutagenesis, through physical (e.g., gamma rays and X-rays) and chemical (e.g., EMS and DES) mutagens, offers an effective strategy to generate novel genetic variation rarely achieved through traditional breeding. Globally, mutation breeding approaches have led to the successful development of improved varieties with enhanced yield potential in pulse crops. Current breakthroughs in genomics have further expedited the identification and selection of beneficial mutations. The integration of mutation breeding with modern genomic approaches offers substantial potential for mitigating issues related to climate change, enhancing pulse crop productivity, and ensuring nutritional security. This chapter details the contribution of mutation breeding to the improvement of yield and quality traits in pulse crops.

Keywords

Conventional breeding · Mutagenesis · Pulses · Yield · Quality traits

1 Introduction

Pulses are edible seeds, well known for their nutritional richness and environmental benefits. They contain high levels of protein, dietary fiber, vitamins, and essential minerals, rendering them essential to nutritionally balanced diets worldwide. In addition to their nutritional value, pulses enhance soil properties through atmospheric nitrogen fixation, thereby reducing dependence on synthetic fertilizer. The important pulse crops include chickpea, lentil, mungbean, pigeon pea, cowpea, and black gram, each contributing significantly to global food security. Despite their importance, pulse yields have remained relatively stagnant compared to cereals, highlighting the need for genetic improvement. Efforts to improve yield and quality in pulses are critical for meeting future food and nutritional demands (Kumar et al. 2020). However, the progress of genetic improvement in pulse crops remains slow. Despite continuous and systematic efforts, conventional breeding has achieved only limited success owing to several constraints. Mutation breeding is a non-conventional breeding method employed to enhance genetic variability and improve crops via the use of mutagens (Kharkwal et al. 2004; Pandit et al. 2021). This approach is particularly valuable for self-

A. Karthikeyan
Subtropical Horticulture Research Institute, Jeju National University, Jeju, South Korea

M. Akilan
Department of Plant Breeding and Genetics, Agricultural College and Research Institute, Tamil Nadu Agricultural University, Trichy, India

J. K. Lekshmi
Department of Entomology, CentreforPlant Protection Studies, Tamil Nadu Agricultural University, Coimbatore, India

D. Sankaran · N. Senthil (✉)
Department of Plant Molecular Biology and Bioinformatics, Centre for Plant Molecular Biology and Biotechnology, Tamil Nadu Agricultural University, Coimbatore, India
e-mail: senthil_natesan@tnau.ac.in

A. Gunasekaran
Department of Biotechnology, Centre of Excellence in Innovation, Agricultural College and Research Institute, Tamil Nadu Agricultural University, Madurai, India

M. Dhasarathan
Agro-Climate Research Centre, DirectorateofCropManagement, Tamil Nadu Agricultural University, Coimbatore, India

J. Devi (ed.), *Mutation Breeding in Crop Improvement*, Sustainable Landscape Planning and Natural Resources Management,
https://doi.org/10.1007/978-3-032-17911-1_8

pollinating crops likes pulses, where genetic diversity is often limited.

Mutation breeding involves inducing genetic variations through physical or chemical mutagens, leading to novel traits that are challenging to attain by conventional breeding or require extensive time for natural selection (Kavuri et al. 2024). Induced mutations result in both qualitative and quantitative variations within a relatively short period by modifying alleles at characterized and novel loci, as well as altering linkage groups (Stearns et al. 2016; (Jegadeesan and Punniyamoorthy 2023). Novel alleles of existing genes were developed through inducing mutations via radiation, such as X-rays and γ-rays, or chemical mutagens like ethyl methanesulfonate (EMS) and methyl methanesulfonate (MMS). Primarily, this technique induces hereditary changes in treated or irradiated planting materials, from which mutants with desirable characters may arise. These mutants are then selected and developed into new cultivars or used as parent lines in crop improvement programs. The selection of parent material is important for achieving targeted objectives in mutation breeding. The primary goal is to expand region specific cultivars through introduction of new alleles that does not exist in germplasms. To date, mutation breeding approaches have successfully contributed to the development of improved pulse varieties—exhibiting enhanced yield potential. In this chapter, we discuss the importance of mutation breeding for improving yield and quality traits in major pulse crops such as mungbean, black gram, cowpea, common bean, faba bean, lentil, pea, chickpea, pigeon pea, and grass pea.

2 Mutation Breeding

The conventional method of breeding requires high degree of genetic diversity which is artificially induced by various breeding methods. Mutation breeding is one such technique that creates novel mutations and widens the genetic base of crop plants for the improvement of yield and quality traits. Mutation breeding efforts initially focused on inducing random genetic changes, followed by field selection, which led to the development of early mutant varieties with improved yield potential along with enhanced environmental resilience. This approach highlights the deliberate efforts of plant breeders and the specific methods used to produce and apply beneficial variations in the development of elite breeding lines and improved crop varieties. Modern biotechnological approaches have transformed mutation breeding, enhancing the precision of trait identification. Techniques such as Targeting Induced Local Lesions in Genomes (TILLING) have enabled targeted mutagenesis, allowing breeders to develop improved cultivars with greater accuracy. Additionally, advances in genomics and phenomics have accelerated the breeding process by linking specific mutations to desirable traits, facilitating more efficient selection, shortening breeding cycles, and improving trait stability. In comparison to hybridization, mutation breeding requires less time to develop cultivars with desirable traits. Over the decades, mutation breeding has played a crucial role in the development of new cultivar with improved yield, nutritional quality, and stress tolerance across various crops, including pulses. Mutations in plants are caused by various factors. Based on their causes, mutations are classified into two broad categories: spontaneous (natural) mutations and induced (artificial) mutations. Spontaneous mutations, which occur naturally without human intervention, are relatively rare in natural populations. Several factors that increase the frequency of mutations occurring naturally in living organisms are known as mutagens. These mutagens are commonly employed by plant breeders to induce mutations for crop improvement and are broadly classified as physical and chemical mutagens. The schematic representation in Fig. 1 depicts the stepwise process involved in producing stable and superior mutants with desirable traits.

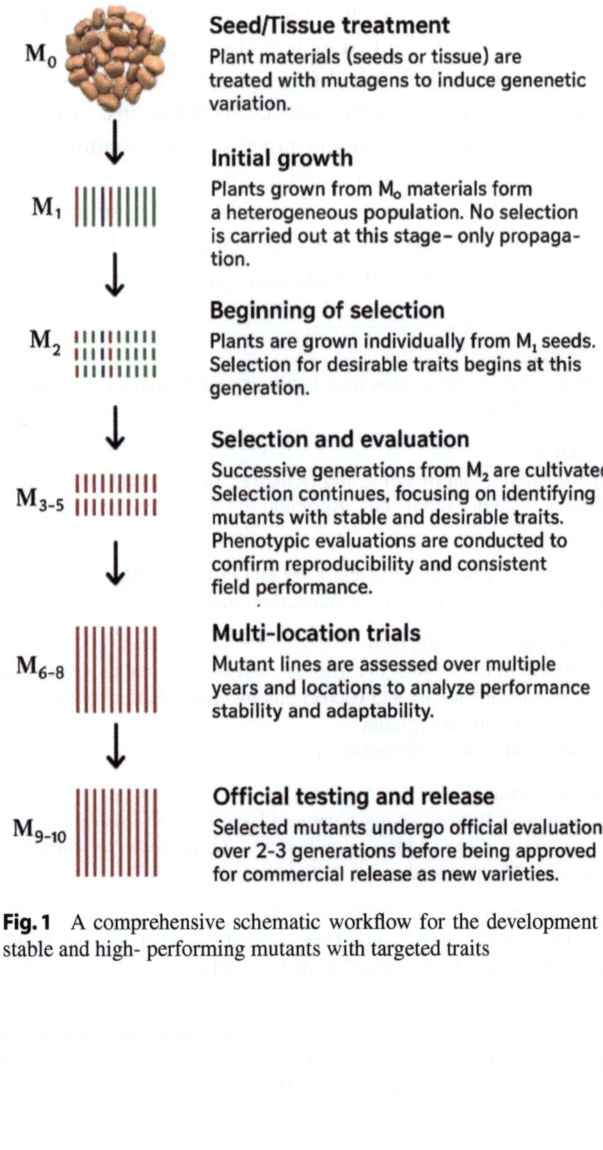

Fig. 1 A comprehensive schematic workflow for the development of stable and high- performing mutants with targeted traits

3 Mutation Breeding for Yield and Quality Improvement in Pulses

The history of mutation breeding in pulse crops dates back to the early twentieth century, with early successes recorded in chickpea, mung bean, lentil, and cowpea. Research efforts intensified during the 1960s and 1970s. A large number of pulse crop varieties have been developed through mutation breeding worldwide, with India contributing the highest number of released varieties. In India, the highest number of mutant varieties has been developed in mung bean, followed by black gram, cowpea, chickpea, pigeon pea, moth bean, lablab, lentil, horse gram, pea, and common bean. Early mutation breeding programs primarily targeted improvements in yield and yield-related traits. A substantial number of studies have subsequently been undertaken to improve yield and quality traits across several major pulse crops (Tables 1, 2 and 3).

3.1 Mungbean

Khattak et al. (2008) employed gamma rays at 0.20, 0.30, and 0.40 KGy to select superior-yielding individuals from

Table 1 Mutation breeding for yield and quality improvement in mungbean, black gram, and cowpea

S. No.	Crop /Species	Generation/ Line s	Direct/Derived from	Method/Dosage	Major Findings	Reference
1	Mungbean (*Vigna radiata*)	M_2 and M_3	NM-1	EMS (0.1, 0.2, and 0.3%), 0.01, 0.02, 0.03% (HZ), and SA	Long pod and bold seeded mutants with higher yield	Wani et al. (2017)
2		M_5	B1, Pusa-9632, and K-851	Gamma rays	Increased pod length and girth were observed. Mutant plants exhibited 13–40% shorter heights compared to controls but developed more branches and larger pods	Sarkar and Kundagrami (2018)
3		M_7	K-851 and Sona mung	Gamma rays (100–400 Gy)	High yielding mutant lines accompanied by high harvest index were identified	Dewanjee and Sarkar (2018)
4		M_2	CO5, CO (Gg) 7, and VBN (Gg) 3	Gamma rays (550 Gy and 600 Gy)	Different types of chlorophyll deficient mutants	Arulselvi et al. (2019)
5	Blackgram (*Vigna mungo*)	M_2	T-9 and Pant U-30	Gamma rays—(100–400 Gy) and EMS (0.1, 0.2, 0.3 and 0.4%)	Mutants exhibiting increased pod number and size showed a significant positive correlation with enhanced plant yield	Goyal et al. (2019)
6		M_5	ADT 3, Co 6, and TU 17–9	Gamma rays—(200–500 Gy)	High-yielding mutants for specific environments were identified	Dhasarathan et al. (2021)
7		M_3	CO6	Combination of Gamma rays and EMS (200 Gy + 20 mM, 200 Gy + 30 mM, 300 Gy + 20 mM, 300 Gy + 30 mM, 400 Gy + 20 mM and 400 Gy + 30 mM EMS)	Mutants with short duration, bold seeds and high yield were isolated	Tamilzharasi et al. (2022)
8	Cowpea (*Vigna unguiculata*)	M_4	Gomati VU-89 and Pusa-578	Gamma rays and SA and combination of both treatments	High-yielding genotypes with broad genetic diversity and significantly enhanced protein and micronutrient content were identified	Raina et al. (2020)
9		M_6	Kaha 1 and Dokki 331	Gamma rays (0, 50, 100, 150, 200, 250, and 300 Gy)	Resulting mutants with improved yield characteristics and nutritional value	Adly et al. (2024)

EMS—Ethyl methanesulfonate, HZ—Hydrazine hydrate, SA—Sodium azide

Table 2 Mutation breeding for yield and quality improvement in common bean and lentil

S. No.	Crop /Species	Generation/ Lines	Direct/Derived from	Method/Dosage	Major findings	Reference
1	Common bean *(Phaseolus vulgaris)*	M_2	Contender and Arka Komal	EMS (0.15, 0.25 and 0.35%)	Several desirable mutants for yield and its contributing traits were identified	Ramandeep et al. (2018)
2		M_3	Nebraska	Gamma rays (10, 20 and 30 kR)	Superior mutants for higher yield and protein content were identified	Masry et al. (2021)
3		M_2	Traditional cultivar	EMS (0.1, 0.2, 0.3 and 0.4%) and SA (0.1, 0.2, 0.3 and 0.4%)	Superior mutants for yield and its contributing traits were identified	Laskar et al. (2024)
4	Lentil *(Lens culinaris)*	M_3	Pant L 406	Gamma rays (100, 200, 300 and 400 Gy)	Obtained "multipodding" (*mp*) mutant	Laskar et al. (2018b)
5						
6		M_3	L-4076	EMS (0.10% to 1.0%)	Obtained increase in pod size, seed count, and higher yield	Shahwar et al. (2024)

EMS—Ethyl methanesulfonate, SA—Sodium azide, DES—Diethyl sulfate

NM 92 and NM 98. Wani et al. (2017) applied three different chemical mutagens to mungbean seeds and identified pod mutations in the M_3 generation, which resulted in enhanced pod length and girth—traits considered beneficial for breeding programs. Sarkar and Kundagrami (2018) focused on mutation breeding in mungbean varieties such as B1, Pusa-9632, and K-851. In the M_3 generation, several pod mutants exhibited increases in both pod length and girth. Although these mutant plants were 13–40% shorter than controls, they exhibited better branching and larger pods while retaining a normal plant phenotype. These improvements highlight their breeding value for yield enhancement through induced mutations. Another study employed gamma irradiation (100– 400 Gy) to induce mutations in mungbean varieties K-851 and Sona mung to induce genetic variation. Traits positively correlated with yield guided selection. Significant improvements were seen in yield-related traits. Four high-yielding mutant families were identified in M_7 generation, showing uniformity and high harvest index. Some lines also showed synchronous maturity and potential for developing small-seeded, high-yielding mungbean varieties (Dewanjee and Sarkar 2018). Arulselvi et al. (2019) detected chlorophyll-related mutations in three mungbean varieties—CO5, CO (Gg) 7, and VBN (Gg) 3—following exposure to 550 and 600 Gy of gamma radiation. Among these treated lines, five distinct categories of chlorophyll-deficient mutants were recognized, demonstrating effective genetic variation induced by mutagenesis.

3.2 Black Gram

Seeds from the blackgram cultivar Vamban-4 were subjected to gamma rays across a range of doses (200–1200 Gy) of ^{60}Co gamma irradiation, and the resulting mutant populations were advanced to M_3 and M_4 generations to isolate stable, high-performing mutants. A key focus was the isolation of a pod mutant with improved agronomic traits in the M_2 generation, demonstrating enhanced yield potential through gamma irradiation (Yasmin et al. 2020). Goyal et al. (2019) applied mutagens—including gamma rays, EMS, and combinations thereof—to urdbean cultivars T-9 and Pant U-30. They observed 16 different morphological mutant types in the M_2 generation. The combined treatments produced the highest mutation rates, with promising mutants selected for further studies based on increased pod number and size with positive yield associations. The same research team also worked on genetic enhancement of yield-related and quantitative traits in blackgram genotypes T-9 and Pant U-30 using varying doses of gamma radiation and EMS. From the M_3 generation, six high-yielding mutant lines were identified, displaying significant increases in genotypic and phenotypic coefficients of variation (GCV and PCV), heritability, and genetic advance. Mutants exhibited higher yield traits, pigment content, protein polymorphism, and could serve as valuable germplasm for crop improvement (Goyal et al. 2022). Dhasarthan et al. (2021) isolated the novel blackgram mutants with improved yield traits using gamma irradiation and assess their stability across environments. Mutants from cultivars ADT 3, CO 6, and TU 17-9 were selected from M_3–M_5 generations. Key yield traits showed strong associations, and GGE/AMMI analyses identified stable, environment-specific genotypes valuable for future breeding. Another study applied gamma rays and EMS to induce mutations in cultivar CO 6, and mutants with short duration, bold seeds and high yield were isolated. These agronomically superior mutants are useful as donors and are potential resources for blackgram improvement (Tamilzharasi et al. 2022).

Table 3 Mutation breeding for yield and quality improvement in pea, chick pea and pigeon pea

S. No.	Crop / Species	Generation/ Lines	Direct/Derived from	Method/Dosage	Major findings	Reference
1	Pea *(Pisum sativum)*	–	Pea Cv-Climax	Gamma rays (0.8, 1.6, 2.4 and 3.6 kGy)	An increased number of seeds per pod, higher 1000-seed weight, and greater dry biomass were observed	Majeed et al. (2016)
2		M_2	S10	HA and SA (0.1, 0.15, 0.20, 0.25)	Significant variation was observed in traits like germination, flowering, and pod characteristics. Sodium Azide, especially at 0.20% and 0.25%, was more effective than HA in inducing genetic variability and phenotypic alterations	Bansod et al. (2022)
3		M_2	Dodoni, Early Onward, and Rondo	100 Gy	Variations observed in traits like plant height, dry biomass, and pod length, but not in growth rate or seed yield	Sarri et al. (2024)
4 5	Chickpea *(Cicer arietinum)*	Pusa-256-A, Pusa-256-B, Pusa- 256-C, Pusa-256- D, BG-1053-A, BG-1053-B	Pusa-256 and BG- 1053	Gamma rays (300 Gy), EMS (0.2%), EMS (0.3%), gamma rays + EMS (100 Gy + 0.2%), gamma rays (200 Gy) and EMS (0.3%) respectively	High yielding mutants	Kozgar (2014)
6		M_2	C-235	Gamma rays (100, 200 and 300 Gy), HZ (0.1, 0.2 and 0.3%), SA (0.01, 0.02 and 0.03%) and combination treatments of HZ and SA (0.1% HZ + 0.02% SA, 0.2% HZ + 0.02% SA and 0.3% HZ + 0.02% SA)	Superior mutants for several yield attributing traits were identified	Raina et al. (2017)
7		Pusa-212A, Pusa- 212B and Pusa 212C	Pusa-212	0.01%, 0.02% EMS and 0.02% SA for 6 h	High yielding mutants	Wani (2017)
8 9	Pigeon pea *(Cajanus cajan)*	M_1	BSMR 736	Gamma rays (5, 10, 15, 20, and 25kR)	Mutants with superior physiological traits and high protein content were identified	Desai and Rao (2014)
10		M_2	CO-7	Gamma rays (15, 20, and 25 kR) and EMS (20, 25, and 30 mM)	Bold seeded and high yielding mutants were identified	Ariraman et al. (2016)
11		M_4	AL 201	Gamma rays (300 and 400 Gy)	Mutants with altered yield and its contributing traits were identified	Gaur et al. (2018)

HA—Hydroxyl Amine, SA—Sodium azide, EMS—Ethyl methanesulfonate, HZ—Hydrazine hydrate, NMU—N-nitroso-N-methyl urea, NG—N-ethyl-N-nitro-N-nitrosoguanidin

3.3 Cowpea

EMS and gamma irradiation have been extensively applied to induce genetic variation in agro-morphological traits of cowpea. Moderate EMS concentrations (0.25%–0.30%) and gamma radiation doses ranging from 200 to 300 Gy were particularly effective in producing beneficial mutations. Raina et al. (2020) applied sodium azide and gamma rays to induce variability in cowpea genotypes Gomati VU-89 and Pusa-578, successfully selecting 11 high-yielding M_4 mutant lines. Biochemical and molecular characterization using SDS-PAGE, SSR, and CBDP markers revealed significant genetic divergence and enhanced protein, micronutrient, chlorophyll, and nitrate reductase levels in mutant lines. Gomati VU-89-G and Pusa-578-C emerged as the most genetically diverse and nutritionally superior lines, offering valuable genetic resources for future cowpea improvement. Induced mutations in Egyptian cowpea varieties Kaha 1 and Dokki 331 were generated using gamma rays (0–300 Gy), with optimal effects at 150 Gy and 300 Gy, respectively. Kaha 1 showed enhanced yield traits, while Dokki 331 exhibited pod and seed color changes with improved nutritional content. Genetic diversity analysis using 20 microsatellite markers revealed considerable polymorphism, confirming successful mutation induction and potential for varietal improvement (Adly et al. 2024).

3.4 Common Bean

Earlier, common bean breeding efforts focused mainly on seed yield and plant morphological alterations for functional studies. Later, most crop improvement programmes targeted the nutritional aspects of common bean, including enhancing protein and micronutrient accumulation and reducing anti-nutritional factors. Chemical mutagenesis, and in particular EMS, was widely used for inducing mutations in common bean. Treatment with EMS as a chemical mutagen resulted in the generation of 3,000 M_2 mutants, many of which showed promise for enhancing diverse agronomic traits (Porch et al. 2009). Several EMS-induced mutant populations were screened to isolate a stable and homozygous low-phytate mutant (lpa-280-10), which showed a 90% reduction in phytate and a 25% reduction in raffinose family oligosaccharides (RFOs) compared to its wild type (Campion et al. 2009). Several desirable mutants for yield and its contributing traits were identified by Ramandeep et al. (2018) in the M_2 generation following EMS treatment. Similarly, gamma ray treatment resulted in the identification of superior mutants with enhanced yield and protein content, as reported by Masry et al. (2021). Recently, Laskar et al. (2024) used EMS and sodium azide to induce mutations in a traditional cultivar and isolated mutants with altered expression of yield and its associated traits. Phytic acid is major anti-nutritional factors present in various crops that limits the availability of micronutrients.

3.5 Lentil

In lentil, limited attempts have been made to induce mutations. Gamma irradiation of lentil varieties "Masoor-85" and "ICARDA-8" produced mutants with improved earliness, plant height, and 100-grain weight. Mutants AEL 12/30/91 and AEL 49/20/91 showed yield increases up to 60.12%, indicating their potential as promising genotypes for future lentil improvement. MMS, alone and with dimethyl sulfoxide (DMSO), induced polygenic variation in lentil. DMSO influenced MMS mutagenicity, possibly through synergism or uptake variation. Moderate MMS + 2% DMSO reduced damage and enhanced desirable mutants, altering trait associations and aiding lentil yield improvement (Amin et al. 2015). Six high-yielding lentil mutants generated from hydrazine hydrate and gamma ray treatments showed enhanced yield and nutrition traits. In particular, mutant lines DPL 62-B and Pant L406-A were most divergent. These lines offer novel gene combinations, supporting genetic improvement, hybridization, and trait stabilization in lentil breeding programs (Laskar et al. 2018a). Gamma irradiation of lentil cultivar Pant L 406 induced a novel "multipodding" (mp) mutant with altered inflorescence traits. Despite reduced seed weight, the mp mutant showed stable morphological changes. It holds potential for genetic studies and breeding of high-yielding lentil cultivars (Laskar et al. 2018b). EMS mutagenesis in lentil variety L-4076 using concentrations from 0.10% to 1.0% induced a novel "bigger pod" (bp) mutant at 0.10% EMS. The bp mutant showed significantly larger pods (6–7 seeds), higher yield, and improved traits, supporting genetic studies and breeding of elite lentil cultivars through genomics-assisted approaches targeting pod architecture and yield stability (Shahwar et al. 2024).

3.6 Pea

Different mutagens, *viz.*, gamma-rays, fast neutrons, EMS, ethylene imine (EI), DES, and nitroso methyl urethane (NMU) were used to develop mutants to enrich genetic stock in pea and among these EMS and gamma rays were found to be highly. Majeed et al. (2016) studied the effects of gamma irradiation doses ranging from 0.8 to 3.6 kGy on the growth and yield performance of the edible pea cultivar "Climax". Their results showed that exposure to 1.6 kGy significantly enhanced the number of seeds per pod, 1000-seed weight, and dry biomass. Bansod et al. (2022) evaluated the effects of Sodium Azide and Hydroxyl Amine on yield and seed quality traits in pea. Significant variation was observed in M_2 for traits like germination, flowering, and pod characteristics. Sodium Azide, especially at 0.20% and 0.25%, was more effective than Hydroxyl Amine in inducing genetic variability and phenotypic alterations. Pandey et al. (2022) used EMS and gamma rays in pea variety "CDC Amarillo" for optimizing the concentration of the mutagen. In addition, with the advancement of targeted mutagenesis, they also optimized the protocol for mesophyll protoplast extraction, PEG-mediated transformation and gene editing of a *LOX* gene using CRISPR/Cas system. These findings widen the scope for pea crop improvement. In another study, three pea varieties ("Dodoni," "Early Onward," and "Rondo") were irradiated with 100 Gy gamma rays and grown under field conditions. Significant differences between M_1 mutant and control plants were observed in traits like plant height, dry biomass, and pod length, but not in growth rate or seed yield. Varietal responses varied, with "Dodoni" showing increased biomass. Results suggest gamma irradiation induces genetic variability, offering valuable traits for future breeding and omics-based improvement (Sarri et al. 2024).

3.7 Chickpea

The lower productivity of chickpea is mainly due to its narrow genetic base, low yield potential, and disease susceptibility. Hence, achieving a higher yield has been one of the major breeding objectives focused on by researchers to improve chickpea productivity. It is a complex trait governed by many genes and influenced by various external factors. Bold-seeded chickpea is one of the important seed quality traits that fetch a premium price in the market. Several induced mutation studies have isolated bold-seeded mutants in chickpea. Induced mutations in four different chickpea genotypes resulted in a significant increase in fatty acid composition. This included an increase in both saturated and unsaturated fatty acids, contributing to enhanced frost tolerance. Chickpea possesses high-quality protein with the highest protein efficiency ratio among legumes. However, improving protein content without yield penalty remains challenging through conventional breeding methods. Induced mutagenesis using physical (gamma rays and fast neutrons) and chemical (EMS and NMU) mutagens in chickpea varieties has resulted in mutants with increased protein content compared to their wild types, without a yield penalty (Kharkwal 1998). Radiation processing of several legumes, including chickpea, followed by germination, resulted in rapid degradation of RFOs (Machaiah and Pednekar 2002), indicating the potential for reducing RFO content through mutation breeding.

3.8 Pigeon Pea

Mutagenesis in pigeon pea has primarily targeted traits such as high yield and bold seed size, leading to the release of several improved varieties. Desai and Rao (2014) used irradiation treatment to improve several traits, including protein content, and found that treated plants exhibited higher protein content than the control. A study was conducted by treating pigeon pea accessions with EMS, sodium azide, and gamma radiation to analyze their response to antinutritional factors (Chougule et al. 2004). They identified mutants in the M_6 generation with altered expression of amylase inhibitors, lectins, and RFOs. Chlorophyll, morphological, and yield-related mutants were observed in Pigeon pea Var CO-7 after gamma ray and EMS treatments (Ariraman et al. 2016). Gamma rays at 300 Gy and 400 Gy induced mutations in pigeonpea cultivar AL 201, resulting in identification of high-yield mutants (Gaur et al. 2018).

4 Conclusion

Mutation breeding has proven to be a powerful approach for improving yield and quality traits in pulse crops. Over the past several decades, numerous high-yielding and quality-improved mutant varieties have been developed and released worldwide, particularly in India. These efforts have significantly contributed to food security and agricultural sustainability. Advances in induced mutagenesis have enabled the creation of genetic variability otherwise limited in self-pollinated pulse crops. However, despite these successes, challenges such as narrow genetic bases and complex trait inheritance persist. Future research should integrate mutation breeding with modern genomics and molecular breeding approaches to accelerate trait discovery and cultivar development. The application of TILLING, genome editing, and high-throughput phenotyping can enhance precision and efficiency. Greater focus on improving yield and quality traits along with nutritional quality, and climate resilience through mutation breeding is essential. Additionally, multi-location trials and participatory breeding approaches will be important for ensuring the adaptability and acceptance of mutant cultivars. Strengthening international collaboration and training programs will further expand the global impact of mutation breeding in pulses.

References

Adly M, Ezzat A, Awad A, El-Fiki A (2024) Mutations induced in some Egyptian cowpea varieties with yield characteristics and high nutritional value using gamma rays and evaluation by microsatellite markers. Int J Radiat Biol 100(8):1117–1125. https://doi.org/10.1080/09553002.2024.2374912

Amin R, Laskar RA, Khan S (2015) Assessment of genetic response and character association for yield and yield components in Lentil (*Lens culinaris* L.) population developed through chemical mutagenesis. Cogent Food Agric 1(1):1000715. https://doi.org/10.1080/23311932.2014.1000715

Ariraman M, Bharathi T, Dhanavel D (2016) Induced mutagenic effectiveness and efficiency studies on Pigeon pea (*Cajanus cajan* (L.) Millsp). Food Biol 4. https://doi.org/10.19071/fb.2015.v4.3040

Arulselvi S, Suresh S, Dhole VJ (2019) Occurrence of chlorophyll deficient mutants in the mutated populations of greengram (*Vigna radiata* (L.) Wilczek). Electron J Plant Breed 10:303–306. https://doi.org/10.5958/0975-928X.2019.00038.3

Bansod SV, Gupta JP, Pandey BK, Awachar SR (2022) Effect of Chemically induced Mutation on Yield Contributing and Seed quality Attribute in Pea (*Pisum sativum* L.). Biol Forum 14(3):1497–1500. https://www.researchtrend.net/bfij/pdf/254%20Effect%20of%20Chemically%20induced%20Mutation%20on%20Yield%20Contributing%20and%20Seed%20quality%20Attribute%20in%20Pea%20_Pisum%20sativum%20L._%20Jai%20Prakash%20Gupta.pdf

Campion B, Sparvoli F, Doria E, Tagliabue G, Galasso I, Fileppi M, Bollini R, Nielsen E (2009) Isolation and characterisation of an lpa (low phytic acid) mutant in common bean (*Phaseolus vulgaris* L.). Theor Appl Genet 118:1211–1221. https://doi.org/10.1007/s00122-009-0975-8

Chougule N, Giri A, Hivrale V, Chhabda P, Kachole M (2004) Identification of amylase inhibitor deficient mutants in pigeonpea (*Cajanus cajan* (L.) Millisp.). Biochem Genet 42:165–180. https://doi.org/10.1023/b:bigi.0000026632.17713.47

Das TR, Baisakh B (2013) Efficacy of physical and chemical mutagenic treatments for developing superior mutants in greengram (*Vigna radiata*). Ind J Agric Sci 83:595–598. https://epubs.icar.org.in/index.php/IJAgS/article/view/30319

Desai AS, Rao S (2014) Effect of gamma radiation on germination and physiological aspects of pigeon pea (*Cajanus cajan* (L.) millsp) seedlings. Int J Res Appl Nat Soc Sci 2:47–52 https://oaji.net/articles/2014/491-1407224238.pdf

Dewanjee S, Sarkar KK (2018) Evaluation of performance of induced mutants in mungbean [*Vigna radiata* (L.) Wilczek]. Legume Res 41(2):213–217. https://doi.org/10.18805/lr.v0iOF.9098

Dhasarathan M, Geetha S, Karthikeyan A, Sassikumar D, Meenakshiganesan N (2021) Development of novel blackgram (*Vigna mungo* (L.) Hepper) mutants and deciphering genotype × environment interaction for yield-related traits of mutants. Agronomy 11(7):1287. https://doi.org/10.3390/agronomy11071287

Gaur AK, Singh I, Singh S, Sharma S (2018) Genetic analysis of M_4 mutant lines of pigeonpea (*Cajanus cajan* (L.) Millspaugh) developed through gamma irradiation. J Food Legumes 31:71–74 https://pub.isprd.in/index.php/jfl/article/view/114/73

Goyal S, Wani MR, Laskar RA, Raina A, Amin R, Khan S (2019) Induction of morphological mutations and mutant phenotyping in black gram [*Vigna mungo* (L.) Hepper] using gamma rays and EMS. Vegetos 32:464–472. https://doi.org/10.1007/s42535-019-00057-w

Goyal S, Wani MR, Raina A, Laskar RA, Khan S (2022) Quantitative assessments on induced high yielding mutant lines in urdbean [*Vigna mungo* (L.) hepper]. Legume Sci 4(2): e125. https://doi.org/10.1002/leg3.125

Jegadeesan S, Punniyamoorthy D (2023) Potential of mutation breeding in genetic improvement of pulse crops. In: Mutation breeding for sustainable food production and climate resilience. Springer Nature Singapore, Singapore, pp 445–485. https://doi.org/10.1007/978-981-16-9720-3_15

Kavuri N, Alavilli H, Manthari R, Lee BH, Mangrauthia SK (2024) Transforming farming: mutational breeding as a sustainable solution for crop improvement in the 21st century. J Plant Biochem Biotechno 1–8. https://doi.org/10.1007/s13562-024-00920-1

Kharkwal MC, Pandey RN, Pawar SE (2004) Mutation breeding for crop improvement. Plant Breed: Mendel Mol Approaches 601-645. https://doi.org/10.1007/978-94-007-1040-5_26

Kharkwal M (1998) Induced mutations for improvement of protein in chickpea (*Cicer arietinum* L.). Indian J Genet Plant Breed 58:61–68 https://www.isgpb.org/journal/index.php/IJGPB/article/view/2393/1133

Khattak GSS, Saeed I, Shah, SA (2008) Breeding high yielding and disease resistant mungbean (*Vigna radiata* (L.) Wilczek) genotypes. Pak J Bot 40:1411–1417. https://pakbs.org/pjbot/PDFs/40(4)/PJB40(4)1411.pdf

Kozgar I (2014) Mutation breeding in chickpea: perspectives and prospects for food security. De Gruyter Open Ltd, Warsaw/berlin. https://doi.org/10.2478/9788376560717

Kumar S, Gupta P, Choukri H, Siddique KH (2020) Efficient breeding of pulse crops. Accel Plant Breed 3:1–30. https://doi.org/10.1007/978-3-030-47306-8_1

Laskar RA, Laskar AA, Raina A, Khan S, Younus H (2018a) Induced mutation analysis with biochemical and molecular characterization of high yielding lentil mutant lines. Int J Biol Macromol 109:167–179. https://doi.org/10.1016/j.ijbiomac.2017.12.067

Laskar RA, Wani MR, Raina A, Amin R, Khan S (2018b) Morphological characterization of gamma rays induced multipodding mutant (mp) in lentil cultivar Pant L 406. Int J Radiat Biol 94(11):1049–1053. https://doi.org/10.1080/09553002.2018.1511927

Laskar RA, Dowarah B, Tamang D, Das S, Borah P, Raina A (2024) Improving French bean yield potential through induced mutagenesis using EMS and SA. Front Hortic 2:1288720. https://doi.org/10.3389/fhort.2023.1288720

Machaiah JP, Pednekar M (2002) Carbohydrate composition of low dose radiation- processed legumes and reduction in flatulence factors. Food Chem 79:293–301. https://doi.org/10.1016/S0308-8146(02)00142-5

Majeed A, Muhammad Z, Ullah R (2016) Growth and yield response of field pea (*Pisum sativum* L.) to gamma irradiation stress. Plant Breed Seed Sci 74:27–35. https://doi.org/10.1515/plass-2016-0012

Masry A, Fayad A, Rakha M (2021) Improving seed yield and protein content of common beans (Phaseolus Vulgaris L.) through mutation breeding. Menoufia J Plant Prod 6:443–460. https://doi.org/10.21608/mjppf.2021.198686

Pandey PK, Bhowmik P, Kagale S (2022) Optimized methods for random and targeted mutagenesis in field pea (*Pisum sativum* L.). Front Plant Sci.13:995542. https://doi.org/10.3389/fpls.2022.995542

Pandit R, Bhusal B, Regmi R, Neupane P, Bhattarai K, Maharjan B, Poudel MR (2021) Mutation breeding for crop improvement: a review. Rev Food Agric 2(1):31–35. https://doi.org/10.26480/rfna.01.2021.31.35

Porch TG, Blair MW, Lariguet P, Galeano C, Pankhurst CE, Broughton WJ (2009) Generation of a mutant population for TILLING common bean genotype BAT 93. J Am Soc Hortic Sci 134:348–355. https://doi.org/10.21273/JASHS.134.3.348

Raina A, Laskar RA, Khursheed S, Khan S, Parveen K, Amin R (2017) Induce physical and chemical mutagenesis for improvement of yield attributing traits and their correlation analysis in chickpea. Int Lett Nat Sci 61:14–22. https://doi.org/10.18052/www.scipress.com/ILNS.61.14

Raina A, Laskar RA, Tantray YR, Khursheed S, Wani MR, Khan S (2020) Characterization of induced high yielding cowpea mutant lines using physiological, biochemical and molecular markers. Sci Rep 10(1):3687. https://doi.org/10.1038/s41598-020-60601-6

Ramandeep DT, Dhall R, Gill B (2018) Effect of mutagen-ethyl methane sulphonate on yield increasing parameters of french bean (*Phaseolus vulgaris* L.). Genetika 50:199–207. https://doi.org/10.2298/GENSR1801199R

Sarkar M, Kundagrami S (2018) Selection of high yielding, extra short duration lines of mungbean derived through gamma radiation. Indian J Genet 78:233–241. https://doi.org/10.5958/0975-6906.2018.00030.5

Sarri E, Styliani-Maria S, Anastasios K, Nasya T, Eleni MA, Eleni T (2024) Effect of γ-irradiation on the growth and yield response of three varieties of pea (*Pisum* spp.) Agronomy 14(8):1695. https://doi.org/10.3390/agronomy14081695

Shahwar D, Ansari MYK, Khatoon B, Park Y (2024). Phenotypic characterization of ethyl methanesulfonate (EMS) induced bigger pod (bp) with multiple seed mutant in lentil (*Lens culinaris* Medik.). Biocatal Agric Biotechnol, 59: 103259. https://doi.org/10.1016/j.bcab.2024.103259

Stearns FW, Fenster CB (2016) The effect of induced mutations on quantitative traits in *Arabidopsis thaliana*: Natural versus artificial conditions. Ecol Evol 6(23):8366–8374. https://doi.org/10.1002/ece3.2558

Tamilzharasi M, Dharmalingam K, Venkatesan T, Jegadeesan S, Palaniappan J (2022) Mutagenic effectiveness and efficiency of gamma rays and combinations with EMS in the induction of macro mutations in blackgram (*Vigna mungo* (L.) Hepper). Appl Radiat Isot 188:110382. https://doi.org/10.1016/j.apradiso.2022.110382

Wani MR (2017) High yielding mutants in chickpea (*Cicer arietinum* L.). Res Crops 18:718–721. https://doi.org/10.5958/2348-7542.2017.00118.8

Wani MR, Dar AR, Tak A, Amin I, Shah NH, Rehman R, Khan S (2017) Chemo- induced pod and seed mutants in mungbean (*Vigna radiata* L. Wilczek). SAARC J Agric 15:57–67. https://doi.org/10.3329/sja.v15i2.35161

Yasmin K, Arulbalachandran D, Dilipan E, Vanmathi S (2020) Characterization of ^{60}CO γ-ray induced pod trait of blackgram-A promising yield mutants. Int J Radiat Biol 96(7):929–936. https://doi.org/10.1080/09553002.2020.1748738

Application of Gamma Radiation for the Genetic Improvement of Crop Plants in Democratic Republic of the Congo

Freddy Otono Bulubulu, Bruno Munzemba Mbukula, Victor Silulundi Matuta, Moïse Itunga Muya, Calvin Ciuma Ilunga, Sébastien Ndiku Luyindula, Jacques Ne Bambi Lutaladio, Alexandre Ntumbula Mbaya, and Isaac Kofi Bimpong

Abstract

The Democratic Republic of the Congo was one of the first African countries to acquire Africa's first research reactor. For many years, this facility has enabled the country to carry out research into plant improvement using radiation-induced mutations to solve the problems of food and nutritional insecurity. New varieties of maize with high levels of lysine and histidine were disseminated to the population. Later, the DRC used gamma rays emitted by the caesium-137 and cobalt-60 irradiators respectively to develop mutants of rice, groundnuts, soybean and maize. Radiation-induced mutations have also been associated with in vitro culture in the improvement of cassava and the production of secondary metabolites with biological activity for treating certain diseases such as malaria. But the challenges in this area remain even greater for the DRC, given the size of the country and the growth and poverty of its population.

Keywords

Gamma radiation · Mutagenesis · Plant breeding

F. O. Bulubulu (✉) · M. I. Muya · C. C. Ilunga · S. N. Luyindula · A. N. Mbaya
Section de Biotechnologie, Département de Radiobiologie, Commissariat Général à l'Energie Atomique/Centre Régional d'Etudes Nucléaires de Kinshasa (CGEA/CREN- K.), Kinshasa, Democratic Republic of Congo
e-mail: freddy.bulubulu.Otono@cgea-rdc.org

M. I. Muya
e-mail: moise.muya.itunga@cgea-rdc.org

C. C. Ilunga
e-mail: calvin.ilunga.ciuma@cgea-rdc.org

S. N. Luyindula
e-mail: sebastien.luyindula.ndiku@cgea-rdc.org

A. N. Mbaya
e-mail: alexandre.mbaya.ntumbula@cgea-rdc.org

B. M. Mbukula · V. S. Matuta · J. N. B. Lutaladio
Institut National pour l'Etude et la Recherche Agronomiques (INERA), Kinshasa, Democratic Republic of Congo
e-mail: brumbukula@gmail.com

V. S. Matuta
e-mail: vickymatuta50@gmail.com

J. N. B. Lutaladio
e-mail: nb.lutaladio@gmail.com

I. K. Bimpong
Plant Breeding and Genetics Section, Joint FAO/IAEA Division of Nuclear Techniques in Food and Agriculture Vienna, Vienna, Austria
e-mail: i.bimpong@iaea.org

1 Introduction

The Democratic Republic of the Congo (DRC) is one of sub-Saharan Africa's largest and most populous countries. Rich in natural resources, the DRC's economy is essentially based on agriculture, with a potential arable land area estimated at over 80 million hectares. However, despite the DRC's enormous potential, the population remains poor and lives with pronounced food and nutritional insecurity due to a number of factors, including various stresses, both biotic and abiotic, poor access to quality inputs such as seeds and fertilisers and to funding, lack of infrastructure.

Traditional plant improvement techniques involving a multi-generational process, including hybridisation, often take five years or more to eliminate undesirable characteristics and develop the desired traits. Nuclear techniques, in particular radio-induced mutations, on the other hand, are capable of considerably speeding up the selection process. These mutations are induced by physical agents including non-ionising radiation such as ultraviolet light, and different types of ionising radiation such as X- and gamma rays, alpha and beta particles, protons and neutrons (Mba et al. 2010, FAO/IAEA 2020).

J. Devi (ed.), *Mutation Breeding in Crop Improvement*, Sustainable Landscape Planning and Natural Resources Management,
https://doi.org/10.1007/978-3-032-17911-1_9

Radiation-induced mutations are a powerful method to introduce genetic variability into crops, improving and making it possible to create plant varieties with desirable characteristics such as better resistance to disease, greater tolerance of difficult environmental conditions or higher yields (Ntsomboh et al. 2023).

These radiation-induced mutation techniques can be combined with biotechnology techniques such as in vitro culture and molecular techniques (including the application of molecular marker techniques for the identification and detection of mutations in the DNA sequence) to increase mutation efficiency (FAO/IAEA 2020).

In the DRC, research into plant improvement through radiation-induced mutations, mainly carried out at the General Atomic Energy Commission (CGEA), is also being applied at the National Institute for Agronomic Studies and Research (INERA) thanks to technical cooperation projects between the DRC government and the International Atomic Energy Agency (IAEA). These have focused not only on the genetic improvement of food plants, but also on medicinal plants.

2 Mutation Induction Techniques

Mutations, or heritable changes that occur in the genetic material of a living organism, can be natural or induced. In other words, according to Lundqvist et al. (2012), originally, mutations can be caused by errors in DNA replication or as results of exposure to natural environmental radiation that leads to DNA modification. Induced mutations also occur as a result of physical, chemical products and/or biological mutagens. Nuclear radiation and all sources of radioactivity, including ultraviolet light and certain types of ionising radiation, can act as physical agents; alkylating agents (ethyl methanesulphonate, NEU (nitrosoethyl urea), MNU (N-methyl N-nitrosourea) and EI (ethylenimine) and other chemicals such as the colchicine (FAO/IAEA 2020). Chromosomal integration based on Agrobacterium and transposon is considered to be a biological mutagen.

Ionising radiation includes radioactive particles such as alpha and beta particles, and certain electromagnetic waves such as X-rays and gamma rays. These types of radiation are called 'ionising' because they have sufficient energy above 10 eV to detach electrons from atoms and create ions. Gamma rays are emitted by the disintegration of an unstable nucleus in an atom. They are the most widely used. With a shorter wavelength, gamma rays have more energy per photon than X-rays. They can induce random mutations in the DNA of plants, leading to genetic variations that can result in desirable characteristics. One of the advantages of gamma-ray mutagenesis is its ability to induce a high frequency of mutations throughout the genome, thus providing a broad spectrum of genetic variability useful for breeding (Mohsen et al. 2023).

Irradiation can be carried out either directly in the reactor or using cobalt-60 or caesium-137—irradiators.

2.1 Seed Irradiation in the Reactor

One of the main functions of research reactors is to provide sources of irradiation energy for various applications, such as radiotherapy, materials testing and radioisotope production. Nuclear research reactors are capable of generating intense neutron and irradiation fluxes for several specific uses (IAEA 2014). Some of the typical irradiation energy sources used in these reactors include.

(a) **Thermal Neutrons**: Reactors use thermal neutrons to carry out irradiation, diffusion, capture or neutron activation experiments. Thermal neutrons are essential for the production of radioisotopes, as well as for research in nuclear physics and materials science.
(b) **Fast Neutrons**: Some research reactors can also produce fast neutrons, which are used for materials testing, in particular for studies into the resistance of materials under irradiation, such as those used in commercial nuclear reactors.
(c) **Sources of Gamma Radiation**: By irradiating samples with neutrons, gamma radiation can be emitted. This radiation is used in radiotherapy, sterilisation, medical imaging and agriculture.
(d) **Radioisotope Production**: Reactors are often used to produce radioisotopes used in medicine (for example, technetium-99 m, used in scans) or in industrial applications. Irradiation of targets in these reactors generates these isotopes.

2.1.1 Irradiation of Seeds with Gamma Rays Emitted by an Irradiator

Gamma rays are highly energetic electromagnetic radiation produced by natural radioactive isotopes and artificially using cyclotrons. They have a shorter wavelength but are highly penetrating (Bharath et al. 2024). Cobalt-60 (^{60}Co) and Caesium-137 (^{137}Cs) are the main sources of gamma rays used in plant selection by mutation. They are highly penetrating. With a half-life of 30.17 years, caesium-137 emits relatively low-energy gamma rays of 662 keV (kiloelectron-volts). Cobalt-60 has a much shorter half-life than caesium-137, at around 5.26 years. On the other hand, it emits higher-energy gamma rays, ranging from 1.17 to 1.33 MeV (megaelectronvolts), giving it a greater penetration capacity than Caesium-137 (FAO/IAEA 2020).

Seed irradiation with a ^{137}Cs or ^{60}Co irradiator is a specific method based on the use of gamma rays to induce genetic mutations in seeds.

(a) **Principle**

Gamma rays emitted by ^{60}Co or ^{137}Cs induce beneficial or neutral changes in the DNA of seeds. Favourable mutations are then selected for the development of new plant varieties.

The process involves exposing the seeds to specific, controlled doses of gamma rays emitted by an irradiator with the aim of inducing genetic mutations. The doses vary depending on the plant species, the purpose of the experiment and the results sought.

(b) **Main Stages in the Seed Irradiation Process**

The main stages in the seed irradiation process include

- Seed selection according to variety and quality to ensure optimum germination after irradiation.
- Preparing the seeds for irradiation: the seeds are cleaned to ensure they are in optimum condition or to avoid interference during irradiation.
- Seeds are packaged in bags to ensure uniform exposure to radiation and avoid external contamination.
- Irradiation of the Seeds. The radiation dose (expressed in Gray, Gy) varies according to the type of mutation desired. A low dose may induce changes in the colour, shape or size of the seeds, while a higher dose may lead to changes in agronomic traits or disease resistance, including chromosome breaks.
- The length of time seeds are exposed depends on the dose of radiation and the type of seed.

(c) **Effect of Gamma Irradiation on Plants**

Gamma rays have considerable effects on the physiological, morphological and biochemical processes of plants. The direct (physical) action of the radiation results in damage to the molecules; the indirect (chemical) action takes place via highly reactive free radicals derived from ionised water molecules (Lagoda et al. 2012).

- **Chemical Effects of Gamma Radiation**

  Gamma radiation causes the dissociation of water molecules present in the environment. This leads to the formation of both unstable and reactive positive radical ions and free electrons. In the presence of molecular oxygen (a bi-radical), the free radicals form peroxy radicals. In the solid state, radiation-induced radicals are stable. This is the case for plant seeds with a low water content. The higher water content of the irradiated target, the greater the damage (FAO/IAEA 2020).

- **Effects of Gamma Irradiation on Morphological Characteristics**:

  The interaction between gamma irradiation and atoms or molecules induces free radicals in cells capable of damaging or modifying the main cellular components present in the cell. They modify the morphogenetic, anatomical, biochemical and physiological characteristics of plants as a function of gamma irradiation doses (Yasmin et al 2022).

- **Effects of Gamma Irradiation on Biochemical Characteristics**:

  Gamma rays induce biochemical changes and break the link between chains, the cross-linking of DNA molecules and protein molecules. In plants, the effects of gamma irradiation are determined on the basis of changes in amino acids, which are essential for human nutrition. Increased gamma radiation affects the quality of reducing sugar, starch content, carotenoids and other organic elements in the plant cell (Yasmin et al. 2022).

(d) **Radio-Sensitivity Test**

One of the first steps in mutagenesis treatment is to estimate the most appropriate mutagenic treatment. This involves determining radio sensitivity and the dose resulting in a 50% reduction in vegetative growth (seedling or epicotyl height) (RD50) or 50% reduction in lethality (LD/LD50). The aim is to determine effective doses of gamma irradiation capable of inducing mutations.

Radio sensitivity can vary according to species, cultivar and genotype, with the physiological conditions of plants and organs, as well as with the types of mutagenic treatments (physical or chemical) to be used (FAO/IAEA 2020).

(e) **Improvement by Mutations in Plants Propagated by Seed**

Techniques for improving seed-propagated species by mutation are different from those for vegetatively propagated species. In seed-propagated species, the processes for developing mutant populations begin with the selection of the parent genotype to be mutated, i.e. the production of M0 seeds which are subsequently irradiated to produce M1 seeds and the manipulation of mutant generations, with the emphasis on M1 to M3 generations for mutant selection (FAO/IAEA 2020).

(f) **Improvement by Mutation of Propagated Species Vegetatively**

Given the complexity of hybridising vegetatively propagated plants such as cassava, potato, banana, sugarcane, rubber and tea, mutation induction is an appropriate method for improving these crops. Their manipulation is complex and each of the targets requires specialised methods for mutation induction. Biotechnology techniques, particularly *in vitro* tissue culture techniques, are effectively used for mutational selection. These techniques include *in vitro* micropropagation to generate populations for mutation induction, mutant screening and the development of mutant lines (FAO/IAEA 2020).

In vitro plant tissue culture, a process by which fragments of plant material (explants) are isolated and grown aseptically *in vitro*, is used. As every part of the plant is made up of cells of the same origin, any part (buds, leaves, flowers, roots, etc.) can be used for cultivation.

In vitro culture has proved to be a major tool for accelerating plant breeding. As meristems are known to be the site of repeated cell divisions that prevent viral propagation, their culture makes it possible to eliminate viruses and regenerate healthy seedlings (Tresh and Cooter 2005).

Combined with mutation induction, plant tissue culture increases the overall efficacy of mutagenic treatments, enables the creation of new genetic variations, facilitates the manipulation of large plant populations, early *in vitro* selection and offers a means of easily cloning selected variants. Tissue culture also makes it possible to manipulate large mutagenised populations on a laboratory scale, enabling the development and implementation of effective and reliable methodologies for screening responses to biotic and abiotic stresses (FAO/IAEA 2020).

Micropropagation, the technique of proliferating apical or axillary buds on newly developed stems capable of providing buds for subsequent cycles of cultivation and multiplication, has become an industrial tool for the mass production of plants of economic interest, including ornamental plants, flowering and fruiting trees and vegetatively propagated crops. The combination of micropropagation and mutation induction has proved to be of great economic interest (Jain and Spencer 2006).

In addition, *in vitro* mutagenesis has overcome the main limitations of conventional mutagenesis and is a powerful tool for the multiplication of any new trait of interest obtained. The direct formation of adventitious buds and somatic embryogenesis allow chimeras to be dissolved and facilitate the development of homohistant mutants (Jankowicz-Cieslak et al. 2017).

3 Applications of Mutations Plant Breeding in Democratic Republic of the Congo (DRC)

3.1 DRC at a Glance

Located in the heart of Africa, the Democratic Republic of the Congo (DRC) covers an area of 2,345,410 km^2, making it the 2nd largest country on the African continent. It is the fourth most populous country in Africa. The DRC boasts a variety of ecosystems and natural habitats of exceptional biological diversity, making it one of the 17 most biodiverse countries in the world.

There are two types of climate such as the equatorial climate, hot and humid all year round, in the central zone crossed by the equator and the tropical climate, hot all year round, with a dry season and a rainy season in the regions to the north and south of the equator.

DRC has almost 80 million hectares of arable land and 4 million hectares of irrigable land, of which only 1% is cultivated. Agriculture employs over 70% of the working population, mainly in rural areas. According to the FAO (2024), more than 90% of producers limit themselves to subsistence production on small family farms covering an average area of 1 to 2 hectares for rain-fed farming and 0.5 to 1 hectare for irrigated farming. Women are key players in agri-food systems and play a crucial role in agricultural and rural development. Nearly 27 million people, a quarter of the population of the DRC, are facing crisis conditions or acute emergency food insecurity, fuelled by poor harvests, poor access to quality inputs and displacement due to violence, diseases and the collapse of infrastructure. The low productivity of crops is due to various stresses, both biotic and abiotic, and to the effects of climate change, characterised by the poor distribution of rainfall in different areas, the infertility of certain soils, etc.

Nuclear techniques, particularly radiation-induced mutations, represent an opportunity to improve crops and combat food and nutritional insecurity in the DRC. The use of nuclear techniques can create variability in crops, enhance their resistance to water stress, yield and disease resistance, and create a wealth of genetic variability including desirable changes in crop quality and nutrient content.

3.2 Background on Applications of Radiation-Induced Mutation Technology in the DRC

Thanks to its uranium resources, the DRC has made a significant contribution to the growth in the use of nuclear energy worldwide. So, with a view to making the most of the impact of nuclear energy for its own development, the DRC was the first African country to instal a 50 kW Triga Mark I research reactor in 1959. A second Triga Mark II reactor (Fig. 1) was inaugurated in March 1972. On 20 November 1974, this reactor was boosted to over 1,600 MW, making it the most powerful research reactor on the African continent. It contained the neutron source in a 600 mCi radium-beryllium source. This activity corresponds to an emission of 0.992×10^7 neutrons per second.

Fig. 1 CGEA/CREN-K's TRIGA II research reactor

The reactor offers a number of basic and advanced irradiation facilities for a wide range of experiments, including seed irradiation as part of plant genetic improvement.

For many years, the General Commission for Atomic Energy, Regional Centre for Nuclear Studies of Kinshasa (CGEA/CREN-K) has been carrying out research into plant improvement through radiation-induced mutations.

Since the installation of the TRIGA MARK I reactor in 1959, the first induced mutations research began in 1969, with experiments on the radio sensitivity of groundnut, soybeans, maize and other seeds.

In 1977, the Lisa conservatome irradiator with a caesium 137 source was installed at CGEA/CREN-K (Fig. 2).

This irradiator, manufactured in mas 1977, had a starting dose of 4,400 Curie. Since then, seed irradiation for genetic improvement has been carried out using this equipment. However, as the half-life of this CGEA caesium irradiator decreased, seeds were sent to the IAEA laboratory at Seisberdoff/Vienna or to Morocco for irradiation with Cobalt-60.

Subsequently, around 2005, radiation-induced mutation techniques were combined with biotechnology techniques, in particular in vitro culture techniques, to improve medicinal plants for the production of high-value secondary metabolites.

Fig. 2 CGEA/CREN-K's Caesium 137 irradiator

From 2010, these techniques were applied to vegetatively propagated plants such cassava.

Since 2019, at the instigation of the International Atomic Energy Agency (IAEA), CGEA researchers have been working with those at the National Institute for Agronomic Studies and Research (INERA) to improve food plants using radiation-induced mutation and biotechnology techniques. Like the CGEA, INERA is an institution under the supervision of the Ministry of Research and Technological Innovation. This institution has research centres and stations all over the country, with infrastructures (experimental sites) that can facilitate field breeding activities. Thanks to this partnership with the CGEA, INERA's human capacities have been on mutation-based selection techniques.

3.3 Radiation-Induced Mutagenesis of Propagated-Seed Plants

3.3.1 Seed Irradiation at the TRIGA Reactor

The TRIGA (Training and Research Reactor for Isotope production General Atomic) reactor is a type of light water nuclear reactor designed for research applications, including the irradiation of biological materials such as seeds. It is capable of delivering high-intensity neutron fluxes, which is ideal for inducing genetic mutations by irradiation.

(a) **Principle of Seed Irradiation**

- Seeds are placed in the reactor's cooling pool or in a specific area for irradiation.

   - Irradiation with neutrons or gamma rays modifies the DNA of the seeds, generating mutations. These mutations can be studied for specific effects, such as disease resistance, improved growth or other agronomic traits of interest.

(b) **Advantages of the TRIGA Reactor for Irradiation**

   The TRIGA reactor has the following characteristics:
   - **High Neutron Flux**: TRIGA Mark I is designed to provide a high neutron flux, ideal for seed irradiation.
   - **Safety**: The TRIGA reactor is designed to be safe and easy to control, even at low power, which is crucial for biological experiments.
   - **Accessibility**: TRIGA is a relatively compact reactor, allowing flexible use in academic and scientific research.

(c) **Agricultural Applications of TRIGA Reactor in the DRC**.

   Seed irradiation in the TRIGA reactor is a process used to modify the genetic characteristics of seeds, which can have applications in agronomic research, crop improvement and the production of seeds resistant to disease or extreme environmental conditions.

The first work to induce genetic mutations by irradiation in the DRC involved irradiating seeds in the TRIGA MARK I reactor, the first reactor at the Regional Centre for Nuclear Studies in Kinshasa (CREN-K).

To this end, soybean seeds (*Glycine max* Merril) of TSHIBASHI variety no. 6 were irradiated with thermal neutrons from the TRIGA MARK I reactor at a flux of 3.5 $\times$ 10^{11}n.th/cm^2/sec and a dose of 2100 rads. After irradiation, the seeds were sown. At generation M3, green seeds (a phenotype present in the parents) and brown seeds appeared on the same plant. To form M4 generation, the brown seeds were sown separately from the green seeds. At M5, the brown seeds gave only brown seeds and the green seeds gave only green seeds. After respective multiplications of these two lines up to advanced generations, the mutants obtained did not dominate the initial variety in terms of field performance or lipid, protein, fatty acid and amino acid content.

Work on genetic improvement in the DRC continued after the acquisition of the TRIGA MK II reactor (Fig. 1). In order to study the radiosensitivity of the reproductive organs of cassava (*Manihot esculenta*), cassava seeds and cuttings were irradiated in the irradiation ring of the TRIGA MK II reactor in a mixed Neutron/Gamma flux. The optimum exposure time to the mixed flux was 3 s for the cuttings. However, the results indicated that the recovery rate of the cuttings decreased with increasing exposure time. Seed irradiation showed a sublethal effect on germination (Dimundu et al. 1987).

By irradiating seeds with thermal neutrons, the DRC has a number of varieties developed at CGEA/CREN-K. (Table 1).

3.3.2 Irradiation of Seeds with Gamma Rays from an Irradiator

- **Irradiation of Grains with Caesium 137**:

In the DRC, the Lisa conservatome irradiator with a caesium 137 source (Fig. 2) installed at the CGEA/CREN-K was used as a tool in work on radiation-induced crop mutations.

In the context of legume improvement, *Vigna vexillata* (L.) seeds were irradiated with gamma radiation at increasing doses of 4, 7, 10, 14 and 20 Krds using this irradiator. The percentage of seed germination decreased as the dose intensity increased. Doses above 14 Krds had a considerable effect on seed germination. Major morphological deformations, gigantism and dwarfism were observed.

The same applies to groundnut seeds, which were treated with various doses of gamma rays (100, 200, 400 and 600 Gy) from the Lisa 1 irradiator. The 100 Gy dose resulted in a greater increase in yield and other morpho-agronomic parameters. The number of pods per plant increased by up to 37%. For the number of seeds per plant, there was a significant increase of 62% at this dose. Higher doses of gamma rays (400 and 600 Gy) significantly reduced plant growth and grain yield (Tshilenge et al. 2013).

The effect of different doses of gamma irradiation from the Lisa 1 conservatome irradiator on different morpho-agronomic characteristics of soybean was determined in work carried out by Mudibu et al. (2012). The results revealed a significant increase in grain yields and their components in M2 populations of the three soybean varieties.

Some applications of radiation-induced mutations carried out in the DRC have also involved medicinal plants, with the aim of improving the production of secondary metabolites with high biological activity.

Seeds of *Phyllanthus odontadenius* L., a Phyllanthaceae whose plants are commonly used in the treatment of malaria in the Democratic Republic of Congo, were irradiated with gamma radiation (Cs-137) at doses ranging from 0 to 225 Gy with steps of 25 Gy. Assessment of the in vitro anti-plasmodial activity of aqueous extracts of the aerial part of M2 plants on clinical isolates of *Plasmodium falciparum* and on the chloroquine-resistant *P. falciparum* K1 strain was determined using the microscopic method, the isotopic micro-test method and the ELISA assay. The in vitro anti-plasmodial activities of the irradiated extracts were improved or exceeded by 147.57% to 660% compared to the control plant.

And more, plant cell and tissue cultures are an attractive alternative to medicinal plants for the production of valuable secondary metabolites. They offer a defined production system that ensures a continuous supply of products, uniform quality and satisfactory yield. New compounds, normally absent or in low concentration in the mother plant, can be produced by these biotechnology techniques. In addition, radiation-induced mutations can transform the cell, confer-

Table 1 Irradiation of seeds with thermal neutrons from the TRIGA I or TRIGA II reactor

N°	Culture	Optimum irradiation doses	Improved traits	Source	References
01	Soybean	350, 700 et 1050 rads	2 mutants with high levels of lysine and histidine (BOTULA II) already in widespread use	Neutrons thermiques (TRIGA I)	Botula (1986)
02	Groundnut		4 mutants with high protein levels		
03	Rice		2 drought-resistant short-shoot mutants		
04	Maize		A number of mutants, which form the basis for the creation of new lines, are part of hybrid formulae		

ring unusual (new) abilities. Combined with in vitro culture, they are capable of increasing the extent of genetic variation and consequently improving the biosynthesis of secondary metabolites. This approach has been used in the DRC to improve secondary metabolite productivity in *Phyllanthus niruri* L., *Cassia alata* L. and *Jatropha curcas* L. using gamma irradiation combined with in vitro*culture.*

Phyllanthus niruri is a seasonal herb with antiviral, antioxidant, anti-inflammatory and hepatoprotective properties (Harish and Shivanandappa 2006; Okoli et al. 2010; Navneet et al. 2017; Li et al. 2017; Lavina and Namrata 2024). In vitro culture of P. niruri stem explants irradiated with doses of 1, 5 and 7 Gy of gamma radiation on MS medium enriched with AIB/BAP and coconut milk (*C. nucifera*) showed that the rate of callogenesis decreased with the irradiation dose. Alkaloids, catechic tannins, saponins and terpenoids were identified in extracts from the callus of irradiated explants, in contrast to extracts from the mother plant. Production of secondary metabolites was abundant in aqueous extracts from irradiated explant callus compared with those from non-irradiated explant callus and the mother plant. It was higher in extracts from explant callus irradiated at 5 Gy. The sensitivity of the *Plasmodium falciparum* strain was highest (100%) for extracts of explant callus irradiated at 1 Gy, at concentrations of 300 and 3 μg/ml. The mother plant extract gave the lowest percentage of inhibition (65.25%).

For *Cassia alata* (Caesalpiniaceae), leaf explants were irradiated with doses of 1, 2.5, 5, 7.5 and 10 Gy of gamma radiation and grown on MS medium. The results showed that callogenesis in *C. alata* decreased with increasing irradiation dose. The yield of secondary metabolites was higher in the callus of explants irradiated at 5 Gy (25.37% for chloroform extracts and 35.9% for aqueous extracts) than in the callus of explants irradiated at other doses or not and the mother plant. Antibacterial activity was observed on *Shigella flexneri*, *Staphylococcus aureus* and *Escherichia coli*, and antifungal activity on *Candida albicans*.

As for *Jatropha curcas* (Euphorbiaceae), a plant cultivated mainly for the production of oil for industrial use and for medicinal applications, petiolar explants were irradiated with doses ranging from 5 to 100 Gy of gamma rays emitted by a Caesium-137 source and grown in vitro for 60 days. Callogenesis was disrupted in a decreasing manner as the gamma irradiation dose increased. The yield of water-soluble principles extracted from the sample irradiated at 5 Gy (29.3%) was higher than that of the non-irradiated control (17.2%). The phytochemical screening revealed saponins, alkaloids, amino compounds, steroids and/or terpenoids in appreciable quantities in both the non-irradiated controls and those irradiated at 5 Gy. Antibacterial activity against *Proteus mirabilis* and *Escherichia coli* strains was highly appreciable. It is dependent on one or more identified phytochemical groups (Nyembo et al. 2005).

- **Cobalt-60 Irradiation**:

However, in order to embark on a genuine radiation-induced mutation selection programme, radiosensitivity tests need to be carried out to determine the optimum dose to be used for mutagenesis (Mba et al. 2010). As part of the technique cooperation between the IAEA and the DRC government, the work carried out by CGEA and INERA researchers under technical cooperation projects COD5030 and COD5031 focused on the genetic improvement of soybean, maize and common bean. Seeds from these three crops were sent to the IAEA laboratory at Seisberdof for cobalt-60 irradiation (Fig. 3).

Also, taking advantage of their stay in Indonesia and Morocco as part of their training courses, two researchers from these two institutions irradiated the seeds of the soya and common bean varieties respectively in order to determine their optimum irradiation doses. In order to study the effects of different doses of gamma radiation from a Cobalt-60 isotopic source on seed germination and early growth, *Jatropha curcas* L. seeds were also sent to the joint FAO/IAEA laboratory for irradiation. The results on the radiosensitivity of these crops are presented in Table 2.

In order to improve soybean production and nutritional quality as part of the fight against food and nutritional insecurity, seeds of two soybean varieties, VUANGI and TGX1830-7F, were irradiated with gamma rays emitted by the Cobalt-60 irradiator at a dose of 250 Gy. In the fourth generation (M4), 145 mutant, segregating and stable lines were selected and

Fig. 3 Cobalt-60 irradiator at the Joint FAO/IAEA Plant Breeding and Genetics Laboratory, Seibersdorf-Austria. Seibersdorf-Austria 2020 *Source* FAO/IAEA

characterised in order to select mutants that were productive, early maturing, resistant to lodging, diseases and pests (Fig. 4).

To capture sufficient information to enable accurate identification of G × E interactions, the genotype, environment and genotype-environment effects of 30 mutant genotypes were assessed at three experimental stations. Analysis of variance showed a highly significant difference ($p < 0.001$) both between sites (environment) and between genotypes in terms of seed yield. For the sites, the mean yield ranged from 719 ± 260 to 3148 ± 849 kg.ha^{-1}. Ten genotypes were identified as globally stable. A participatory varietal selection (PVS) session was organised with the involvement of male and female soybean growers. It emerged that 6 varieties were positively assessed by men and/or women. Four varieties were tested for their distinctive characteristics, uniformity and varietal stability (DUS) and agronomic and technological value (VATE) under the supervision of the National Seed Service of the Ministry of Agriculture for release: **TGM0376 (Lutaladio)**, **TGM0260 (Luyindula)**, **TGM0454 (Diamuini1)** and **TGM0454-1 (Diamuini2)** (Fig. 5).

The biochemical and nutritional characterisation of these mutants was carried out by determining their macronutrient and micronutrient content, including iron and zinc. The same applies to the content of anti-nutritional factors.

With a view to producing new maize varieties tolerant to water stress in the context of climate change, seeds of Katanga and Bukidibukidi varieties were irradiated with doses of 350 Gy and 340 Gy respectively of gamma radiation emitted by the IAEA's Cobalt-60 irradiator at Seisberdoff (Vienna). After growing 9,526 M2 seeds, more than 600 mutant lines were selected in M3. Of these lines, 428 were early, 168 late and 4 lines with specific traits such as yellow stripes on the leaves, pink-coloured grains, yellowish-coloured grains and very early lines (75 days). After two growing seasons of the early lines in the dry season, the putative mutants had an average yield varying between 1958 kg.ha^{-1} and 4004 kg.ha^{-1} compared with 2095 kg.ha^{-1} and 3504 kg.ha^{-1} respectively for the Bukidibukidi and Katanga parents. Three Egyptian varieties used as a control for water stress gave yields ranging from 2335 to 3913 kg.ha^{-1}. A local water-stress tolerant variety also used as a control gave a yield of 2335 kg.ha^{-1}. The genotype x environment effects of the selected mutants were assessed at three sites. The average yield of the mutants at the three sites ranged from 4,428 to 8,910 kg.ha^{-1}. Yields were 4,968 kg.ha^{-1} for the Bukidibukidi parents and 5,279 kg.ha^{-1} for Katanga. The

Table 2 Radiosensitivity of certain crops irradiated with cobalt-60

Plant	Radiation doses (Gy)	Genotype	LD_{50}/GR_{50} (Gy)	References
Jatropha curcas	100, 200, 300		254	Nyembo et al. (2018)
Soybean	100, 200, 300, 400, 500, 600, 700, 800, 900, 1000	VUANGI	287	
		TGX1830-7F	313	
		LOCAL Indo	466	
Bean (Phaseolus vulgaris)	80, 160, 240, 320, 560, 800	Bille-Bille (BL019)	175	
		Blanc de Goma (BG019)	58	
		Commando 2 (CM019)	135	
		HM21-7 (HM-21)	121	

LD50: Lethal dose 50
GR_{50}: 50% growth reduction of seedling height.

Fig. 4 Effects of gamma irradiation (Co-60) on soybeans

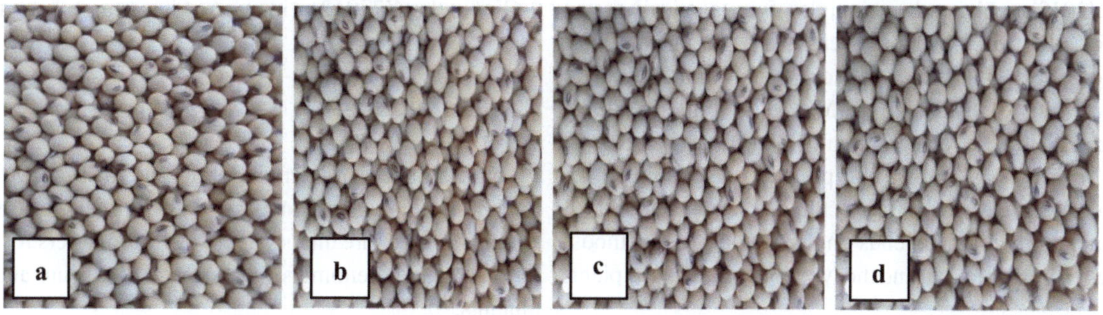

Fig. 5 Approved soybean mutants: **a** Diamuini 1; **b** Diamuini 2; **c** Lutaladio; **d** Luyindula

Fig. 6 Maize mutants

local control gave a yield of 7,027 kg.ha^{-1}. A total of six high-yielding varieties were selected for PVS testing (Fig. 6).

Variability was also observed in the macronutrient and micronutrient content of these mutants.

3.4 In Vitro Mutagenesis of Cassava (*Manihot Esculenta* Crantz)

Cassava is a staple food and a fundamental crop in the Democratic Republic of Congo because of its importance for food security. It is susceptible to a number of diseases and pests that can affect yields and crop quality. These include African cassava mosaic, downy mildew and insect pests (Hillocks et al. 2002). Although cassava can be propagated by sexual reproduction, it is mainly grown vegetatively (by cuttings) because of its self-incompatibility and the difficulties of pollination. However, cassava cultivation is subject to a number of major challenges, including reduced genetic diversity, variable quality of cuttings, the spread of diseases and pests and the limits of genetic improvement through traditional breeding. But vegetative propagation of cassava by traditional selection is slow and limited. Hence biotechnological techniques are used to enhance the genetic diversity of cassava.

Geier (2012) explains the principle of mutagenesis associated with in vitro cultivation of cassava. After irradiation, the explant (M_1V_0) carries mutated and non-mutated cells. When the explant is subcultured (M_1V_1), a chimeric structure appears. Dissociation of the chimeras, which are cells of different origins made up of wild-type and a range of mutants, involves consecutive cycles of sub-culturing. And the number of sub-cultures required depends on the species, the methods used to regenerate the plant and the types of mutagenised plant material.

To this end, Bulubulu et al. (2015) conducted molecular studies revealing the presence of African cassava mosaic virus (ACMV) in the local cassava variety, Boma. With the aim of obtaining virus-free vitroplants, meristems from two local cassava cultivars (*Manihot esculenta* Crantz), Mpelo-Nlongi and Boma, were extracted and grown in two types of culture medium: the first consisted of a Murashige and Skoog basic medium supplemented with salts, vitamins, sucrose (20 $g.L^{-1}$) and growth regulators, in particular Naphthalene acetic acid (NAA, 10 μM), Benzylaminopurine (BAP, 0.66 μM) and gibberellic acid (GA3, 0.1 μM) with 80 mg of adenine sulphate per litre of solution. The second medium was Murashige and Skoog's basal medium without growth regulator. The pH of the media was adjusted to 5.7. The cultures were placed in a phytotron at a temperature of 26 °C under a light of 2000 lx with a photoperiod of 16 h of light and 8 h of darkness at a relative humidity of 70%. Healthy seedlings of the Boma variety obtained by in vitro culture of meristems extracted from virus-infected plants were exposed to gamma irradiation with Cobalt-60 at variable doses (5, 10, 15, 20, 25 and 30 Gy) with the aim of determining the optimum doses for inducing useful genetic changes. The dose of 6 Gy leading to an average damage of 50% was recorded as the optimum dose. The latter, calculated using the linear regression equation, indicates that stem length, fresh weight, number of leaves and nodes of in vitro growing seedlings were reduced by 42, 13, 19 and 19% respectively compared with controls (Diamuini et al. 2015). Mass irradiation was carried out with doses of 3, 6 and 9 Gy. These irradiated vitroplants (M_1V_0) were cultured in vitro and the explants of the newly formed stem apices were subcultured (M_1V_1). From the resulting plantlets, successive cycles of sub-cultures were performed up to advanced generations (M_1V_5). The putative mutants obtained were characterised using TILLING (Targeting Induced Local Lesion In Genomes) reverse genetics techniques. The latter uses traditional mutagenesis followed by the discovery of mutations by identifying deleterious lesions in specifically targeted genes (Colbert et al. 2001). Preliminary tests to determine the density of induced mutations in the plant population were carried out using primers amplifying 1.5 kb fragments of genes involved in starch biosynthesis. Low mutation frequencies were recorded at a dose of 9 Gy for the X77012 primer encoding starch biosynthesis.

4 Perspectives

With the aim of characterising traits and facilitating the identification of valuable genetic variants in maize and soybean mutants, studies are underway using SSR markers for drought in maize and genomics (GWAS) of soybean and maize mutants.

In addition, thanks to technical cooperation with the IAEA and the 'Atoms 4 food' project, a joint IAEA-FAO initiative, the DRC is continuing to develop new varieties of crops including beans, rice, yams and cowpeas, to provide the population with good quality seeds with improved traits capable of improving their income.

5 Conclusion

Despite its considerable resources, DRC is still characterised by precarious food security due to declining soil fertility, insufficient dissemination of improved varieties, losses due to pests and diseases and the effects of climate change. The results obtained showed that irradiating seeds in a reactor or using a Cs-137 or Co-60 irradiator created genetic variability in the different crops used. Thanks to selection techniques based on radiation-induced mutations, the DRC has produced new varieties of some staple foods such as maize, soybean and rice. Combined with biotechnology, radio-mutation techniques have enabled the development of healthy, starch rich cassava plants. The results obtained with TILLING make it possible to select cassava mutants of the M_1V_4 generation at an early stage, without prior phenotype studies. The application of the in vitro culture technique combined with that of radio-induced mutations has also made it possible to improve the production of secondary metabolites with high biological activity. But DRC is still facing challenges linked not only to food insecurity but also to the source of irradiation, given the prolonged stoppage of its research reactor and the shipment of seeds to Vienna for irradiation. Restarting the Trico II reactor or acquiring a cobalt-60 irradiator could enable the DRC to produce new crop varieties with interesting traits and solve the problem of food insecurity.

Acknowledgements We would like to thank the IAEA for funding projects ZAI5029, COD5030 and COD5031, which enabled us to carry out the research that led to the production of cassava, soybean and maize mutants.

References

Bharath RA, Prathmesh SP, Sarsu F, Suprasanna P (2024) Induced mutagenesis using gamma rays: biological features and applications in crop improvement. OBM Genet 8(2):233. https://doi.org/10.21926/obm.genet.2402233

Botula MBL (1986) Contribution du département de Génétique et Amélioration des plantes à la création de nouvelles variétés des plantes et des semences à très haut rendement. In Recherches Nucléaires et Développement du Zaïre: 25 ans d'activités. Ed. Prof. Malu-wa-Kalenga. Presses du Commissariat Général à l'Energie Atomique. 129 – 135.

Bulubulu OF, Diamuini NA, Kikakedimau NR, Mbaya NA, Mutambel H, Lumande K, Luyindula N, Rufflard G, Lepoivre P (2015) PCR and ELISA detection of cassava mosaic virus in a Congolese cassava landrace. Int J Biotechnol Food Sci 3(1):10–16. http://www.sciencewebpublishing.net/ijbfs

Colbert T, Till BJ, Tompa R, Reynolds S, Steine MN, Yeung AT, McCallum CM, Comai L, Henikoff S (2001) High-throughput screening for induced point mutations. Plant Physiol 126:480–484. https://doi.org/10.1104/pp.126.2.480

Diamuini N, Luyindula N, Bulubulu O, Kikakedimau N, Chikelu M, Bradley JT (2015) *In vitro* Gamma radiosensitivity test in Congolese cassava, *Manihot esculenta* Crantz accession. Acad J Biotechnol 3(3):001–005. https://doi.org/10.15413/ajb.2014.0111

Dimundu P, Kalonji M, Botula M, Kembola K, Mahungu N et Kalumvueziko L (1987) Radiosensibilité des organes de reproduction du manioc (Manihot esculenta) exposés au flux mixte Neutron/Gamma. Rev. Zaïr. Sci. Nucl. Vol 8 n° 2. 33 – 38.

FAO (2024) Analyse IPC de l'insécurité alimentaire chronique en République Démocratique du Congo. IPC, p 117. www.ipcinfo.org

FAO/IAEA (2020) Manuel d'amélioration des plantes par mutation-Troisième édition. Sous la supervision de M.M. Spencer-Lopez, B.P. Forster et L. Jankuloski. Vienne, FAO. https://doi.org/10.4060/i9285fr

Geier T (2012) Chimeras: properties and dissociation in vegetatively propagated plants. Plant Mutation Breeding and Biotechnology. CABI, Wallingford, UK: 191–201. https://doi.org/10.1079/9781780640853.0191

Harish R, Shivanandappa T (2006) Antioxidant activity and hepatoprotective potential of *Phyllanthus niruri*. Food Chem 95:180–185. https://doi.org/10.1016/j.foodchem.2004.11.049

Hillocks RJ, Thresh JM, Bellotti A (2002) Cassava: biology, production and utilization, CABI Publishing, pp 2002—2332. https://doi.org/10.1079/9780851995243.0000

IAEA (2014) Applications of Research Reactors. IAEA Nuclear Energy Series No. NP-T-5.3. Vienna, p 97. https://www-pub.iaea.org/MTCD/Publications/PDF/Pub1627_web.pdf

Jain SM, Spencer MM (2006) Biotechnology and mutagenesis in improving ornamental plants. Volume 1, pp 589-600. In: Floriculture and ornamental biotechnology: Advances and tropical issues, J.A. Teixeira da Silva (Editor) Global Science Books, ISSN 1749-0294 (Paper), 1749-0308 (online)

Jankowicz-Cieslak J, Mba C, Till, BJ (2017) Mutagenesis for Crop Breeding and Functional Genomics. In: Jankowicz-Cieslak, J., Tai, T., Kumlehn, J., Till, B. (eds) Biotechnologies for Plant Mutation Breeding. Springer, Cham. https://doi.org/10.1007/978-3-319-45021-6_1

Lagoda P, Shu Q, Forster B, Nakagawa H (2012) Effects of radiation on living cells and plants. Plant Mutation Breeding and Biotechnology. CABI Publishing, Wallingford: 123–134. https://doi.org/10.1079/9781780640853.0123

Lavina S, Namrata K (2024) Treatment of Urolithiasis with medicinal plant Phyllanthus niruri. Int J Pharm Res Appl 9(3): 632–637. www.ijprajournal.com ISSN: 2456–4494.

Li Y, Li X, Wang JK, Kuang Y, Qi MX (2017) Anti-hepatitis B viral activity of *Phyllanthus niruri* L (Phyllanthaceae) in HepG2/C3A and SK-HEP-1 cells. Trop J Pharm Res 16(8):1873–1879. https://doi.org/10.4314/tjpr.v16i8.17

Lundqvist U, Franckowiak J, Forster B (2012) Mutation categories. Plant mutation breeding and biotechnology. CABI, FAO, Oxfordshire, UK: 47–55. https://doi.org/10.1079/9781780640853.0047

Mba C, Afza R, Bado S, Jain SM (2010) Induced mutagenesis in plants using physical and chemical agents. In: Davey MR, Anthony P (eds) Plant cell culture: essential methods. Wiley, Toronto, Canada, pp 111–130. https://doi.org/10.1002/9780470686522.ch7

Mohsen G, Soliman SS, Mahgoub EI, Ismail TA, Mansour E, Alwutayd KM, Safhi FA, Abd El-Moneim D, Al shamrani R, Atallah OO, Shehata WF, Hassanin AA (2023) Gamma-rays induced mutations increase soybean oil and protein contents. PeerJ 11:e16395. https://doi.org/10.7717/peerj.16395

Mudibu J, Nkongolo KKC, Kalonji MA, Kizungu RV (2012) Effect of gamma irradiation on morpho-agronomic characteristics of soybeans (Glycine max L.). Am J Plant Sci 3(3):331–337. https://doi.org/10.4236/ajps.2012.33039

Navneet Kumar U, Afroze A, Pradeepti C, Shailendra K, Naik KK, Sharma M, Swati P, Sunil T and Sameer S (2017). Development and Validation of a New Chromatographic Method for the Simultaneous Estimation of Serratiopeptidase, Aceclofenac and Paracetamol by RP-HPLC. Pharm Anal Chem, 3:2. https://doi.org/10.4172/2471-2698.1000122

Ntsomboh NG, Fokam PE, Likeng-Li-Ngue BC, Mbi KT, Zambou AH, Mafouasson HN, Martin BJ et al (2023) Gamma ray induced mutagenesis for crop improvement: applications, advancements, and challenges. Gamma Rays—Current Insights. IntechOpen. https://doi.org/10.5772/intechopen.100299

Nyembo K, Luyindula N, Onyembe PML, Makosso N, Muambi N (2005) Effets de rayonnements gamma sur la production des métabolites secondaires par les tissus de *Jatropha curcas* L. en culture *in vitro*. Rev Cong Sci Nucl 21:79–84. https://www.osti.gov/etdeweb/biblio/21553159

Nyembo BK, Mbaya AN, Ilunga CC, Muambi JL, Tshilenge LL (2018) Effects of 60Co gamma radiation doses on seed germination of Jatropha curcas L Int J Agron Agri R 13(5):46–52. ISSN: 2223–7054, 2225–3610. http://www.innspub.net/

Okoli CO, Ibiam AF, Ezike AC, Akah PA, Okoye TC (2010) Evaluation of antidiabetic potentials of *Phyllanthus niruri* in alloxan diabetic rats. Afr J Biotechnol 9(2):248–259. http://www.academicjournals.org/AJB

Tresh JM, Cooter RJ (2005) Strategies for controlling cassava mosaic disease in Africa. Plant Pathol 54:587–614. https://doi.org/10.1111/j.1365-3059.2005.01282.x

Tshilenge LL, Kalonji MA, Nkongolo K C et Kizungu R (2013) Effet de l'irradiation gamma sur les caractéristiques morpho-agronomiques de l'arachide (*Arachis hypogaea* L.). Am J Plant Sci 4(11):2186–2192. https://doi.org/10.4236/AJPS.2013.411271

Yasmin K, Chandran AD (2022) Gamma irradiation effects on crop plants. Res J Biotechnol 17:126–135. https://doi.org/10.25303/1708rjbt1260135

Advancing Rice Cultivation Through Mutation Breeding for Sustainable Agriculture

Arulbalachandran Dhanarajan, Karthiksaran Chinnasamy, Balamurugan Venkatachalam, Yasmin Kasim, Ramachandran Marimuthu, Kishore Kumar Ashok, Nedumaran Thillaigovindhan, Soundarya Vasudevan, and S. Velu

Abstract

Mutation breeding is a pivotal technique in agricultural genetic development, focused on creating elite plant varieties while preserving their desirable traits. It involves the induction of mutations, screening for potential mutants and testing for mutant verification. This method has been widely used since the 1950s to improve crop productivity and resilience. When physical and chemical mutagen, like radiation and chemicals, cause mutations, the plants that result have better traits, like being resistant to disease and having a high yield. Ion beam irradiation, UV radiation and chemical mutagens like EMS and MNU have been central to creating diverse plant varieties. This breeding strategy has contributed to significant agricultural advances, particularly in developing high-yielding, disease-resistant crops like rice and wheat. Mutation breeding also helps make farming more sustainable by lowering the need for chemicals and making crops better able to handle climate change and other environmental stresses. Molecular marker-assisted selection (MAS), TILLING, and CRISPR-Cas9 are some of the techniques that have made mutation breeding even better, allowing for more precise and effective improvements. These advancements hold promise for boosting global food security by producing resilient crops capable of thriving in diverse and challenging environments, aligning with the United Nations' Sustainable Development Goals.

Keywords

Mutation breeding · Physical mutagen · Chemical mutagen · TILLING · CRISPR-Cas9

A. Dhanarajan (✉) · K. Chinnasamy · B. Venkatachalam · Y. Kasim · R. Marimuthu · S. Vasudevan
Division of Crop Molecular Breeding and Stress Physiology Laboratory, Department of Botany, School of Life Sciences, Periyar University, Salem, Tamil Nadu, India
e-mail: arul78bot@gmail.com; mutabreed@periyaruniversity.ac.in

Y. Kasim
PG and Research Department of Botany, Kandaswami Kandars College, Velur, Namakkal, Tamil Nadu, India

R. Marimuthu
PG Department of Botany, Sri Vidya Mandir Arts and Science College (Autonomous), Katteri, Uthangarai, Krishnagiri, Tamil Nadu, India

K. K. Ashok
Integrative Agriculture Department, College of Agriculture and Veterinary Medicine, UAE University, Al Ain, United Arab Emirates

N. Thillaigovindhan
Department of Plant Biology and Plant Biotechnology, Sir Theagaraya College, Old Washermanpet, Chennai, Tamil Nadu, India

S. Velu
Department of Botany, Thiru. A. Govindasamy Government Arts College, Tindivanam, Villupuram, Tamil Nadu, India

1 Introduction

Rice is one of the staple food crops, and an important portion of the global human population faces increasing pressure from climate change and the growing demand for food. Mutation breeding serves as an effective method to increase rice production and promote sustainable agriculture. This method may produce better cultivars with desired characteristics, including increased yields, greater stress tolerance, and improved nutritional content, by causing genetic variants through induced mutagenesis (Zafar et al. 2020). Mutation breeding utilizes physical or chemical mutagens to generate genetic variety within a plant species. This generated variation may then be evaluated for advantageous characteristics, resulting in the creation of enhanced crop varieties. Especially during the Green Revolution, this method has been crucial in creating high-yielding rice cultivars that have transformed rice production. Mutation breeding is essential for sustainable agriculture since it improves resource usage efficiency by producing varieties that need less amounts of

J. Devi (ed.), *Mutation Breeding in Crop Improvement*, Sustainable Landscape Planning and Natural Resources Management,
https://doi.org/10.1007/978-3-032-17911-1_10

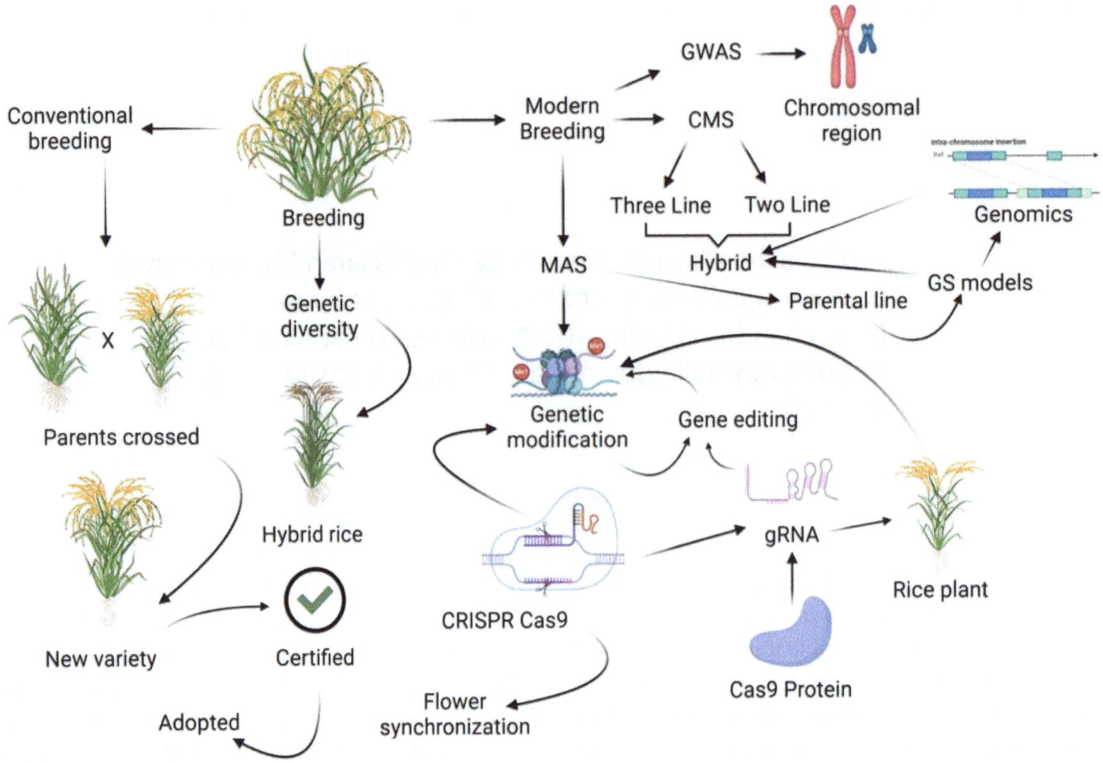

Fig. 1 Schematic representation of advancing rice cultivation through mutation breeding for sustainable agriculture

water, fertilizer, or pesticides, so aiding in resource conservation and diminishing the environmental impact of agriculture. To increase arable area and decrease the requirement for freshwater irrigation, salt-tolerant rice cultivars created by mutation breeding, for instance, may flourish in salty environments. The abiotic stressors including drought, salt, and severe temperatures are becoming more frequent and intense due to climate change. Rice varieties that can resist these difficulties may be developed with the use of mutation breeding, guaranteeing consistent yields even in the face of adversity (Yadav et al. 2021). Enhancing nutritional quality: Enhancing the nutritional profile of rice may bolster food security and public health. By altering other nutritional characteristics or raising micronutrient levels, mutation breeding may make rice a more complete food supply. Mutation breeding may expedite the creation of novel kinds more effectively than conventional breeding techniques, especially when integrated with contemporary genomic technologies and breeding may be streamlined by using methods like MutMap which can quickly find mutations linked to desired features. The research has been remarkably conducted on the adaptation of rice to climate change. Significantly, rice's root system plays a critical role in its capacity to endure osmotic stress, a typical result of climate change. Genome modification technologies like CRISPR have also accelerated the speed and precision of rice breeding by means of genome alteration, hence facilitating the creation of stress-tolerant types with better yields under difficult climatic circumstances (Zafar et al. 2020). The future of mutant breeding in rice entails the use of modern genomic technologies such as CRISPR/Cas. These techniques provide enhanced accuracy and efficiency in generating and choosing favorable mutations, hence expediting the generation of superior rice varieties. Utilizing these technologies allows us to improve rice agriculture and foster a more sustainable and food-secure future. Schematic representation of Advancing rice cultivation through mutation breeding for sustainable agriculture is illustrated in Fig. 1.

2 Fundamentals of Mutation Breeding

The term mutation breeding was coined by Freisleben and Lein. It is referred to intentionally induced mutant lines created for agricultural enhancement. Breeders now widely use mutation breeding to create leading lines of breeding and cultivated varieties, all while maintaining their quality attributes. Mutation breeding is a crucial aspect of agricultural genetic development and has proven crucial in ensuring global food and nutrition security. In the future, there will likely be 9 to 11 billion people on the planet by 2050; crop genetic innovation is crucial to maintaining future food security (Fróna et al. 2019). When an organism's genetic material alters at a specific location, it is called a mutation. De Vries defined a mutation as an unanticipated alteration in the genetic material that is passed down from parent to offspring and is not brought on by segregation or recombination. De Vries coined the term "sudden" to differentiate between alterations that are explained by the normal process of recombination and those that are subtle. But newer techniques have shown that mutations happen in changes, mainly at the genotypic level. This leads to small phenotypic changes that might not be noticeable at first.

Mutation breeding demonstrates the three primary concepts. The official mutant variety is released when mutation induction, screening for prospective mutants, and mutant testing (mutant verification) are completed. A physical, chemical, and biological mutagen exposes plant propagules to initiate the first phase, mutation induction. "Mutant screening," the second step, is the process of selecting desirable features from a significant number and the number of treated mutants. The selected mutants in this procedure are referred to as "putative mutants," which implies that they are not true mutants. A chosen mutant individual, for instance, could show resilience to stresses both biotic and abiotic, but in reality, the person is not infected as the virus, insects, and environmental conditions are not there.

Physical mutagens like X-rays, UV and fast thermal neutrons and gamma rays are used to produce most of these mutant kinds. Breeders breed nearly 80% of these mutant varieties from seeds. FAO reports that worldwide farmers cultivate over half of these mutant species as cereals. Mutation breeding is a crucial aspect of agricultural genetic development and has proven crucial in ensuring global nutrition and food security. Typically, the mutagenic agent is applied to planting materials like seeds and vegetative propagules to cause mutations in plants. The specific plant material, its physiology, and its genetic composition all affect the radiation dosage selection. Mutants are the result of mutagenesis. A person is considered a mutant if they have one or more genetic mutations, which can be detected by molecular techniques or phenotyping. Since the discovery of ionizing radiation, which can change an organism's genetic makeup, spontaneous mutations were the only known way for new genetic diversity to exist in plants and animals. This discovery was especially important for farming. Soon after this finding, mutant breeding was based on ionizing radiation. The technique of treating seeds with physical or chemical mutagens creates mutants, which can lead to new varieties with better qualities, like higher yields and resistance to diseases. This process is known as mutagenesis induction. However, creating mutant lines requires more time and work. Induced mutagenesis can alternatively be classified as chemical or physical mutagenesis, depending on the kind of mutagenic agent utilized. The kind of ions, accelerator, ion radiation dosage, species, and developmental stage all influenced the plants' reactions to the ion beam. Double-strand breaks with a high frequency and broad spectrum are produced by ion beam irradiation. Ion beam irradiation can result in significant DNA alterations, including point-line alterations in half of the mutants and large deletions, translocations, and inversions in another half. Although electron beam radiation is thought to be a less accurate technique, it works on plants in a manner similar to that of ion beam radiation. Neutrons are categorized as ionizing radiation, just like gamma rays. Neutrons are not charged particles like gamma rays or other electromagnetic radiation. When neutrons come into touch with materials, they can knock atomic nuclei out of their usual locations, which can lead to ionization and atom excitation. Neutrons can cause mutations by deleting or altering a nucleotide, breaking chemical connections inside the DNA molecule, or doing both. As a result, it is essential to give the radiation at the appropriate dose, which is determined by the exposure duration and radiation intensity. Mutation breeding is one of the finest ways to improve a variety of agronomic and economically important features, and plant breeders have been adopting it a lot lately. It has been successfully applied to increase the yield of crop plants. In mutation breeding, increasing the dosage of a mutagen is believed to increase the frequency of mutations, but it also reduces the survival rate and regeneration capacity (Khan et al. 2009). The first commercially available mutant variety was *Nicotiana tabacum*, which was developed in 1934. Later, some breeders successfully produced mutants of *Saccharum officinarum* that were resistant to illness. The *S. officinarum* variant Co S 687 was created using a chemical mutagen and gamma radiation. It is reported that this kind is immune to red rot illness and multiple breeders have successfully created disease-resistant *S. officinarum* mutants. By using the gamma radiation and a chemical mutagen, the red rot disease could not affect the strain of *S. officinarum* called S 687, f. Rice blast disease is caused by *Pyricularia grisea*, an anamorph of *Magnaporthe grisea*, which reduces crop yield and affects Necrotic lesions on leaves, nodes, seedlings, and panicles are a result of *Oryza sativa*'s global production. Several attempts have been made to use an induced mutation approach to create a disease-resistant variant of *O. sativa*. A new commercial variety with a unique agronomic trait namely, resistance to blast was developed as a result of a mutation that produced a new gene combination that had not before existed in the germplasm pool (*Pyricularia oryzae*).

3 Types of Mutations

A plant's genetic makeup and traits can be impacted by point, chromosomal, somatic, induced, fasciation, and transposable element changes. These mutations are crucial for the development and breeding of plants and offer genetic diversity for agricultural improvements. Induced mutations using mutagenic agents, such as chemicals or gamma rays, are generally used in mutant breeding to produce new plant types with desired traits. These advancements improve crop resilience and yield and enable the cultivation of plant crops that can better withstand environmental stresses. As researchers continue to explore the genetic landscape of plants, the potential for innovative solutions to global food security challenges becomes increasingly promising.

3.1 Mutagenic Agents

Mutagenic agents are essential to genetics and plant breeding because they can result in mutations that improve crop traits and increase genetic diversity (Fig. 2). New plant types that are more resilient to environmental stressors, disease, and the nutritional demands of a developing country are produced as a result of this process. Breeders can select for particular qualities thanks to the genetic changes these agents produce, which eventually increases agricultural productivity (Atia et al. 2024). Mutagenic compounds used to protect crops fall into three major categories. In addition to ways of action, plant breeding employs chemical, biological, and physical techniques. These methods can introduce mutations that enhance desirable characteristics, such as increased yield or improved resistance to pests. By understanding how these agents function, researchers can develop more effective strategies to ensure food security and sustainability in agriculture. Sustainability in agriculture is becoming increasingly crucial as global populations rise and environmental pressures mount. Continued research into mutagenic agents not only promises to boost crop resilience but also aligns with the goal of minimizing chemical inputs, thereby fostering an eco-friendlier approach to farming practices. This approach not only enhances agricultural productivity but also contributes to the preservation of biodiversity and the health of ecosystems. As we explore innovative techniques and technologies, the potential for creating resilient food systems that can withstand climate challenges becomes more attainable. By integrating traditional knowledge with modern science, we can develop strategies that not only improve yields but also ensure the sustainability of our agricultural landscapes.

3.1.1 Physical Mutagenesis

Physical mutagens are believed to be a primary source of mutations and have been an important component since 1920. Physical mutagens include ionizing radiation, including gamma rays, ion beams, fast neutrons, X-rays and α and β rays. Other nonionizing radiations, like as ultraviolet (UV) light, are also used as mutagens. UV light, a nonionizing radiation with a poor penetration capability, often causes dimer formation and DNA base deamination, whereas physical mutagenesis causes deletions in DNA fragments that result in gene-silenced or knockout mutants.

Gamma and X-rays are 10^{-7} cm. Due to their deep penetration and sparing ionization; these radiations share similar physical and biological consequences. Particulate ionizing radiation includes α rays, β rays, and neutrons. Many plant species, including *O. sativa* have undergone physical mutagenesis to generate mutants. The first use of X-rays was replaced by gamma rays from radioactive sources such as ^{60}Co and ^{137}Cs, which were more widely available to developing nations. Fast neutrons from nuclear reactors can also be used,

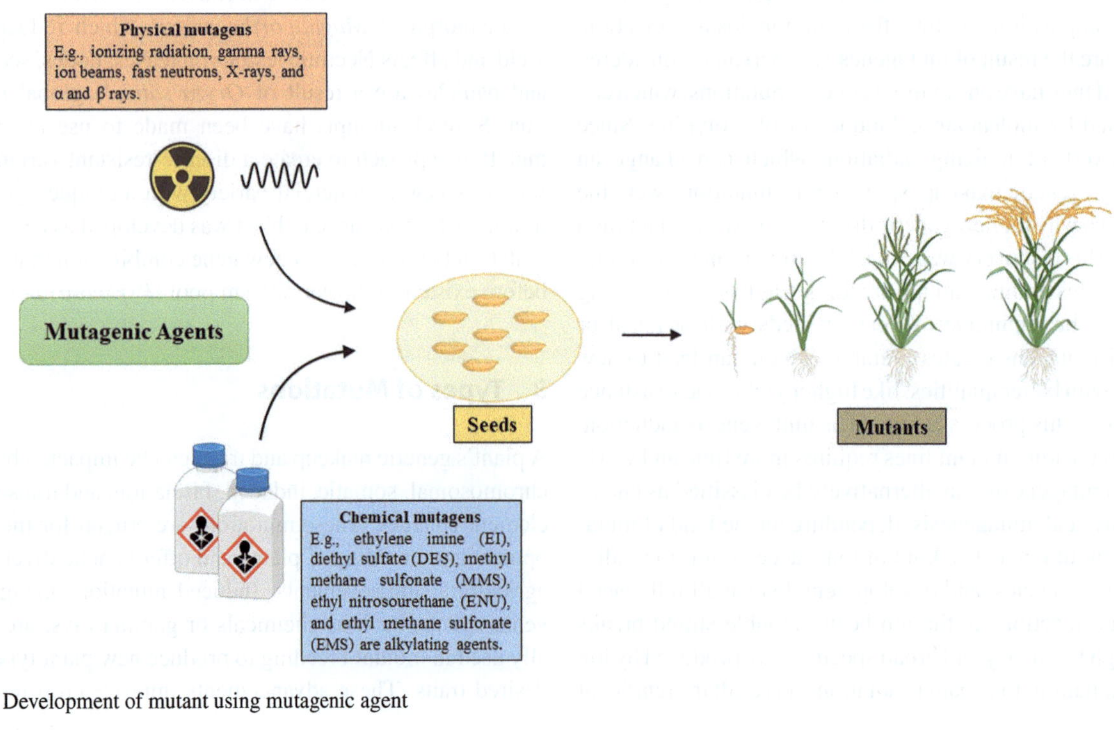

Fig. 2 Development of mutant using mutagenic agent

especially through the Joint FAO/IAEA Division in Vienna. Ionizing radiation can cause DNA strands to cross-link, which can be more harmful than chemical mutagens. UV radiation is one of the physical mutagens that can cause point mutations because it is nonionizing, has a limited penetration rate, and effectively produces purine or pyrimidine dimers. Pollen in late or early uninucleate stages can benefit from exposure to UV light. Concerning radiation doses and conditions for irradiating key agricultural species' seeds and buds. Experiments are required to determine treatment conditions for minor or novel species. Radiation is affected by external factors such as temperature, oxygen content, moisture content, and storage conditions. These factors can significantly influence the efficacy of radiation treatments and the subsequent mutation rates in the target organisms. Further research into optimizing these variables is essential to enhance the application of UV radiation in agricultural practices, particularly for species that have not been extensively studied. This optimization could lead to improved crop resilience and productivity, ultimately benefiting sustainable farming methods. By understanding how these external factors interact with radiation, researchers can develop tailored protocols for a wider range of plant species, paving the way for innovations in agricultural biotechnology (Xu et al. 2020).

3.1.2 Chemical Mutagenesis

The initial efforts to use chemical mutagens to cause mutations in biological systems occurred around the start of the twentieth century. Because chemical mutagens are so good at causing mutations, many individuals utilize them in addition to physical mutagens. Chemical mutagens have the ability to alter a single nucleotide polymorphism (SNP) or a single base pair (bp). This is different from radiation, which can delete large chunks of DNA, rearrange chromosomes, or move genes around. Many chemical mutagens, including ethylene imine (EI), diethyl sulfate (DES), methyl methane sulfonate (MMS), ethyl nitrosourethane (ENU), and ethyl methane sulfonate (EMS), are alkylating agents. Azide (sodium azide), base analogs (5-bromouracil, 5-chlorouracil, 2-aminopurine), and nitrous acid are examples of several chemical mutagens. Numerous gene alterations can result in deletions, frame shift mutations, and transition mutations from AT to GC and GC to AT. EMS has been utilized as an alkylating agent in plants among chemical mutagens. It targets O-atoms like O6-guanine and O2-thymine to a lesser degree, as well as highly nucleophilic sites like the N7 atom of guanine and N3-adenine. It causes mispairing with thymine (T) by alkylating guanine (G) to produce O-6-ethylguanine. During replication, DNA polymerases frequently confuse T with O-6-ethylguanine. Random mutations in the G/C to A/T transition occur if this damage is not repaired. Alkylating compounds are among the chemical mutagens that result in a variety of chromosomal abnormalities, including disturbances to meiosis and mitosis. Chemical mutagens have been employed in mutant breeding projects to create novel kinds and variants with characteristics including disease resistance, shorter plants, and greater yields. BPT 2231 rice varieties treated with gamma rays and electromagnetic stimulation (EMS) exhibited both bold-grain and fine-grain variations, along with a high nutritious content. Future high-yielding hybrids can be created via heterosis breeding with these types. EMS involves soaking the rice seed in a solution with a known EMS concentration, usually 0.2 to 2.0% is determined by the sensitivity or lethal curve of the genotype used. After using EMS 1.5% for two hours, they were able to make 340 families of mutant rice plants. This indicated that the dosage and length of exposure were important for making mutants. There was a notable variation between the rice families. EMS treatment also caused genetic variability in Basmati rice. Rice may undergo alterations as a result of the 0.5% to 1.0% dosages given for six hours, according to the research. A greater number of non-lethal point mutations are produced by EMS mutagenesis, hence a relatively small population of plants roughly 10,000 is needed to saturate the genome with mutations. In diploid organisms, one mutation every 294 kilobases might result from EMS therapy (1.5%). Traits associated with abiotic stress tolerance were also identified. Among these were increased root weight, volume, and length; these traits are associated with drought tolerance. Resistance to heat and drought is connected to a decrease in ROS formation, which is a result of both stresses. Furthermore, mutants that retained more chlorophyll were also more heat resistant. The applied mutagen in rice is N-methyl-N-nitrosourea (MNU). The monofunctional alkylating agent MNU has a strong reaction with the oxygen atoms in the DNA molecule. Alkyl groups are added to nitrogenous bases, mainly guanine and cytosine, as the initial stage of mutation. O6-alkylguanine doesn't just happen; it's affected by the structure of chromatin, the shape of DNA, and the guanine 5' and 3' flanking sequences. When 5 adenine and 3 cytosine flank 5 guanine, the likelihood of forming O6-alkylguanine increases tenfold. Research suggests that MNU mutagenesis performs more effectively in developing cells compared to dry rice seeds. Mutations may result from the ionic substance sodium azide. By producing an organic metabolite that resembles L-azidoalanine, it does this. After entering the cell nucleus, this chemical interacts with the DNA to cause point mutations in the genome. Similar sequences were found at the mutation locus as 5'-GGR-3' by Tai et al. (2016). Between 1.4 and 2.9 mutations per Mb was the found mutation rate. The SA treatment yielded important findings towards rice improvement. The Suweon 542 is a mutant rice with potential applications in rice flour production, showing floury grains and small, less degraded starch particles. In this mutant, the recessive locus flo7(t) accounts for 92.2% of the difference in grain flour. Red rice grains treated with

SA also exhibited higher levels of procyanidins, γ-oryzanol, vitamin E, and iron accumulation. Researchers have observed an increase in iron and zinc levels in polished grains, along with higher amounts of starch and amylose. They have also noted the emergence of new aromatic rice. SA treatment has also produced lines that are resistant to abiotic stress. It has been found that the salt hypersensitive 1 (shs1) mutant affects rice's antioxidant metabolism and sodium homeostasis when it is under salt stress (Lin et al. 2016).

4 Benefits of Mutation Breeding in Rice Cultivation

In India, 33% of the GDP (gross domestic product) and 62% of the workers are employed in agriculture. Approximately 62% of Indian workers are employed in agriculture, which also accounts for 33% of the nation's GDP (gross domestic product). One of the nation's biggest problems after gaining independence in 1947 was meeting the food needs and demands of the 1.4 billion people living there at the time. Later, a huge population boom further exacerbated the strain on agriculture. One of India's most vital food crops is rice. India is one of the world's largest producers of rice. This nation cultivates more rice than any other. India generated 120 million tons for 2021. A substantial proportion of the carbs, fibers, vitamins, and proteins needed in the Indian diet may be found in rice crops, which are also rich in other nutrients. In a similar vein, rice production increased from 35 MT to 89.5 MT. According to their phenotype, these crops were semi-dwarf. Given that just a small portion of land is suitable for cultivation, India must work to boost yields per acre from the land that is already under cultivation (currently, 60.43%). The mid-1960s saw the successful introduction of several high-yielding rice and wheat cultivars, and research using artificial embryo development, test-tube plant culture, and intra-ovarian pollination led to the achievement of PTC in India. With the use of these plant breeding techniques, India's food output increased dramatically. The production of C4-fixing rice cultivars, hexaploidy wheat, and the autotetraploid origin of *Solanum tuberosum* were all investigated during this time period; these endeavors are now referred to as the Green Revolution. Mutation breeding is used to create several high-yielding hybrid crops in India.

The semi-dwarf rice cultivars were created by fusing Taichung Native-1 and IR-8. The Indian rice belt is home to the semi-dwarf rice types Jaya and Ratna. In India, 200 varieties were developed using the mutation breeding process. Radiation therapy causes mutations in some qualities that manifest in the anticipated time frame. For example, X-ray-induced mutation results in semi-dwarf rice, and the rice variety is exposed to radiation. The brown hopper-resistant, high-yielding variety Atomita-2 is a product of Pelite-1. Ò-irradiation is used to make Reimei, high-yield rice. The purpose of the multiple cross study was to introduce novel character kinds or varieties into the crop of the future. Rice ADT-37 was created by fusing *Oryza japonica* with *Oryza indica* (Predieri 2001).

5 Breeding for Disease Resistance Crops

Most crops in tropical regions of the world are vulnerable to nematodes and bacteria, and crop productivity and yield are impacted by their infection. Consequently, crop output could be impacted by 25% or sometimes the entire amount. In this situation, disease-resistant plants can be produced by tissue culture, trans gene cloning, and breeding studies, increasing the availability of food. Nearly every plant has resistance genes, or "R-genes," for immunity and disease resistance. Pesticide exposure on agricultural land, crop rotation, tillage, disease-free seeds, etc., can all help stop the spread of disease among crops. Pure crop lines should be employed to avoid intrinsic disease pressures. Because cross-pollinated crops are known to have a high degree of inbreeding depression, sterile crops may be produced. Self-pollination may assist plants overcome this by probably reducing the inbreeding depression.

In order to integrate cell genomes and enhance immunity to tolerate diseases, laboratory studies on plants employ genetic engineering, backcrossing, inbreeding, mutation breeding, interspecific hybridization, and hybrid breeding. Genetic engineering, somaclonal variations, and other mutagens that induce plant mutations are further breeding methods. Mutations alter signaling and metabolic order, which alters the ordering of peptides or nucleotides. This causes a variety to exhibit new traits that are not apparent in the parent generation. By allowing them to acquire new traits, mutation breeding improves the quality of the progeny. There are other varieties of atomic gardening, such as tulips and snapdragons. In order to take control of the host's cellular machinery, plant infections usually destroy the metabolic networks. UV light inactivates some disease-specific genes of infections; these genes are knocked out, or KO, to assess the pathogen's pathogenicity. Chemi-mutagens are alkylating chemicals that generate point or transition mutations in purines, especially guanine, in DNA sequences. One example of such an agent is EMS (ethylmethanesulphonate). For instance, X-rays caused cotton plant mutant breeding in India, while Ò-rays created Calrose-76 rice in the US, the world's first high-yielding rice (Xu et al. 2020).

Insect resistance can be attributed to morphological, biochemical, or physiological characteristics of host crops. Certain physical traits of plants can discourage pests from attacking them. Plants with venation and regular leaves attract insects, while those with rough or hairy leaves repel them.

6 Sustainability and Environmental Benefits

Mutation breeding is crucial for enhancing agricultural sustainability and reducing environmental impact. By creating genetic variations in crops, this technique facilitates the development of improved plant varieties that support eco-friendly farming practices. Due to the increasing global human population, the demand for food is escalating, making it essential to adopt innovative agricultural methods that are both productive and sustainable. Mutation breeding provides an effective solution by producing resilient crops capable of thriving in challenging environmental conditions while simultaneously decreasing reliance on harmful agricultural inputs.

6.1 Decrease in Chemical Inputs

Developing crops with enhanced resilience to herbicides and diseases is one of the main environmental benefits of mutant breeding. This genetic enhancement minimizes the necessity for chemical pesticides and herbicides, thereby reducing the environmental impact of agricultural practices. Overuse of these chemicals can result in soil degradation, water contamination, and the decline of beneficial insects and microorganisms. By lowering dependence on chemical inputs, mutation breeding promotes healthier ecosystems, maintains soil fertility, and safeguards water bodies from toxic runoff. Additionally, farmers experience economic benefits through reduced costs for chemical treatments, making agriculture more cost-effective and sustainable. Mutation breeding technique was used to produce the disease resistant varieties are used to the agriculture field to reduce the chemicals usage. Induced the disease-resistance gene in the particular varieties, and it's used to reduce the chemical use for the agriculture field. The resistance gene induces the rice crop to reduce the disease viability and avoid the loss of yield in the rice crop (Van Esse et al. 2020).

6.2 Adaptability and Resilience to Climate Change

Developing crops that can withstand harsh weather conditions is essential as climate change puts the world's food security at greater risk. In this process, mutation breeding is essential because it allows for the development of plant types that are more adaptable to environmental stresses like flooding, heat, salinity, and drought. By requiring less irrigation, these hardy crops help save limited and valuable water supplies. Furthermore, mutant breeding helps to lower greenhouse gas emissions linked to the manufacture and use of synthetic fertilizers by reducing the need for them. We can ensure consistent food production despite erratic weather patterns by growing climate-resilient crops, which will eventually advance long-term agricultural sustainability. For food security in a time of environmental instability, this creative strategy is crucial. With climate change posing significant challenges, mutation breeding has been instrumental in developing drought, heat, and salt tolerant crops. Recent studies have successfully used gamma irradiation to enhance the tolerance of rice and wheat to extreme environmental conditions (Chaudhary et al. 2019).

6.3 Higher Yields and Resource Efficiency

To fulfil the demands of a growing population without increasing agricultural acreage, crop yield must rise. Through the improvement of features including nutrition efficiency, stress tolerance, and faster growth, mutation breeding is a valuable strategy for increasing yield potential. Mutation breeding contributes to the preservation of important ecosystems by increasing productivity on already-existing cropland and halting habitat loss and deforestation. Furthermore, high-yield crops use less labor, fertilizer, water, and other resources, increasing farming efficiency and reducing environmental impact. Mutation breeding guarantees that food production maintains up with growing demand while preserving natural resources for future generations through sustainable intensification. This strategy promotes long-term environmental sustainability in addition to enhancing food security. Mutation breeding has contributed to biofortification, improving the nutritional content of the staple food crops. For example, the development of vitamin A-enriched rice and iron-fortified wheat through induced mutagenesis has played a crucial role in combating malnutrition.

6.4 Biodiversity Conservation

Mutation breeding is essential for conserving biodiversity by infusing new traits into both traditional and underutilized plant species. This method significantly enhances crop genetic diversity, which in turn fortifies agricultural resilience against environmental stresses and disease outbreaks. Maintaining genetic variation is vital for ensuring that crops can adapt to shifting conditions, thereby reducing the risks of crop failure and bolstering food security. In addition, by enhancing the growth potential of underutilized crops, mutation breeding not only supports local farming practices but also promotes sustainable agricultural methods. These initiatives harmonize with natural ecosystems, fostering long-term agricultural sustainability while ensuring a diverse and varied food supply for coming generations. Mutation breeding helps create a more stable and sustainable food system by promoting genetic variety and preserving regional customs (Frison et al. 2011).

7 Contribution to Sustainable Development Goals (SDGs) in Agriculture

The Sustainable Development Goals (SDGs) of the UN, especially those pertaining to responsible consumption and production, life on land, climate action, and zero hunger, are greatly advanced via mutation breeding. By enhancing crop productivity and resilience, this technique effectively combats hunger and malnutrition, ensuring food security for an expanding global population. Moreover, mutation breeding promotes sustainable agricultural practices by decreasing dependence on chemical inputs like pesticides and fertilizers while optimizing resource efficiency. This contributes to reducing environmental degradation and encourages responsible consumption and production patterns. By increasing crop tolerance to climate-related stresses such as drought, heat, and salinity, mutation breeding aids in both climate adaptation and mitigation efforts. Additionally, this approach is crucial for preserving biodiversity and protecting ecosystems by encouraging the cultivation of diverse and underutilized crops. By merging agricultural innovation with environmental stewardship, mutation breeding emerges as an essential tool for achieving global sustainability and ensuring long-term food security (Raina et al. 2016).

8 Recent Advancements and Innovative Techniques

Mutation breeding has developed as an influential tool for crop production and development, allowing scientists to grow plant varieties with enhanced characteristics such as increased yield, resistance to pests and diseases, and tolerance to environmental stresses. Unlike traditional breeding methods, mutation breeding induces genetic variations through physical or chemical mutagens, accelerating the development of desirable traits. Recent advancements in mutation breeding, particularly through the integration of biotechnological tools, have significantly increased its efficiency and precision.

8.1 Marker-Assisted Selection (MAS) Breeding

In mutant breeding, molecular markers are essential for quickening the selection process. Higher precision in identifying desired mutations is made possible by methods like simple sequence repeats (SSR) and single nucleotide polymorphism (SNP) markers.

MAS is a trait selection strategy that uses molecular markers. To identify genetic markers linked to any traits or phenotypes, the initial stage of MAS employs both genotypes and phenotypes. The impact of these markers on traits and phenotypes is then calculated and integrated into genetic assessment and selection techniques. Plant breeding becomes more efficient as a result of adding the desired genes, or QTLs, into the plant genome. The molecular markers may also help enhance the efficiency of target gene introgression and recovery from the recurrent parent genome through backcrossing. Affordability, a high degree of parental polymorphism, professional marker assay, dependability, and superior DNA quality are also considered to be the five most important criteria for marker choice. The effective use of MAS breeding across a broad range of crops, such as rice, barley, maize, sorghum, and wheat, has been documented by several studies (Gouda et al. 2020).

A number of essential germplasm lines were effectively utilized in breeding, resulting in varieties that hold significant importance in the market. Along with the conventional breeding plan, MAS created and produced rice compounds and published variants with key traits. The majority of genes utilized in MAS and variety production, including those linked to disease resistance (*Xa23, Xa21, Xa4, Pi-1, Pi-2, Pi-25, Pi-33, R-sb2t, R-sbzt,* etc.), sterility gene (*Rf5*), and date heading, as well as genes associated with grain consistency (*Lgc1, Wx, Frg, badh2*). Through the application of marker-aided backcross [MABC], high-amylase content was combined with kernel quality attributes such as weight, width/length ratio, basmati fragrance, and bacterial blight resistance [MABC]. Through the application of marker-assisted selection (MAS) documented in Thailand, submergence resistance types associated with the Sub1 gene have been enhanced, and their suitability for deep-water paddy production in eastern India has been assessed. To determine the function of Sub1 in various genetic settings, the actions of MABC-enhanced resistance types were assessed. The Sub1 loci were efficiently and precisely inserted by MABC into the rain-fed lowland rice variety BR11, recognized as the premier variety in Bangladesh. By the end of 2015, there were about 100 different genes identified. Due to their broad resistance, 14 genes have been added to this list: "*Pi1, Pi2, Pi9, Pi20[t], Pi33, Pi39, Pi40[t], Pi47, Pi48, Pi54rh, Pi56, Piz, Piz-t,* and *Pigm*." By providing hybrid and inbred variants, these markers aid in the development of rice in Asian nations, increasing its percentage from 11 to 34%. By developing resistance to bacterial blight from hybrid/inbred varieties, this affects an anticipated increase in grain production of 0.8 million metric tons ["worth US$20.5 million"] per cropping season of paddy rice in China's main countries of Indonesia, the Philippines, and India (Leung et al. 2003).

8.2 TILLING (Targeting Induced Local Lesions in Genomes)

One accurate technique of reverse genetics is called *TILLING* (targeting induced local lesions in genomes), which uses chemicals to create mutations and advanced screening methods to find those gene changes. Chemical mutagens like ethyl methane sulfonate (EMS), which cause random point mutations throughout the genome, are first applied to plant seeds in this procedure. The generated mutant populations are then screened for target gene mutations using methods such as next-generation sequencing or enzymatic mismatch cleavage founded on polymerase chain reaction (PCR). This method is generally recognized in agricultural research as a non-GMO methodology and allows researchers to find useful genetic variants without introducing foreign DNA.

The capacity of *TILLING* to recognize and capitalize on inherent changes in crop genomes is one of its main advantages, which makes it an affordable technique for improving traits. Rice varieties with greater drought tolerance and wheat lines with enhanced resilience to fungal diseases have both been successfully developed using this strategy. Sustainable agriculture is aided by *TILLING*, which allows breeders to create crops that are more tolerant to environmental challenges by leveraging the genetic variety already present. This creative method promotes initiatives to guarantee food security in the face of changing climate circumstances in addition to enhancing crop performance. Furthermore, this method has developed into *Eco TILLING*, which finds advantageous alleles for use in breeding initiatives by utilizing the natural genomic variety found in populations. It is one of the techniques in contemporary mutant breeding because of its efficiency and adaptability. It enables targeted gene changes that enhance crop resilience, nutritional value, and yield stability in the face of changing environmental circumstances. Food security and agricultural breakthroughs are greatly aided by this technique, which is especially effective at producing functional genetic variants in staple crops like rice and wheat.

8.3 CRISPR-Cas9 and Gene Editing in Mutation Breeding

Rice breeding has been transformed by the gene-editing technique CRISPR-Cas9, which enables precise alterations at particular loci. In contrast to traditional mutagenesis, this technique creates random genetic alterations that allow for the creation of enhanced rice cultivars with desired characteristics. CRISPR-Cas9 introduces a double-strand break by directing the Cas9 enzyme to a particular sequence of DNA using guide RNA. Rapid development of rice varieties with higher nutritional profiles, increased tolerant of pests, and increased drought tolerance has resulted from this exact method. Knocking down genes linked to tolerance to bacterial blight is one recent use (Zhou et al. 2022). To produce better rice varieties, CRISPR-Cas9 is being included into mutant breeding techniques. This enables breeders to maximize nutritional value, stress tolerance, and yield. However, there are still major obstacles to wider adoption, including ethical and regulatory issues. It is anticipated that CRISPR-Cas9's function in mutant breeding will grow as it develops, offering more effective, economical, and sustainable solutions for the world's rice production. For focused enhancements, CRISPR-induced mutations can be used in conjunction with traditional mutagenesis to replicate naturally existing variants (Zhang et al. 2022).

These days, rice assortments that employ genome-altering technologies contain novel and varied genotypes of yield-related variables. The CRISPR-Cas9 framework's potential to increase rice crop productivity and grain quality. CRISPR-Cas9 may be used to modify the genes Gn1a, DEP1, GS3, and IPA1, which control traits, plant and panicle architecture, the number of grains per panicle, and grain size (Shan et al. 2014).

Currently, a number of studies are being conducted to use CRISPR-Cas9 to further increase grain output. Grain size is one of the most important variables influencing the nutritional value and possible production of rice grains. Breeders and molecular scientists are particularly interested in it since it can boost yield (Wang et al. 2015). The GS2Emutant gained function as a result of Wang et al. (2022) targeting the miR396 location in the CS2 gene, which controls a number of critical features such seed breaking, grain size, grain quality, nitrogen utilization efficiency, and abiotic stress response. In terms of grain size and yield, the mutant had several advantageous trait expressions; also, the thousand grain weight increased by 23.5%, which led to a 10.4% rise in yield.

Researchers have also successfully edited genes involved in starch biosynthesis, improving rice quality and digestibility for better human consumption (Xu et al. 2020). susceptibility of improved rice cultivars. Salinity is one of the most important abiotic stresses affecting rice production worldwide. Cultivating salinity-tolerant cultivars is the most environmentally friendly and cost-effective method for managing salinity. For target-site genome editing, CRISPR-Cas9 systems have become increasingly popular in recent years to develop the salinity tolerance crop varieties. CRISPR-Cas9 was used to make specific changes in the OsRR22 gene, which helped the japonica rice variety WPB106 better withstand salt (Zhang et al. 2019). Creating specific changes in the *OsPUB7* gene using CRISPR-Cas9 to develop rice that can

Table 1 Integration of biotechnology in mutation breeding for sustainable agriculture

Crop	Integration of biotechnology in mutation breeding	Traits	References
Rice	MAS	Tolerance to bacterial blight	Leung et al. (2003)
	MAS	Blast disease resistance	-
	CRISPR-Cas9	Tolerance to bacterial blight, drought tolerance	Zhou et al. (2022)
		Enhance plant and panicle architecture, Improve quantity of grains per panicle, and grain size	Shan et al. (2014)
		Improve grain Yeild	Sedeek et al. (2019)
		Increase yield	Wang et al. (2015)
		Increase grain size and yield	Wang et al. (2022)
		Rice quality	Xu et al. (2020)
		Salinity tolerance	Zhang et al. (2019)
		Drought tolerance, resistance to bacterial leaf blight disease	Kim et al. (2019)

better handle drought and other environmental stresses in the future. Kim et al. (2019) reported that Os8N3 EBEs to reduce the activity of this gene and help the Kitaake (*Oryza sativa* L. sp. Japonica) variety resist bacterial leaf blight disease. *Xanthomonas oryzae pv. Oryzae*-induced bacterial blight in Basmati varieties resulted in significant yield losses (Table 1).

9 Economic Importance

Rice is one of the vital food sources for over half of the global population and faces increasing threats from climate change, insect infestations, diseases, and rising population numbers. The productivity and sustainability of agriculture have significantly increased as a result of the introduction of better crop types. Due to their improved yields, quality, and less dependency on fungicides and pesticides, high-yielding cultivars have seen an increase in production levels and lands devoted to them. Short-duration and drought-tolerant crops have maximized land use efficiency and made water conservation easier. Cropping systems have become more intensive due to changes in maturity and photoperiod responses, increasing their efficiency. The market value of agricultural goods, including oil, starch, malt, has grown due to improvements in processing quality. Specialty crops, nutritional improvements, and attractive features have all been developed to satisfy consumer desires. Targeted phenotypic enhancements have been made easier by induced mutations, which have also produced a number of other agronomic, financial, and environmental advantages, highlighting their crucial role in the invention of contemporary agriculture (Ahloowalia et al. 2004) (Fig. 3; Table 2).

9.1 Yield Improvement and Economic Benefits

The mutation breeding method used produced the high yield and grain production, and it's used to produce the biotic and abiotic resistant varieties of crops (Table 2). In addition to improving food security, this creative strategy supports sustainable farming methods by encouraging biodiversity and lowering dependency on chemical inputs. Farmers are better prepared to handle issues like climate change, pests, and

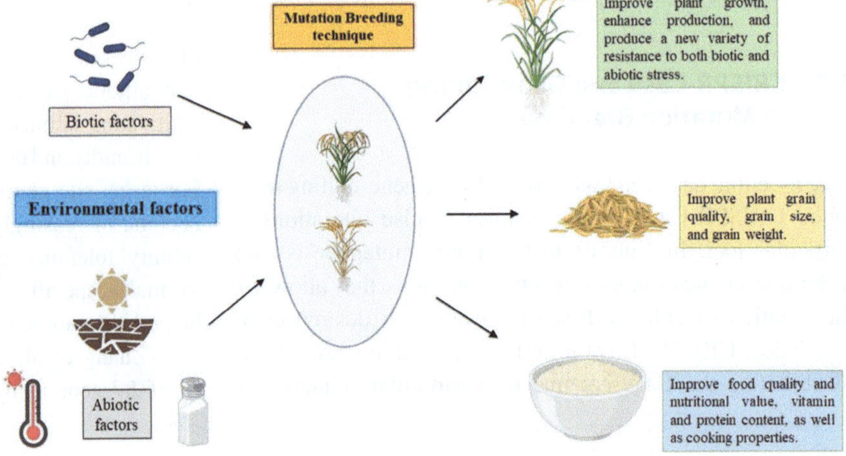

Fig. 3 Economic importance of mutation breeding in rice for sustainable agriculture

Table 2 Development of crops through mutation breeding

Crop	Mutagen	Mutant variety	Trait	References
Rice	Gamma rays	PNR	High-yield, early maturation and low stature	Chakrabarti (1995)
	Gamma ray, chemical mutagens	IR64	Salinity tolerance, Blast disease resistance, Tungro disease resistance, Drought tolerance, Herbicide tolerance	Nakhoda et al. (2012), Madamba et al. (2009), Zenna et al. (2008), Cairns et al. (2009), Chhapekar et al. (2015)
	MAS	IR64	Resistance to bacterial blight	Djedatin et al. (2016)
	Heavy-ion beam	M89	Salt-tolerant, high-grain yield	Zhang et al. (2022)

diseases by creating crops that can tolerate harsh climatic circumstances, which eventually results in a more resilient agricultural system. Higher yields translate into increased farm revenues, improved rural livelihoods, and enhanced economic stability in rice-dependent economies, particularly in Asia and Africa. Rice has seen the publication of 434 mutant cultivars with enhanced traits, including early maturity, disease and cold resistance, better grain production, semi-dwarf height, and enhanced grain quality. Twelve were caused by other radiation sources, seven by fast neutrons, sixteen by X-rays, and 225 (56%) by gamma rays. Sixty-seven percent of them were acquired as direct mutants. Through the induction of genetic variants that produce high-yielding cultivars, mutation breeding has played a significant role in increasing rice output. The rice variety "PNR" has not only contributed to greater food security but also improved the livelihoods of farmers by providing them with a more reliable crop. India, for instance, was created using gamma irradiation and showed notable early maturation and low stature, boosting farmer incomes and enhancing national food security. Because this kind of variety produces a lot of crops per acre, farmers can optimize their harvests while using the least number of resources and land possible (Chakrabarti 1995). Grain quality is a crucial determinant of rice's economic value, affecting consumer preferences and market prices. Mutation breeding has been utilized to improve rice quality traits, including grain size, cooking properties, aroma, and nutritional content. The famous Basmati rice varieties, such as "Basmati 385," were improved through mutation techniques to enhance elongation and aroma, thereby increasing their export value. These mutant breeding methods are used to improve the high-yield crop by changing its genetic makeup. Economic value can be raised by putting into practice ideas that focus on optimizing production procedures and reducing waste. Companies can increase their profitability and productivity by looking at and changing these aspects. Investing in research and development can also lead to the improvement of innovative techniques that further increase crop resilience and adaptation. Ultimately, a comprehensive approach that combines mutant breeding with sustainable farming practices will promote long-term agricultural success. Such improvements boost trade competitiveness, allowing rice-producing countries to access premium markets and enhance foreign exchange earnings.

9.2 Resistance to Biotic and Abiotic Stresses

Rice production is frequently challenged by various environmental stresses like biotic and abiotic stresses, including pests, diseases, drought, salinity, and extreme temperatures (Table 2). These stresses result in significant yield losses, threatening food supply stability. Mutation breeding has developed as an effective method to developing stress-resistant rice varieties. By inducing mutations in genes responsible for stress tolerance, researchers have successfully enhanced the resilience of rice to environmental adversities. Biotic stresses such as bacterial blight, blast disease, and insect infestations cause substantial losses in rice production. Mutation breeding has been instrumental in developing resistant rice varieties by modifying genes involved in plant defense mechanisms and Abiotic stresses, including drought, salinity, and high temperatures, pose important challenges to rice cultivation. Mutation breeding has played a pivotal role in developing rice varieties that can thrive under such conditions. The International Rice Research Institute (IRRI) has used induced mutation techniques extensively to create a number of mutants in the widely cultivated rice variety IR64. Numerous genes governing important agronomic properties in rice have been identified thanks in large part to this mutant collection (Wu et al. 2005). Some of these mutants have advantageous phenotypic characteristics that may be applied directly to breeding programs for rice. Notable examples include resistance to salinity (Nakhoda et al. 2012), blast disease (Madamba et al. 2009), tungro disease (Zenna et al. 2008), and drought (Cairns et al. 2009). There are presently no commercially grown genetically modified rice varieties, and transgenic uses in rice are still restricted despite advances in genetic engineering. Agrobacterium-mediated transformation procedures for IR64. By introducing advantageous genes into IR64, this technique has made it possible to create transgenic lines with enhanced nutritional content, including increased endosperm accumulation of iron (Fe) and zinc (Zn). Further increasing its potential uses in contemporary rice breeding and farming, transformation efforts have also produced IR64 lines that demonstrate herbicide tolerance (Chhapekar et al. 2015). The growth of rice varieties resistant to sheath blight remains a significant challenge due to the limited availability of naturally resistant germplasm. Because traditional breeding techniques rely on the genetic variety already present in rice populations, they have not been able to adequately address this problem. As a result, genetic engineering has become a potent technique for introducing genes linked to sheath blight resistance into rice, providing a more efficient means of creating resistant cultivars. The identification and isolation of important genes linked to sheath blight resistance has advanced significantly in recent years. For example, it has been discovered that the IDD14 and IDD13 genes activate PIN1a, which is essential for increasing rice tolerance of Sheath Blight. Breeders may incorporate these genes into superior rice types by utilizing genetic engineering technology, which might lessen the destructive effects of Sheath Blight on rice production worldwide. In addition to providing tolerance to African strains of *Xanthomonas oryzae*. Multiple associated QTLs was identified (Djedatin et al. 2016), IR64 carries the major resistance gene Xa4, which confers resistance to bacterial blight (BB) caused by *Xanthomonas oryzae pv. oryzae* and may have agronomic benefits beyond disease resistance (Xu et al. 2012). Reducing production costs, guaranteeing steady farmer incomes, and keeping rice prices accessible for consumers all depend on minimizing yield losses brought on by pests and environmental stressors. One of the most detrimental abiotic stressors affecting rice farming is soil salinization, which can cause yield reductions of 20% to complete crop failure (Haque et al. 2021). Production of rice is seriously threatened by the growing occurrence of salty and alkaline soils, particularly in areas where arable land is becoming scarcer. Mutation breeding has been recognized as an effective method for developing rice varieties that exhibit greater resistance to saline-alkaline environments, thereby addressing this challenge. The primary objective of mutation breeding is to introduce genetic changes that enhance resilience to saline-alkaline stress while also producing high-yielding elite varieties with desirable agronomic traits (Ahloowalia et al. 2004). The ability to create salt-tolerant rice lines has been shown by recent developments in mutant breeding. Notably, Zhang et al. (2022) isolated saline-alkaline-tolerant mutant lines by effectively inducing mutations by heavy-ion beam irradiation. This innovation greatly improves the economic feasibility of farmers' agricultural operations by enabling them to cultivate rice in previously unsuitable places for cultivation in saline-alkaline soils. By increasing the range of arable area available for agriculture, such creative breeding techniques not only improve crop output in difficult environmental conditions but also support sustainable agricultural practices. Modern genetic engineering technologies combined with creative breeding techniques like mutation breeding can help researchers and breeders create rice varieties that are more resistant to biotic and abiotic stresses. These developments have a great deal of potential to stabilize rice markets, improve food security, and encourage sustainable rice cultivation in light of climate change and other changing environmental concerns.

10 Challenges and Limitations

Developing rice varieties to address issues like resistance to pests, environmental stress, and yield gains has been made possible via the use of mutation breeding. However, the method comes with several challenges and limitations that hinder its widespread application. These challenges arise from the inherent nature of mutation breeding, as well as the broader agricultural and socio-economic context in which rice is cultivated.

10.1 Genetic Variability and Mutation Unpredictability

Mutation breeding relies on inducing genetic changes through physical (e.g., radiation) or chemical mutagens, and the resulting mutations are often random. Mutations are unexpected and can cause a wide range of genetic alterations, many of which may not result in desirable features. For example, a mutation might produce undesirable phenotypes that negatively affect plant growth or yield. Furthermore, the genetic variability resulting from mutation breeding can be vast, complicating the identification of beneficial mutations. This unpredictability means that while some mutations may enhance rice characteristics like disease resistance or drought tolerance, others could impair plant health or growth. It can therefore take a lot of time and effort to screen for advantageous traits (Chaudhary et al. 2019).

10.2 Long Timeframe for Developing New Varieties

One of the major drawbacks of mutation breeding is the time needed to develop new types of rice varieties. The process involves several stages, including inducing mutations, selecting mutated plants, evaluating their characteristics, and stabilizing new varieties through multiple generations (Lenaerts et al. 2019). This prolonged timeline makes mutation breeding less attractive compared to other modern breeding methods, such as genetic modification or genomic selection, which can yield faster results. Rice breeders often need to wait for multiple generations of plants to assess whether a mutation has led to an advantageous trait. Moreover, because mutations can have recessive effects, it may take several rounds of breeding before the beneficial mutation becomes stable and prominent in subsequent generations.

10.3 Limited Scope of Induced Mutations

While a variety of genetic alterations may be introduced by mutation breeding, the range of attributes that can be enhanced is constrained. Many characteristics, such complicated tolerance to illnesses and pests, are regulated by many genes, and mutagen-induced mutations could not significantly affect these polygenic features. This limitation makes it difficult to address multifactorial traits like high yield or multi-stress tolerance through mutation breeding alone. Additionally, induced mutations are more likely to affect single-gene traits, whereas complex traits require multi-genic interactions, which mutation breeding might not adequately address. In these situations, molecular breeding methods like genetic manipulation or marker-assisted selection (MAS) are frequently more successful.

10.4 Environmental and Ecological Concerns

The use of mutagens in mutation breeding, particularly radiation and chemical mutagens, raises concerns regarding their environmental impact. While radiation mutagenesis is widely used in agriculture, the introduction of irradiated seeds into the environment has raised questions about the long-term ecological consequences, such as potential changes in soil microbiota or water contamination. Furthermore, the safety of chemical mutagens for non-target organisms and their persistence in the environment requires careful regulation. In regions with stricter environmental policies, mutation breeding may be subject to regulatory limitations that slow its implementation. These concerns necessitate the development of safer and more sustainable mutation breeding techniques.

10.5 Lack of Genomic Information and Marker Resources

In certain areas, the use of genetic markers in mutant breeding is still in its infancy, despite rice being a model plant for genomics. To find advantageous mutations, mutation breeding necessitates effective screening methods, which can be expensive and time-consuming. The effectiveness of mutant breeding programs is limited by the complexity of the selection process caused by the absence of complete genomic data and readily available markers for certain phenotypes. Not all rice breeders have access to cutting-edge technology like next-generation sequencing (NGS), particularly in impoverished nations where mutant breeding may be most advantageous. The extensive use of mutant breeding in rice agriculture is hampered by this digital divide.

10.6 Economic and Resource Constraints

Mutation breeding is an expensive and resource-intensive process. The need for specialized facilities for irradiation or chemical treatment, along with the manpower required for screening and evaluating thousands of mutations, increases the cost of developing new rice varieties. For resource-poor farmers in developing countries, these costs can be prohibitive. Furthermore, mutation breeding may not be the most cost-effective approach when compared to other breeding techniques like hybridization or genetic engineering, which may offer quicker and more direct improvements.

11 Conclusion

Mutation breeding plays an essential role in advancing agricultural genetics, serving as a vital tool for developing improved crop varieties with desired traits, such as enhanced yield, disease resistance, and resilience to environmental stress. By harnessing the power of mutagenic agents, such as radiation and chemicals, breeders have successfully created new genetic variations that enhance the sustainability and productivity of crops. This technique not only promotes food security but also reduces the environmental impact of agriculture by lowering the need for harmful chemical inputs. Additionally, mutation breeding supports biodiversity conservation and helps crops adapt to changing climate conditions, offering a promising path toward more eco-friendly and economically viable farming practices. As research continues to evolve, combining traditional breeding techniques with modern technologies like CRISPR-Cas9 and TILLING, mutation breeding will remain a cornerstone of agricultural innovation, ensuring the resilience of crops and the long-term sustainability of global food systems.

References

Ahloowalia BS, Maluszynski M, Nichterlein K (2004) Global impact of mutation- derived varieties. Euphytica 135:187–204. https://doi.org/10.1023/B:EUPH.0000014914.85465.4f

Atia M, Jiang W, Sedeek K, Butt H, Mahfouz M (2024) Crop bioengineering via gene editing: reshaping the future of agriculture. Plant Cell Rep 43(4):98. https://doi.org/10.1007/s00299-024-03183-1

Cairns JE, Acuna TLB, Simborio FA, Dimayuga G, Praba ML, Leung H, Torres R, Lafitte HR (2009) Identification of deletion mutants with improved performance under water-limited environments in rice (*Oryza sativa* L.) Field Crop Res 114:159–168. https://doi.org/10.1016/j.fcr.2009.07.019

Chakrabarti SN (1995) Mutation breeding in India with particular reference to PNR rice varieties. J Nuclear Agric Biol 24:73–82. https://hal.science/hal-05143726v1

Chaudhary J, Alisha A, Bhatt V, Chandanshive S, Kumar N, Mir Z, Kumar A, Yadav SK, Shivaraj SM, Sonah H, Deshmukh R (2019) Mutation breeding in tomato: advances, applicability and challenges. Plants 8(5):128. https://doi.org/10.3390/plants8050128

Chhapekar S, Raghavendrarao S, Pavan G, Ramakrishna C, Singh VK, Phanindra MLV, Dhandapani G, Sreevathsa R, Kumar PA (2015) Transgenic rice expressing a codon-modified synthetic CP4-EPSPS confers tolerance to broad-spectrum herbicide, glyphosate. Plant Cell Rep 34:721–731. https://doi.org/10.1007/s00299-014-1732-2

Djedatin G, Ndjiondjop MN, Sanni A, Lorieux M, Verdier V, Ghesquiere A (2016) Identification of novel major and minor QTLs associated with *Xanthomonas oryzae pv. oryzae* (African strains) resistance in rice (*Oryza sativa* L.) Rice 9:18. https://doi.org/10.1186/s12284-016-0090-9

Frison EA, Cherfas J, Hodgkin T (2011) Agricultural biodiversity is essential for a sustainable improvement in food and nutrition security. Sustainability 3(1):238–253. https://doi.org/10.3390/su3010238

Fróna D, Szenderák J, Harangi-Rákos M (2019) The challenge of feeding the world. Sustainability 11(20). https://doi.org/10.3390/su11205816

Gouda G, Gupta MK, Donde R, Mohapatra T, Vadde BL (2020) Marker-assisted selection for grain number and yield-related traits of rice (*Oryza sativa* L.). Physiol Mole Biol Plants 26:885–898. https://doi.org/10.1007/s12298-020-00773-7

Haque MA, Rafii MY, Yusoff MM, Ali NS, Yusuff O, Datta DR, Anisuzzaman M, Ikbal MF (2021) Advanced breeding strategies and future perspectives of salinity tolerance in rice. Agronomy 11:1631. https://doi.org/10.3390/agronomy11081631

Khan S, Al-Qurainy F, Anwar F (2009) Sodium Azide: a chemical mutagen for enhancement of agronomic traits of crop plants. Environ We Int J Sci Tech 4:1–21

Kim YA, Moon H, Park CJ (2019) CRISPR/Cas9-targeted mutagenesis of Os8N3 in rice to confer resistance to *Xanthomonas oryzae pv. oryzae*. Rice 12:1–13. https://doi.org/10.1186/s12284-019-0325-7

Lenaerts B, Collard BC, Demont M (2019) Improving global food security through accelerated plant breeding. Plant Sci 287:110207. https://doi.org/10.1016/j.plantsci.2019.110207

Leung H, Zhu Y, Revilla-Molina I, Fan JX, Chen H, Pangga I, Mew TW (2003) Using genetic diversity to achieve sustainable rice disease management. Plant Dis 87(10):1156–1169. https://doi.org/10.1094/PDIS.2003.87.10.1156

Lin KC, Jwo WS, Chandrika NNP, Wu TM, Lai MH, Wang CS (2016) A rice mutant defective in antioxidant-defense system and sodium homeostasis possesses increased sensitivity to salt stress. Biol Plant 60:86–94. https://doi.org/10.1007/s10535-015-0561-7

Madamba MRS, Sugiyama N, Bordeos A, Mauleon R, Satoh K, Baraoidan M, Kikuchi S, Shimamoto K, Leung H (2009) A recessive mutation in rice conferring non-racespecific resistance to bacterial blight and blast. Rice 2:104–114. https://doi.org/10.1007/s12284-009-9027-x

Nakhoda B, Leung H, Mendioro MS, Mohammadi-nejad G, Ismail AM (2012) Isolation characterization and field evaluation of rice (Oryza sativa L, Var. IR64) mutants with altered responses to salt stress. Field Crop Res 127:191–202. https://doi.org/10.1016/j.fcr.2011.11.004

Predieri S (2001) Mutation induction and tissue culture in improving fruits. Plant Cell Tiss Organ Cult 64(2/3):185–210: S2CID 37850239. https://doi.org/10.1023/A:1010623203554

Raina A, Laskar RA, Khursheed S, Amin R, Tantray YR, Parveen K, Khan S (2016) Role of mutation breeding in crop improvement-past, present and future. Asian Res J Agric 2(2):1–13. https://doi.org/10.9734/ARJA/2016/29334

Sedeek KE, Mahas A, Mahfouz M (2019) Plant genome engineering for targeted improvement of crop traits. Front Plant Sci 10:114. https://doi.org/10.3389/fpls.2019.00114

Shan Q, Wang Y, Li J, Gao C (2014) Genome editing in rice and wheat using the CRISPR/Cas system. Nat Protoc 9(10):2395–2410. https://doi.org/10.1038/nprot.2014.157

Tai TH, Chun A, Henry IM, Ngo KJ, Burkart-Wako D (2016) Effectiveness of sodium azide alone compared to sodium azide in combination with methyl nitrosurea for rice mutagenesis Plant Breed. Biotech 4(4):453–461. https://doi.org/10.9787/PBB.2016.4.4.453

Van Esse HP, Reuber TL, van der Does D (2020) Genetic modification to improve disease resistance in crops. New Phytol 225(1):70–86. https://doi.org/10.1111/nph.15967

Wang S, Li S, Liu Q, Wu K, Zhang J, Wang S, Fu X (2015) The OsSPL16-GW7 regulatory module determines grain shape and simultaneously improves rice yield and grain quality. Nat Genet 47(8):949–954. https://doi.org/10.1038/ng.3352

Wang W, Pan Y, Tan C, Li H, Chen Y, Ma C (2022) A new gain-of-function OsGS2/GRF4 allele generated by CRISPR/Cas9 genome editing increases rice grain size and yield. The Crop J 10(4):1207–1212. https://doi.org/10.1016/j.cj.2022.01.004

Wu JL, Wu C, Lei C, Baraoidan M, Bordeos A, Madamba MR, Ramos-Pamplona M, Mauleon R, Portugal A, Ulat VJ, Bruskiewich R, Wang G, Leach J, Khush G, Leung H (2005) Chemical- and irradiation-induced mutants of indica rice IR64 for forward and reverse genetics. Plant Mol Biol 59:85–97. https://doi.org/10.1007/s11103-004-5112-0

Xu J, Jiang J, Dong X, Ali J, Mou T (2012) Introgression of bacterial blight (BB) resistance genes Xa7 and Xa21 into popular restorer line and their hybrids by molecular marker-assisted backcross (MABC) selection scheme. African J Biotech 11(33):8225–8233. https://doi.org/10.5897/AJB12.341

Xu DP, Yao ZE, Pan JB, Feng HY, Guo ZQ, Lu XL (2020) Study on the multiple characteristics of M3 generation of pea mutants obtained by neutron irradiation. Nuclear Sci Techni 31(7):67. https://doi.org/10.1007/s41365-020-00777-8

Yadav S, Sandhu N, Dixit S, Singh VK, Catolos M, Mazumder RR, Kumar A (2021) Genomics-assisted breeding for successful development of multiple-stress-tolerant, climate-smart rice for southern and southeastern Asia. Plant Genome 14(1):e20074. https://doi.org/10.1002/tpg2.20074

Zafar K, Sedeek KE, Rao GS, Khan MZ, Amin I, Kamel R, Mahfouz MM (2020) Genome editing technologies for rice improvement: progress, prospects, and safety concerns. Front Genome Edit 2:5. https://doi.org/10.3389/fgeed.2020.00005

Zenna NS, Cabauatan PQ, Baraoidan M, Leung H, Choi IR (2008) Characterization of a putative rice mutant for reaction to rice tungro disease. Crop Sci 48:480–486. https://doi.org/10.2135/cropsci2007.03.0127

Zhang Y, Polinski MP, Morrison PR, Brauner CJ, Farrell AP, Garver KA (2019) High- load reovirus infections do not imply physiological impairment in salmon. Front Physiol 10. https://doi.org/10.3389/fphys.2019.00114

Zhang X, Yang F, Ma H, Li J (2022) Evaluation of the saline–alkaline tolerance of Rice (*Oryza sativa* L.) mutants induced by heavy-ion beam mutagenesis. Biol 11(1):126. https://doi.org/10.3390/biology11010126

Zhou Y, Xu S, Jiang N, Zhao X, Bai Z, Liu J, Yang Y (2022) Engineering of rice varieties with enhanced resistances to both blast and bacterial blight diseases via CRISPR/Cas9. Plant Biotech J 20(5):876–888. https://doi.org/10.1111/pbi.13766

Mutagenesis for Crop Breeding and Functional Genomics

Rakshana Palaniswamy, Vignesh Mohanavel, Raveendran Muthurajan, and Sudha Manickam

Abstract

Mutagenesis has emerged as a powerful tool in crop breeding that facilitates the creation of elite crop varieties with improved agronomic traits. Innovations in mutagenesis techniques and functional genomics have revolutionized crop improvement strategies. Various mutagenesis approaches, including chemical, physical, biological, site-directed mutagenesis, soma clonal variation and modern genome-editing techniques such as CRISPR/Cas9 are employed to generate genetic diversity for crop breeding applications. Genome level mutations result in changes in chromosome structure (such as inversions, translocations, duplications, and deletions) and chromosome number (polyploidy, aneuploidy, and haploidy), while gene mutations are point mutations in the DNA sequence. Traditional phenotyping methods and advanced genotyping techniques including PCR assay, Molecular markers, next-generation sequencing, and MutMap approaches are used to characterize these mutations. Mutagenesis serves as both forward and reverse genetic approach in crop improvement. This chapter investigates the pivotal role of mutagenesis in crop improvement and its integration with functional genomics to identify and validate candidate genes that govern important traits, including enhanced yield, better nutritional quality, resilience to biotic and abiotic stressors, and other agronomic attributes. We demonstrate how the combination of mutagenesis and genomic technologies is transforming crop breeding practices, using examples from various crop species to improve food security amid evolving environmental challenges.

Keywords

Mutagenesis · Crop improvement · Functional genomics · Agronomic traits · Stress tolerance · Disease resistance · Nutritional content

1 Introduction

Genetic diversity is the foundation for successful plant breeding. Over time, the genetic variability of crops has steadily decreased, mainly because breeders have concentrated on 'elite' cultivars. This loss of genetic diversity has resulted in a bottleneck, prompting the development of various induced mutagenesis techniques for generating new genetic variations. Additionally, the conventional plant breeding process can be challenging and labour-intensive because of long breeding cycles. It is difficult to create new crops because of the decreased genetic variety as well as the decreased vigour and fertility. Mutagenesis refers to the process of inducing heritable mutations in an organism's genetic material, which can lead to changes in traits that may enhance agricultural productivity. Mutagenesis breeding aims to enhance genomic variation while minimizing its impact on the viability of the organism. This method has been crucial for crop improvement, contributing to development of plants with better yield, biotic/abiotic stress resilience, and other desirable characteristics. Mutagenesis can produce new allelic variants in both targeted and random ways. To create new alleles in any organism, random mutagenesis has been employed since the early twentieth century. Depending on the type of mutagen, its dosage and the duration of exposure, the mutations can cause changes from single nucleotide polymorphisms (SNPs) to significant chromosomal alterations at random sites on the genome.

Rakshana Palaniswamy and Vignesh Mohanavel: These authors Contributed Equally to This Work

R. Palaniswamy · V. Mohanavel · R. Muthurajan · S. Manickam (✉)
Department of Plant Biotechnology, Centre for Plant Molecular Biology and Biotechnology, Tamil Nadu Agricultural University, Coimbatore, India
e-mail: sudhatamil@gmail.com

J. Devi (ed.), *Mutation Breeding in Crop Improvement*, Sustainable Landscape Planning and Natural Resources Management,
https://doi.org/10.1007/978-3-032-17911-1_11

In contrast, induced mutagenesis can produce precise nucleotide alterations at predefined locations on the genome. It enhances genomic diversity while maintaining viability, especially when compared to spontaneous mutations. This significantly accelerates crop improvement by providing considerable genetic variation and novel alleles for economically important traits, achieving results faster than traditional breeding methods. The alleles generated through induced mutagenesis serve as valuable genetic resources for improving crops and identifying candidate genes through functional genomics across diverse species. Varieties developed during this process typically be similar to their closest ancestors but differ in one or a few characters. These varieties have been shown to be environmentally sustainable, safe, and non-hazardous. Global plant breeding programs have heavily relied on mutagenesis, resulting in significant genetic variation over the years. By 2025, approximately 3460 mutant varieties across nearly 250 crop species have been registered in FAO/IAEA Mutant Variety Database (https://mvd.iaea.org/). Physical mutagens such as gamma rays, fast neutrons, and X-rays, as well as chemical mutagens like EMS, NEU, and NMU have been predominantly used for mutagenesis. Almost half of these mutants have been developed in three Asian countries: China, Japan, and India (Fig. 1a). Rice exhibits the greatest number of mutant varieties among all crops, followed by barley, chrysanthemum, and common wheat (Fig. 1b). The key traits modified include enhanced yields, better nutritional quality, resilience to biotic and abiotic stressors, and other agronomic advancements. These advancements are crucial for boosting agricultural productivity and improving food

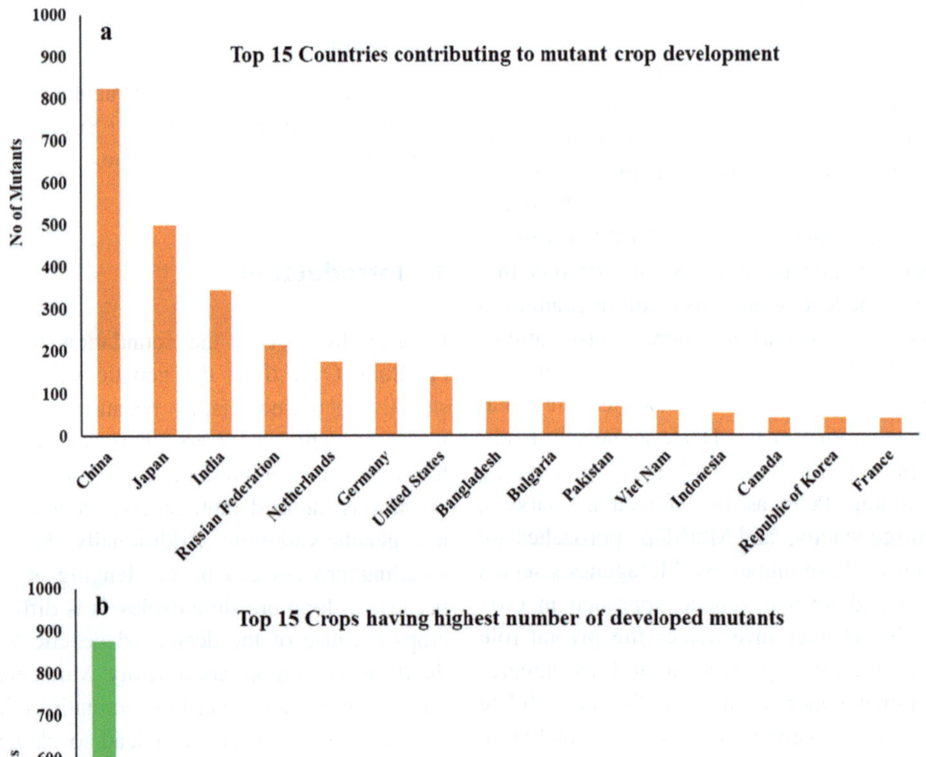

Fig. 1 **a** Bar chart illustrating the contribution of the top 15 countries in mutant crop development **b** Bar chart illustrating the contributions of the top 15 mutant crop species recorded in the database (*Source* FAO/IAEA Mutant Variety Database)

security, particularly in areas where environment factors pose significant challenges to crop production.

2 Mutagenesis Approaches

There are different approaches to mutagenesis that can be applied in crop breeding, each contributing to various aspects of plant improvement. Below are the major types of mutagenesis techniques:

2.1 Chemical Mutagenesis

Chemical mutagenesis is a random mutagenesis method that employs chemicals to create mutations in plant DNA, altering the genetic code and resulting in various traits. Silent and missense mutations are more common mutations, while nonsense mutations are rare types. It can be utilized to identify genes, create mutant alleles, study protein structure, and enhance enzyme activity, selectivity, and stability. Chemical mutagens, such as 5-Bromouracil (5BU), ethyl methanesulfonate (EMS),Benzo(a)pyrene (BaP), 2-Aminopurine, N-ethyl-N-nitrosourea (ENU), and N-methyl-N′-nitro-N-nitrosoguanidine (MNNG) are used to introduce specific base substitutions in the genome (Table 1). In mutation breeding, chemical mutagens are commonly employed to develop new varieties that exhibit enhanced agronomic characteristics like grain size, yield components, pest/disease resistance, and resistance to abiotic stresses.

2.2 Physical Mutagenesis (Radiation Mutagenesis)

Physical mutagenesis, referred to as radiation mutagenesis, is a process that induces mutations in the genetic code of an organism by subjecting it to physical agents such as ionizing radiation (including X-rays, gamma rays), non-ionizing radiation (UV rays) and particle radiation (alpha, beta rays, electrons, fast and thermal neutrons). Ionizing radiations generates highly reactive free radicals that damages DNA molecules, leading to point mutations and large chromosomal rearrangements (Table 1). Mutation breeding uses radiation to induce beneficial mutations in plant seeds, leading to improved crop varieties. Research also explores radiation's

Table 1 Mutagenesis approaches and their mechanism of action

Mutagenesis method	Specificity	Mechanism of action	Mutation frequency	Reference
Chemical mutagenesis				
Ethyl methane sulfonate (EMS)	Random	Alkylation: transitions (G:C to A:T) or transversions (G:C to C:G or G:C to T:A)	Approximately 10 mutations (per Mb)	Till et al. (2007)
Base analogues (5-Bromouracil)	Random	Transition (A:T to G:C)	–	
Sodium azide (Az)	Random	Non-alkylation: transition (G:C to A:T)	Approximately 2.9 mutations (per Mb)	Tai et al. (2016)
Physical mutagenesis				
γ rays	Random	SNP and deletion	7.5 × 10 – 6 to 9.8 × 10 – 6	Li et al. (2016a)
Fast-neutron	Random	Transition (A:T to G:C), large deletions, chromosomal rearrangements	Approximately 78 mutations	Li et al. (2016b)
UV rays	Random	Pyrimidine dimers: transition (C:G to T:A, T:A to C:G), tandem transition (C:C to T:T)	26 to 45-fold	Nakamura et al. (2021)
Biological mutagenesis				
T-DNA insertion	Random	Insertional mutagenesis	>80%	Sallaud et al. (2004)
Transposon insertion	Random	Insertional mutagenesis	>50%	Kolesnik et al. (2004)
Retrotransposon insertion	Random	Insertional mutagenesis	1 insertion (per 100 Kb)	Miyao et al. (2003)
Genome editing (targeted mutagenesis)				
Zinc finger nuclease	Targeted	Small deletions or insertions	7–16%	Zhang et al. (2010)
TALEN	Targeted	small deletions, insertions, and substitution	4–30%	Zhu et al. (2017)
CRISPR/Cas9	Targeted	small deletions, insertions, and substitution	85–100%	Zhu et al. (2017)

effects on living organisms and DNA repair mechanisms. Crops like tomatoes, cotton, and wheat have been developed through this method, showing increased disease resistance, higher yields, and improved quality.

2.3 Biological Mutagenesis

Biological mutagenesis approaches employ natural biological elements, including transposons, retrotransposons, and T-DNA insertions to induce mutations and create genetic variability for crop enhancement (Table 1). Transposon-mediated mutagenesis uses transposon and retrotransposon sequences, while the T-DNA-mediated method utilizes T-DNA sequence (*Agrobacterium tumefaciens*) to introduce random mutations in the crop genome. It enables researchers to study gene function, identify genes linked to specific traits, and enhance desired traits. They are commonly applied in model plants to investigate the functions of genes in relation to stress responses and stress tolerance mechanisms. In entrapment tagging, the insertional mutagen elements are coupled with a tag (gene or enhancer) to discover novel genes and gene regulatory cascades in crops.

2.4 Site-Directed Mutagenesis

Site-directed mutagenesis (SDM) is a highly precise technique used to introduce specific alterations in genes or nucleotide sequences, unlike random mutagenesis which affects the entire genome. This approach is a powerful tool in molecular biology for examining the structural and functional dynamics of proteins and nucleic acids (Bachman 2013). SDM can be accomplished using three primary approaches: (1) vector-based, (2) PCR-based, and (3) nuclease-based methods. The vector-based method involves plasmids or phage vectors to carry the mutation into the target gene. The PCR-based method employs mutated primer pairs, where one primer carries the targeted mutation while the other without mutation, to induce mutations into the gene sequence through polymerase chain reaction (PCR). The nuclease-based method, such as the cre-lox recombinase system, uses site-specific nucleases to induce mutations at precise genomic locations. SDM is used to introduce point mutations, short deletions, or additions in protein sequences. This enables researchers to investigate how individual amino acids contribute to maintaining the structural integrity, functional characteristics, and overall stability of proteins. Common applications of SDM include targeted modifications in protein–protein interactions, post-translational modifications, suppression of enzymatic activity, and generation of RNAi-resistant rescue constructs. SDM minimizing the risk of random, undesired mutations that can arise through traditional breeding methods. This makes SDM an essential tool for the rapid development of improved crop varieties.

2.5 Gene Editing (Targeted Mutagenesis)

Gene-editing tools such as transcription activator-like effector nucleases (TALENs), zinc finger nucleases (ZFNs) and clustered regularly interspace short palindromic repeat (CRISPR)are new plant breeding techniques that employs site-directed nucleases (SDNs) to induce double-stranded breaks (DSB) at targeted sites within the DNA, facilitating precise nucleotide modifications within the genome (Table 1). Following the creation of DSBs, the cell's natural DNA repair mechanisms are leveraged to introduce mutations, which can involve the insertion of new genetic material or the deletion of existing sequences at the break site. It results in generation of novel knock-in and knock-out mutants, which serve as highly valuable resources for breeding programs, as it eliminates the complications associated with linkage drag, a common challenge in conventional breeding. Moreover, genetic improvements can be achieved through the targeted removal of genetic elements that negatively impact gene function, enhancing desired traits. It presents a more precise, efficient, and accelerated approach to improve yield, stress resilience, and nutritional content.

2.6 In Vitro Mutagenesis and Soma Clonal Variation

Soma clonal variation refers to the genetic variation observed among plants that are produced through tissue culture. This variation often results from spontaneous mutations and epigenetic changes that occur during prolonged growth periods and stress affecting the cultured cells. In contrast, *in-vitro* mutagenesis is a controlled procedure employed to induce genetic variations in plant cells using either physical or chemical mutagens. Although both processes can result in genetic changes, soma clonal variation occurs naturally as a consequence of the tissue culture environment, while in vitro mutagenesis is a targeted, induced process (Kumar et al. 2010). Plants regenerated through in vitro cultures, especially when subjected to mutagenesis, may exhibit novel traits including high yield, enhanced stress resistance, variations in fruit size, or changes in nutrient composition.

3 Types of Mutation

Mutations can be broadly classified into three main categories: intergenic mutations that encompass large-scale structural alterations in chromosomes such as inversions, translocations,

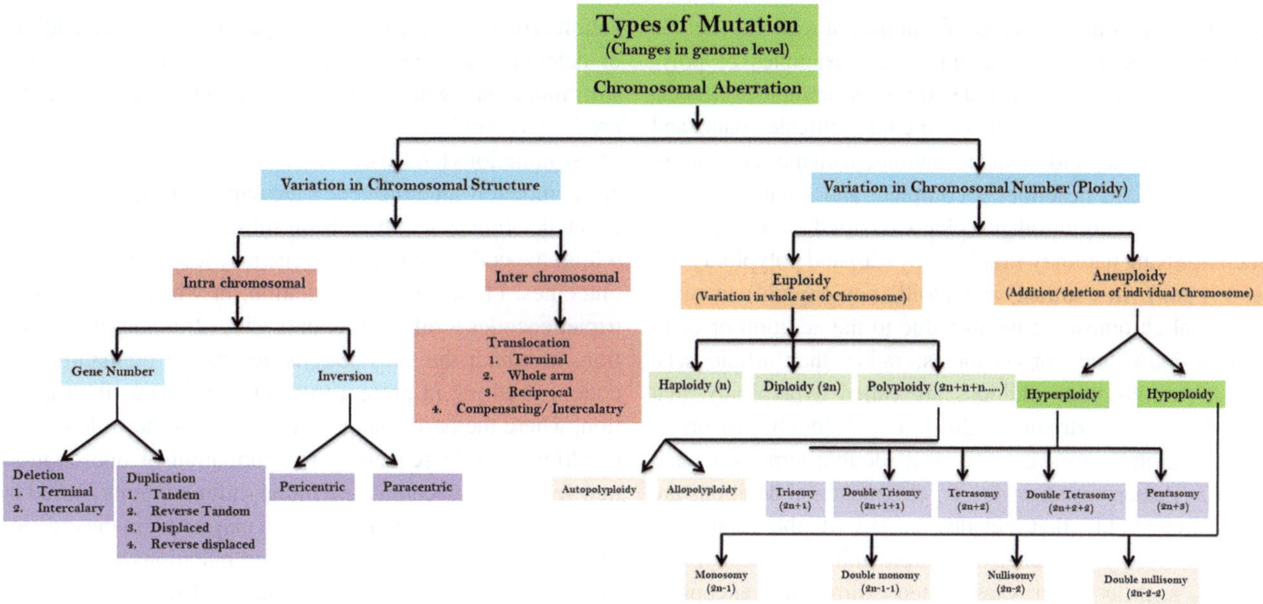

Fig. 2 Schematic illustration of various mutation types (chromosomal aberrations) observed at the genome level

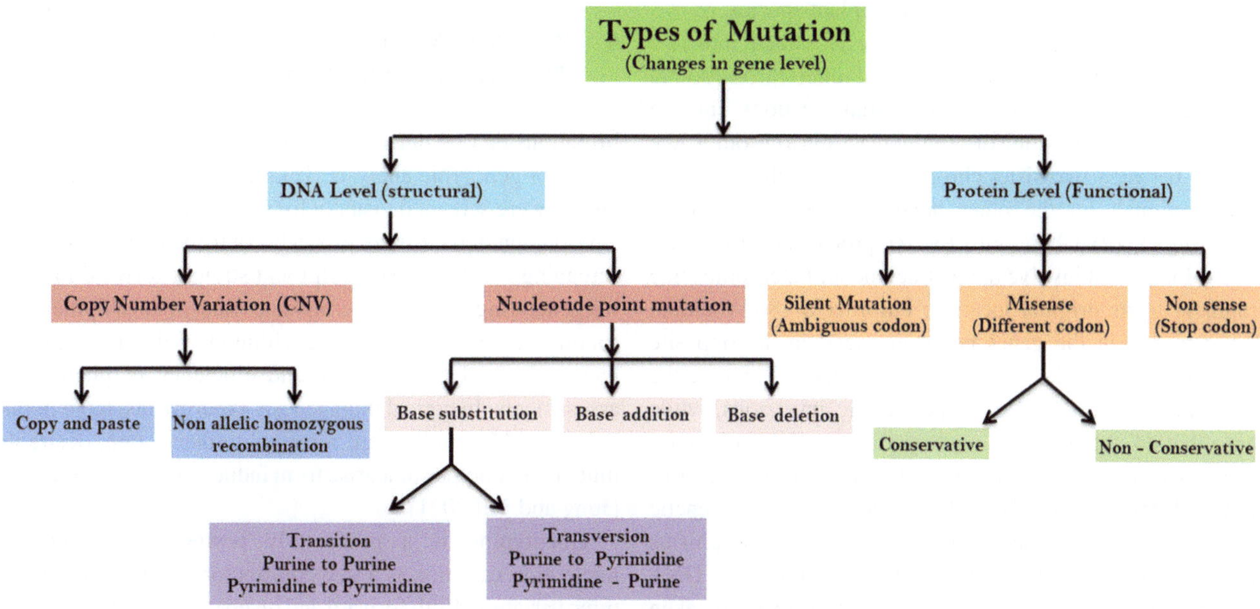

Fig. 3 Schematic illustration of various mutation types observed at the gene level

duplications, and deletions); mutations that affect chromosome number (ploidy), including aneuploidy, euploidy, and haploidy; intragenic mutations, that consist of small-scale point mutations occurring within a gene (Oladosu et al. 2016). Different types of mutations and chromosomal aberrations induced at gene and genome level are illustrated in Figs. 2 and 3.

Ploidy mutations involve changes to the genome's chromosome number, such as the addition or subtraction of entire chromosome sets, or a transition from a diploid (2n) state to a haploid (n) state. These mutations can arise spontaneously or be induced experimentally. Polyploidy, which can result from genome duplication (autopolyploidy) or the addition of genomes from different species (allopolyploidy). Polyploidy has played a vital role in the natural evolutionary development of numerous plant species, and has been routinely employed to develop new species in agriculture. A notable example is triticale, a hybrid that merges the genetic traits of wheat and rye.

Polyploids, whether occurring spontaneously or through induced methods, plays a crucial role in domestication and genetic improvement of crops. Examples of triploids include bananas, watermelons, and apples are examples of polyploid crops; tetraploids include cotton, groundnuts, brassica, potato, and wheat; hexaploids include triticale, oats, and wheat; and octaploids include dahlias, triticale, and strawberries. Euploid describes a complete and balanced set of chromosomes (normal)that is characteristic for each species, encompassing haploidy (n), diploidy (2n) and polyploidy (2n + n + n..). In contrast, aneuploid (non-euploid) involves abnormal chromosome number due to the addition or deletion of individual chromosomes, rather than whole sets. These chromosomal additions are called hyperploidy, that includes 2n + 1 (trisomy), 2n + 1 + 1 (double trisomy), 2n + 2 (Tetrasomy), 2n + 2 + 2 (double tetrasomy) and 2n + 3 (Pentasomy).Conversely, chromosomal deletions are called hypoploidy, that includes 2n–1–1 (double monomy), 2n-2 (nullisomy) and 2n–2–2 (double nullisomy). Polyploid species, such as bread wheat, often exhibit such aneuploid chromosomal configurations. The artificial induction of polyploidy has been instrumental in creating genetic diversity and aiding in gene discovery across various plant species.

Chromosomal translocations can be classified into two primary types: intra-chromosomal and inter-chromosomal rearrangements. 'Intra-chromosomal translocations' involve movement or transposition of a segment within a chromosome arm, or between two sister chromatids. 'Inter-chromosomal translocations', on the other hand, can be classified as, (1) terminal, (2) whole-arm, (3) reciprocal, and (4) intercalary. Inversions involve the reattachment of a chromosome fragment to its original location, but after a 180° rotation. These results in the genes being arranged in the opposite order compared to the wild type. Inversions that involve the centromere are referred to as pericentric inversions, while those that do not are called paracentric inversions. Chromosomal inversions have been key contributors to the process of species differentiation. These changes appear as fixed genetic differences between species or as variable polymorphisms within populations across different taxa. Additionally, DNA sequence duplication is a common mutation that has taken place during the evolutionary history of life forms. Typically, duplications larger than 1 kb in length are classified as segmental duplications.

Base substitutions are a type of gene-level mutation, referring to nucleotide changes where one base is replaced by another. This phenomenon arises when a base-analogue incorporated into the DNA mis-pairs during replication, resulting in substitutions mutations. Transition mutations occur when there is a substitution between two purines (A-G) or pyrimidines (T-C), while transversions involve the substitution of purines with pyrimidines or the reverse (A:G-T:C) (Kharkwal 2012). Although, transitions and transversions are among the simplest nucleotide alterations, they can still lead to mutations that are observable at the phenotypic level. Additionally, replication errors may also result in specific nucleotide addition or deletion mutations. Such alterations can lead to frameshift mutations, where the gene's reading frame is shifted to produce a completely different amino acid during translation. These mutations tend to have more severe effects because they can completely alter the gene's message, starting from the site of the deletion or addition. Base substitutions that take place within the coding region of a gene may lead to the following outcomes: (1) a missense mutation, in which the modified triplet encodes a different amino acid; (2) a nonsense mutation, in which the substitutions generate a premature stop codon, resulting in protein truncation; or (3) a silent mutation, where the base change does not affect the codon or the resulting peptide sequence. The formation of one or more new features may come from frame shift mutations caused by base additions and deletions, which may result in the creation of a completely new protein. Some mutations arise due to the rearrangement of nucleotides in the DNA (Oladosu et al. 2016).

4 Mutation Screening Methods (Mutational Genomics)

Mutations can be detected through either phenotypic observation or genotypic analysis. Traditionally, the mutant populations were screened visually for phenotypic changes. When environmental factors are properly controlled, the phenotypic screening can be a resourceful and straight forward method for generating improved germplasm. Recent developments in robotics and machine learning within the field of phenomics have enhanced the accuracy and efficiency of phenotypic assessments. In typical mutagenesis experiments, large populations of M_2 families are screened to detect rare, desirable mutant phenotypes that arise from induced genetic alterations (Jung and Till 2021).

Genotyping is a particularly powerful approach, as nucleotide variation is a key contributor to heritable phenotypic variation. Both forward and reverse genetics have significantly advanced as a result of the development of efficient methodologies for identifying uncommon, caused mutations. Detection of large-scale genomic variations, including insertions, substitutions, deletions, and chromosomal rearrangements, have progressed from traditional PCR techniques to array-based hybridization methods, and to advanced high-throughput genome sequencing technologies. High-throughput enzymatic mismatch cleavage, capillary electrophoresis, denaturing HPLC, high-resolution melting analysis, and direct DNA sequencing are employed to interpret chemically induced point mutations within TILLING techniques (Jung and Till 2021).

Molecular markers, particularly DNA-based markers, are important tools in genotypic screening. These markers consist of specific DNA sequences linked to traits or mutations. Molecular indicators derived from variations in nucleotide sequences encompass point mutations, insertions, substitutions, deletions, duplications, and rearrangements. When it comes to identifying mutants with desired traits, molecular markers such as Restriction Fragment Length Polymorphisms (RFLPs), Single Nucleotide Polymorphisms (SNPs), Simple Sequence Repeats (SSRs), Inter small RNA polymorphism (iSNAP), Intron length polymorphism (ILP), and Penta-primer amplification refractory mutation system (PARMS) are indispensable. These molecular markers are crucial for identifying mutants with desired characteristics, facilitating efficient selection during mutagenesis screening.

Advancements in next-generation sequencing (NGS) technologies has significantly improved the efficiency and affordability to detect millions of mutations in a short time. For instance, high-throughput techniques like 'MutMap', 'MutMap-Gap', 'MutChromSeq', 'Mut-Ren-Seq', 'QTL-seq', and 'whole-genome sequencing(WGS)-based mutation mapping' have recently been developed. These methods are highly applicable and cost-efficient for identifying desirable genes in a shorter timeframe. WGS constitutes the most comprehensive method for genotypic screening of mutants. In recent years, this approach has become more accessible for small-to medium-sized genomes due to the continued decline in sequencing costs and the increased availability of well-annotated reference genomes. According to Harloff et al. (2012), a representative M_2 population of tetraploid rapeseed, with a genome size of 2258 Mbp/2C, has been sequenced, revealing an exceptionally high mutation density of 130,000 mutations per M_2 plant. Linking the co-segregation of genotype with phenotype allows for the sequencing of whole genomes and gene cloning in plants with smaller genomes. The MutMap approach has been used to clone alleles induced by EMS in rice through a bulked segregant strategy, and has since been modified to facilitate gene cloning without the need for outcrossing (Jankowicz-Cieslak et al. 2017). Mutations in all transcribed sequences can be recovered using exome sequencing for big genomes where WGS is still prohibitively expensive. Hexaploid wheat has been subjected to a capture that includes a probe set spanning 107 Mb. After 2735 mutant lines' exomes were sequenced, almost 10 million mutations in the protein-coding areas were discovered, yielding 35–40 mutations per kb (Krasileva and Vasquez-Gross 2017).

5 Mutagenesis as Tool for Forward Genetics and Reverse Genetics Studies

Mutagenesis is an effective technique for both forward and reverse genetics (Fig. 4), allowing scientists to explore the genetic mechanisms behind important biological processes and apply this knowledge to crop improvement. In forward genetics, candidate genes are linked to their corresponding

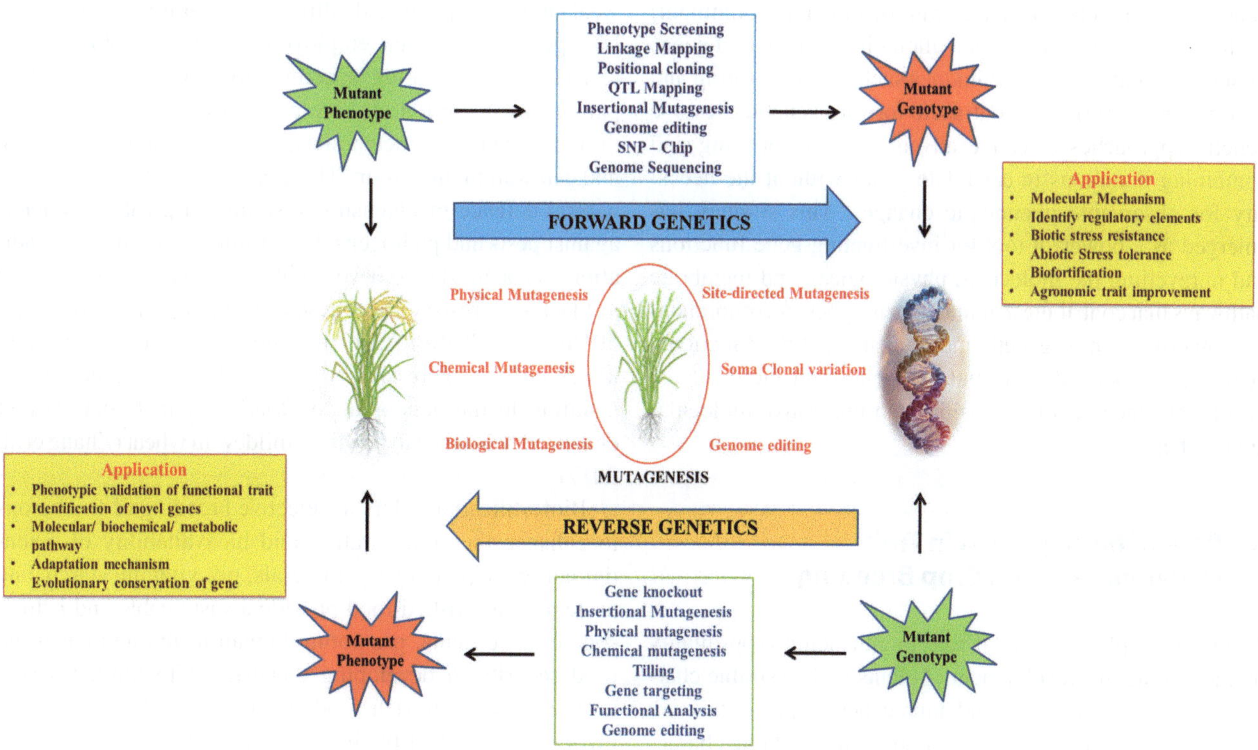

Fig. 4 Schematic illustration of forward and reverse genetic approaches and their applications in crop improvement

phenotypes to determine gene function. In forward genetics, Mutagenesis typically starts with random mutations induced by chemical or physical mutagens. The mutant population is then screened for advantageous phenotypes like disease resistance, changed growth patterns, or altered metabolic pathways. To investigate the underlying mutations in genes/QTLs, researchers use molecular approaches including genetic mapping, QTL mapping, GWAS, positional cloning, fine mapping, and other cutting-edge genomic approaches. These methods provide valuable insights into gene function and regulation, with the process beginning with the identification of a desirable phenotype.

Most mutant varieties developed through forward genetics approach exhibit favourable phenotypes and have been introduced as new varieties, often without a comprehensive understanding of the molecular mechanism causing these traits. In most cases, mutations convert a dominant allele into a recessive one, resulting in a 'loss-of-function', while 'gain-of-function' mutations are infrequently observed. Moreover, forward genetics is primarily effective for modifying monogenic traits, with its utility further limited by the low probability of simultaneously inducing multiple beneficial mutations in unlinked genes within a single plant.

Reverse genetics, on the other hand, focuses on understanding the function of a gene by examining the phenotypic changes that occur due to alterations in a known genetic sequence. Advances in sequencing technologies and accessibility of reference-genome annotations, this approach allows for the investigation of any gene. Through controlled mutagenesis, researchers can create 'gain-of-function' (dominant) or 'loss-of-function' (recessive) alleles in certain target genes by induced mutagenesis techniques such genome editing and transposon insertions. In contrast to forward genetics, reverse genetic approaches offer the advantage of identifying and combining mutations in candidate genes without the necessity for observable phenotypic changes. This strategy has emerged as a powerful tool for investigating gene functions and unravelling the molecular, physiological, and metabolic pathways that contribute to mutant phenotypes. A comprehensive investigation of genetic mechanisms and the functional genomics associated with mutated genes can facilitate the genetic enhancement of crops through innovative molecular approaches.

6 Role of Mutagenesis in Trait Development and Crop Breeding

Mutagenesis plays a pivotal role in enhancing crop quality by facilitating the development of plants with desirable characteristics. Both random and mutagenesis approaches are harnessed to enhance various agronomic and botanic characteristics, improved yield, stress resilience, nutritional quality, and other agronomic advancements. Among various factors, abiotic stresses represent the primary constraints that hinder crop growth and its yield potential. Key abiotic stressors encompass drought, elevated temperatures, chilling stress, soil salinity, alkalinity, flooding, and heavy metal toxicity, which can result in yield reductions of up to 70% in major food crops. Mutagenesis technique was successfully utilized to generate novel genetic variations that enhance tolerance to various abiotic stresses. This approach has facilitated the development of mutant cultivars exhibiting traits associated with stress tolerance, including root-adaptive traits, improved water retention, root architecture, stay green traits, altered metabolism, limited Na ion uptake, salt extrusion, increased expression of heat-shock proteins, ROS scavenging enzymes, and osmo-protectants. A notable example includes a soybean mutant developed through gamma-ray induction that can germinate at 4 °C, compared to the wild-type, which typically germinates at around 8 °C (Khan and Tyagi 2013). A novel stay-green rice mutant with enhanced harvest index and delayed senescence was developed through EMS mutagenesis to improve drought stress resilience (Ramkumar et al. 2019). In rice, T-DNA mediated mutagenesis of '*OsMADS23*' and '*OsPLC4*' genes led to improved salt stress tolerance (Deng et al. 2019; Li et al. 2021).

Biotic stressors, such as pests (insects, herbivore, nematodes, weeds), and disease-causing pathogens (bacteria, fungi, virus, phytoplasma) deprive host plants of essential nutrients and growth factors. These stressors interfere with plant's metabolic processes and physiological functions, severely weakening the plant and ultimately resulting in its death. Mutagenesis has been employed to improve biotic stress resistance by mutating specific target genes involved in pathogen recognition, disease resistance (R) genes, production of secondary metabolites, synthesis of phytoalexins, and cell wall modification. This approach strengthens plant's natural defence mechanisms, resulting in greater resilience against pests and pathogens. For example, gamma ray irradiation has been used to develop blight-resistant lentil, as well as black spot-resistant Japanese pear (Sadiq et al. 2008; Saito 2016). The CRISPR/Cas9 technique used to concurrently knock-out the expression of three '*EDR1*' gene homologs, resulting in the generation of *Taedr1* mutants that exhibit enhanced resistance to powdery mildew in wheat (Zhang et al. 2017).

Biofortification refers to selective breeding of food crops to enhance the concentration and bioavailability of essential nutrients, particularly, minerals, oil, vitamin, starch, and proteins. Biofortified food provide a sustainable and natural source of nutrients, promoting human health and enhanced food security in developing countries with limited access to diverse, nutrient-rich food. Strategies for biofortification involve alteration of the biosynthetic pathways of nutrients, enhancing their transport and storage, increasing bioavailability, minimizing anti-nutritional factors (phytic acid) and other toxic compounds that hinder nutrient absorption. For instance, brassica mutants with improved edible oil quality have been developed by decreasing the levels of toxic substances including glucosinolates and erucic acid (Singh and Verma 2015). Similarly, high oleic and low linoleic acid soybean oil was developed through the precise mutagenesis of the fatty acid desaturase gene families '*FAD2-1A*' and '*FAD2-1B*' utilizing TALEN technology (Haun et al. 2014).

Mutagenesis is a powerful technique utilized to improve both agronomic and botanic traits in plants. This technique can lead to improvements in various characteristics such as plant architecture, seed dormancy and germination, flowering and ripening times, photoperiod-insensitivity, improved yield, resistance to lodging and shattering, accelerated flower and fruit development. These improvements render mutagenesis an essential tool for developing stress-resilient, high-yielding crops adaptable to diverse environmental conditions. Rice grain yield has been significantly improved through the targeted knockout of three key genes, '*GW2*', '*GW5*', and '*TGW6*' utilizing the CRISPR/Cas9 technology (Xu et al. 2016). To simplify the post-harvest processing, naked barley grains (without hull) were created by editing '*Nud*' gene through CRISPR/Cas9 mutagenesis approach (Gasparis et al. 2018). Furthermore, to enhance stress resilience, late-flowering soybean mutants were created by editing the '*GmFT2a*' gene utilizing the CRISPR/Cas9 approach, allowing the plants to escape unfavourable environmental conditions and better adapt to stress (Cai et al. 2018).

7 Role of Mutagenesis in Functional Genomics Studies

Beyond routine crop improvement, induced mutations combined with advanced genomic technologies create a robust and cost-effective functional genomics platform. This platform facilitates the development of molecular markers and aids in the discovery of candidate genes for precise manipulation, thereby contributing to paving the way for more efficient and sustainable agriculture. Additionally, it uncovers valuable insights into gene function, enabling the development of crops with enhanced yield, and stress resilience. One of the first examples of this approach was the discovery of a new 'waxyE1100' allele within the '*granule-bound starch synthase I* (*GBSS-I*) gene (waxy)' through EMS mutagenesis. This novel allele was subsequently used in breeding programs for development of crops with high-amylopectin starch for various industrial and food applications (Muth et al. 2008).

Functional genomics studies rely on both forward and reverse genetic approaches. TILLING, a classical reverse genetics strategy, has been employed to identify and analyse molecular alterations in specific genes, playing a pivotal role in both crop improvement and pathogen defence. This technology offers several key benefits, including its applicability across a wide range of plant species irrespective of their ploidy level or genome sizes. It can detect diverse DNA sequence alterations, such as small deletions, insertions, and microsatellite polymorphisms. EMS was the first chemical mutagen used to mutagenize 'allohexaploid' and 'allotetraploid' wheats. Following TILLING, 1920 individuals were screened, and 246 novel alleles of 'waxy' locus were identified (Slade et al. 2012). More recently, TILLING was used to investigate the functionality of 'heat shock binding protein 1 (HSBP1)'. Mutations in negative regulator 'HSBP1' enhanced heat stress tolerance resistance in tomato plants (Marko et al. 2019).

Insertion mutagenesis is crucial for dissection of novel genes and their stress regulatory modules utilizing transposons, activator/dissociation (Ac/Ds) insertions, retrotransposons, and T-DNA insertion elements. Diverse collection of insertional knockout mutants and mutant libraries were generated to study gene functions. It is frequently employed in the gene trap technique, activation tagging, promoter trapping, and knockout mutant identification. Here, the mutagen helps identify the damaged genes by serving as a molecular tag. The insertion tag and the desired phenotype are associated, which aids in identifying the target gene.

Insertional mutagenesis has been extensively employed to create mutant libraries, facilitating the straight forward identification of tagged genes through PCR-based methods, including thermal asymmetric interlaced PCR (TAIL-PCR) or inverse PCR. For instance, two key genes associated with floral development, '*RID1*' (a key regulator of floral induction) and '*JMJ706*' (which modulates flower formation), have been successfully identified using the T-DNA mutagenesis. Despite its utility, T-DNA mutagenesis is characterized by low-copy number insertions, which necessitates the development of extensive T-DNA libraries that encompass the entire genome (Sun and Zhou 2008). Advancements in NGS technologies addresses the need for the characterization of comprehensive mutant collections and has facilitated systematic studies in functional genomics.

Transposons including the 'non-autonomous DNA-based active rice transposon' (nDart1), 'miniaturePing' (mPing), 'enhancer/suppressor mutator' (En/Spm-dSpm), and 'activator/dissociation' (Ac/Ds) elements have proven to be effective tools for the functional investigation of rice genes. For instance, insertion of the Ds element into '*Oryza sativa no pollen*(*Osnop*)' gene resulted in a male sterile mutant characterized by pollen-free flowers. A gene's biological function as well as that of regulatory modules like enhancers or promoters can be evaluated via entrapment trapping. In rice, stress-responsive genes are identified using GUS-trapped T-DNA tagging lines. Additionally, a 'chlorophyll-deficient mutant (*OschlH*)' exhibiting a 'chlorina' phenotype has been

Table 2 List of publicly available mutant crop databases for functional genomics research

S. No.	Database name	Crop(s)	Mutant source	Data available	Website
1	SoyTEdb	Soybean	Transposons	Genome-wide transposon map	http://www.soytedb.org
2	SALK	Arabidopsis	T-DNA insertion	T-DNA mutant collection, T-DNA flanking sequences, and genotyping primers	http://signal.salk.edu/
3	MaizeGDB	Maize	Ac, Ds, Mu, EMS	Phenotype data, SNP, Allele / Polymorphism	https://www.maizegdb.org/data_center/phenotype
4	OryGenesDB	Rice	T-DNA, Tos17, Ds	Flanking sequence tag (FST), EST, Gramene markers	http://orygenesdb.cirad.fr/
5	Rice Mutant Database (RMD)	Rice	Enhancer trap, Tos17	Flanking sequences, Mutant phenotype, gene expression data	http://rmd.ncpgr.cn/
6	HapRice	Rice	SNP haplotype	Haplotype-based trait dissection	http://qtaro.abr.affrc.go.jp/
7	DPTEdb	Dioecious plants	Transposons	Details of transposable elements in 8 dioecious plant species	http://genedenovoweb.ticp.net:81/DPTEdb/index.php
8	SPTEdb	Salicaceous plants	Transposons	Details of transposable elements in 3 salicaceousplant species	http://genedenovoweb.ticp.net:81/SPTEdb/index.php
9	TOMATOMA	Tomato	EMS and γ-ray	Mutant phenotype and metabolite data	http://tomatoma.nbrp.jp/index.jsp

discovered using a promoter-less GUS reporter gene in rice. The combination of 'GAL4/VP16-UAS' elements with GUS was successfully employed to develop enhancer-trapped lines of japonica rice (Ram et al. 2019).

Recent studies reported two main types of high-throughput CRISPR-mediated knockout screening approaches in plants, (1) small- or medium-scale screening, which targets the mutagenesis of several hundreds of genomic loci; and (2) large-scale genome-wide screening, which is employed to induce mutations in thousands of genomic loci. CRISPR knock-out mutant libraries are available for a wide range of crop plants including tomato, soybean, cotton, maize, rice, tobacco, and rapeseed. The large-scale CRISPR library was used to report genome-wide targeted mutagenesis in rice. Nearly 88,541 sgRNAs were designed to mutate 34,234 functional genes encompassing 83% of the rice genome. The sgRNAs were identified by the authors using barcode-based high-throughput amplicon sequencing on 91,004 T_0 plants. The 1 × sgRNA coverage was correlated with the size of the transgenics collection. Additionally, the authors found several mutants with aberrant phenotypes and reported an editing effectiveness of about 80% (Lu et al. 2017). Another research group generated over 14,000 T_0 plants, achieving over 0.5 × sgRNA coverage, by utilizing 25,604 sgRNAs to mutate over 12,802 genes that are highly expressed in shoot tissues (Meng et al. 2017). In vast and repetitive genomes, CRISPR/Cas9-driven repeat depletion improves high-density, genome-wide genotyping by directing sequencing efforts on single-copy regions. To reduce the repetitive elements in *Lens culinaris* genome with 84% repetitive DNA, a large-scale CRISPR library containing 566,766 gRNAs targeting 2.9 Gbp of repeat elements was recently employed for high-throughput genotyping in complex plant genomes (Rossato et al. 2023).

A wide range of bioinformatics databases with unlimited access to extensive collection of mutants and associated genomic data have been developed to assist researchers working on functional genomics, gene annotation, molecular pathways, and various other facets of crop improvement. Table 2 presents a list of some representative publicly available mutant crop databases for functional genomics research. These databases provide valuable resources for the identifying functional genes, understanding gene regulatory mechanisms, improving traits, validating candidate genes, and advancing crop genetic improvement. They play a pivotal role in understanding the gene functions and their impact on key agronomic traits, thereby enhancing crop breeding and development.

8 Conclusion

Mutagenesis represents a powerful and versatile tool in crop breeding and functional genomics by creating genetic diversity and enabling precise gene function studies. It facilitates discovery of novel genes, the functional validation of candidate genes, understanding of the complex molecular mechanisms, and promotes accelerated development of elite crop varieties with improved agronomic traits. These induced mutations are regarded as non-GMO (with exception of SDN2 and SDN3 genome-edited mutants) and are similar to spontaneous mutations, hence it does not carry intellectual property rights. Hence, the mutant varieties are free from ethical and regulatory concerns related to its use, release, import, or export facilitating its widespread adoption and utilization. Mutagenesis is an attractive option for plant breeders in developing countries where access to complex genetic engineering techniques is restricted. Despite these advancements, challenges persist in ensuring that mutations are stable and safe, without any unintended consequences. Nevertheless, with ongoing research and technological advancements, mutagenesis will remain a vital tool for development of next-generation crops to meet the demands of expanding global population. In conclusion, mutagenesis and functional genomics holds immense promise towards sustainable agriculture, but it requires careful monitoring and management to maximize its benefits while minimizing risks.

References

Bachman J (2013) Site-directed mutagenesis. In: Methods in enzymology, vol 529. Academic Press,, pp 241–248. https://doi.org/10.1016/B978-0-12-418687-3.00019-7

Cai Y, Chen L, Liu X, Guo C, Sun S, Cunxiang W, Jiang B, Han T, Hou W (2018) CRISPR/Cas9-mediated targeted mutagenesis of GmFT2a delays flowering time in soya bean. Plant Biotechnol J 16(1):176–185. https://doi.org/10.1111/pbi.12758

Deng X, Yuan S, Cao H, Lam SM, Shui G, Hong Y, Wang X (2019) Phosphatidylinositol hydrolyzing phospholipase C4 modulates rice response to salt and drought. Plant, Cell Environ 42(2):536–548. https://doi.org/10.1111/pce.13437

FAO/IAEA Mutant variety database. https://mvd.iaea.org/. Accessed 22 Feb 2025

Gasparis S, Kała M, Przyborowski M, Łyżnik LA, Orczyk W, Nadolska-Orczyk A (2018) A simple and efficient CRISPR/Cas9 platform for induction of single and multiple, heritable mutations in barley (Hordeum vulgare L.). Plant Methods 14:1–14. https://doi.org/10.1186/s13007-018-0382-8

Harloff HJ, Lemcke S, Mittasch J, Frolov A, Wu JG, Dreyer F, Leckband G, Jung C (2012) A mutation screening platform for rapeseed (Brassica napus L.) and the detection of sinapine biosynthesis mutants. Theoret Appl Genet 124(2012):957–969. https://doi.org/10.1007/s00122-011-1760-z

Haun W, Coffman A, Clasen BM, Demorest ZL, Lowy A, Ray E, Retterath A, Stoddard T, Juillerat A, Cedrone F, Mathis L (2014) Improved soybean oil quality by targeted mutagenesis of the fatty acid desaturase 2 gene family. Plant Biotechnol J 12(7):934–940. https://doi.org/10.1111/pbi.12201

Jankowicz-Cieslak J, Mba C, Till BJ (2017) Mutagenesis for crop breeding and functional genomics. In: Biotechnologies for plant mutation breeding: protocols. pp 3–18. https://doi.org/10.1007/978-3-319-45021-6_1

Jung C, Till B (2021) Mutagenesis and genome editing in crop improvement: perspectives for the global regulatory landscape. Trends Plant Sci 26(12):1258–1269. https://doi.org/10.1016/j.tplants.2021.08.002

Khan MH, Tyagi SD (2013) A review on induced mutagenesis in soybean. J Cereals Oilseeds 4(2):19–25. https://doi.org/10.5897/JCO10

Kharkwal MC (2012) A brief history of plant mutagenesis. In: Plant mutation breeding and biotechnology. CABI, Wallingford UK, pp 21–30. https://doi.org/10.1079/9781780640853.0021

Kolesnik T, Szeverenyi I, Bachmann D, Kumar CS, Jiang S, Ramamoorthy R, Cai M, Ma ZG, Sundaresan V, Ramachandran S (2004) Establishing an efficient Ac/Ds tagging system in rice: large scale analysis of Ds flanking sequences. Plant J 37(2):301–314. https://doi.org/10.1046/j.1365-313X.2003.01948.x

Krasileva KV, Vasquez-Gross HA, Howell T, Bailey P, Paraiso F, Clissold L, Simmonds J, Ramirez-Gonzalez RH, Wang X, Borrill P, Fosker C (2017) Uncovering hidden variation in polyploid wheat. Proc Natl Acad Sci 114(6):E913–E921. https://doi.org/10.1073/pnas.1619268114

Kumar K, Gill MI, Kaur H, Choudhary OP, Gosal SS (2010) In vitro mutagenesis and somaclonal variation assisted salt tolerance in 'Rough Lemon' (Citrus jambhiri Lush.). Europ J Horticult Sci 75(6):233. https://doi.org/10.1079/ejhs.2010/2005779

Li X, Bo Y, Qi W, Min Q, Zeng R, Xie Z, Huang J (2021) OsMADS23 phosphorylated by SAPK9 confers drought and salt tolerance by regulating ABA biosynthesis in rice. PLoS Genet 17(8):e1009699. https://doi.org/10.1371/journal.pgen.1009699

Li G, Chern M, Jain R, Martin JA, Schackwitz WS, Jiang L, Vega-Sánchez ME, Lipzen AM, Barry KW, Schmutz J, Ronald PC (2016a) Genome-wide sequencing of 41 rice (Oryza sativa L.) mutated lines reveals diverse mutations induced by fast-neutron irradiation. Mol Plant 9(7):1078–1081. https://doi.org/10.1016/j.molp.2016.03.009

Li T, Liu B, Chen CY, Yang B (2016b) TALEN-mediated homologous recombination produces site-directed DNA base change and herbicide-resistant rice. J Genet Genom 43(5):297–305. https://doi.org/10.1016/j.jgg.2016.03.005

Lu Y, Ye X, Guo R, Huang J, Wang W, Tang J, Tan L, Zhu J-K, Chu C, Qian Y (2017) Genome-wide targeted mutagenesis in rice using the CRISPR/Cas9 system. Mol Plant 10(9):1242–1245. https://doi.org/10.1016/j.molp.2017.06.007

Marko D, El-Shershaby A, Carriero F, Summerer S, Petrozza A, Iannacone R, Schleiff E, Fragkostefanakis S (2019) Identification and characterization of a thermotolerant TILLING allele of heat shock binding protein 1 in tomato. Genes 10(7):516. https://doi.org/10.3390/genes10070516

Meng X, Hong Y, Zhang Y, Zhuang F, Song X, Gao S, Gao C, Li J (2017) Construction of a genome-wide mutant library in rice using CRISPR/Cas9. Mol Plant 10(9):1238–1241. https://doi.org/10.1016/j.molp.2017.06.006

Miyao A, Tanaka K, Murata K, Sawaki H, Takeda S, Abe K, Shinozuka Y, Onosato K, Hirochika H (2003) Target site specificity of the Tos17 retrotransposon shows a preference for insertion within genes and against insertion in retrotransposon-rich regions of the genome. Plant Cell 15(8):1771–1780. https://doi.org/10.1105/tpc.012559

Muth J, Hartje S, Twyman RM, Hofferbert HR, Tacke E, Prüfer D (2008) Precision breeding for novel starch variants in potato. Plant Biotechnol J 6(6):576–584. https://doi.org/10.1105/tpc.012559

Nakamura M, Nunoshiba T, Hiratsu K (2021) Detection and analysis of UV-induced mutations in the chromosomal DNA of Arabidopsis. Biochem Biophys Res Commun 554:89–93. https://doi.org/10.1016/j.bbrc.2021.03.087

Oladosu Y, Rafii MY, Abdullah N, Hussin G, Ramli A, Rahim HA, Miah G, Usman M (2016) Principle and application of plant mutagenesis in crop improvement: a review. Biotechnol Biotechnol Equip 30(1):1–16. https://doi.org/10.1080/13102818.2015.1087333

Ram H, Soni P, Salvi P, Gandass N, Sharma A, Kaur A, Sharma TR (2019) Insertional mutagenesis approaches and their use in rice for functional genomics. Plants 8(9):310. https://doi.org/10.3390/plants8090310

Ramkumar MK, Senthil Kumar S, Gaikwad K, Pandey R, Chinnusamy V, Singh NK, Singh AK, Mohapatra T, Sevanthi AM (2019) A novel stay-green mutant of rice with delayed leaf senescence and better harvest index confers drought tolerance. Plants 8(10):375. https://doi.org/10.3390/plants8100375

Rossato M, Marcolungo L, De Antoni L, Lopatriello G, Bellucci E, Cortinovis G, Frascarelli G, Nanni L, Bitocchi E, Di Vittori V, Vincenzi L (2023) CRISPR-Cas9-based repeat depletion for high- throughput genotyping of complex plant genomes. Genome Res 33(5):787–797. https://doi.org/10.1101/gr.277628.122

Sadiq MS, Haidar S, Haq MA, Abbas G (2008) A high yielding and disease resistant mutant of lentil developed through seed irradiation of an exotic germplasm. Can J Pure Appl Sci 2(2):411–416. https://www.cjpas.net/uploads/pdfs/2/2/May-08_14.pdf

Saito T (2016) Advances in Japanese pear breeding in Japan. Breeding Sci 66(1):46–59. https://doi.org/10.1270/jsbbs.66.46

Sallaud C, Gay C, Larmande P, Bès M, Piffanelli P, Piégu B, Droc G, Regad F, Bourgeois E, Meynard D, Périn C (2004) High throughput T-DNA insertion mutagenesis in rice: a first step towards in silico reverse genetics. Plant J 39(3):450–464. https://doi.org/10.1111/j.1365-313X.2004.02145.x

Singh S, Verma AK (2015) A review on efforts of induced mutagenesis for qualitative and quantitative improvement of oilseed brassicas. J Pharmacogn Phytochem 4(2). https://www.phytojournal.com/archives/2015/vol4issue2/PartD/4-3-27.1-541.pdf

Slade AJ, McGuire C, Loeffler D, Mullenberg J, Skinner W, Fazio G, Holm A, Brandt KM, Steine MN, Goodstal JF, Knauf VC (2012) Development of high amylose wheat through TILLING. BMC Plant Biol 12:1–17. https://doi.org/10.1186/1471-2229-12-69

Sun Q, Zhou D-X (2008) Rice jmjC domain-containing gene JMJ706 encodes H3K9 demethylase required for floral organ development. Proc Natl Acad Sci 105(36):13679–13684. https://doi.org/10.1073/pnas.0805901105

Tai TH, Chun A, Henry IM, Ngo KJ, Burkart-Waco D (2016) Effectiveness of sodium azide alone compared to sodium azide in combination with methyl nitrosurea for rice mutagenesis. Plant Breed Biotechnol 4(4):453–461. https://doi.org/10.9787/PBB.2016.4.4.453

Till BJ, Cooper J, Tai TH, Colowit P, Greene EA, Henikoff S, Comai L (2007) Discovery of chemically induced mutations in rice by TILLING. BMC Plant Biol 7:1–12. https://doi.org/10.1186/1471-2229-7-19

Xu R, Yang Y, Qin R, Li H, Qiu C, Li L, Wei P, Yang J (2016) Rapid improvement of grain weight via highly efficient CRISPR/Cas9-mediated multiplex genome editing in rice. J Genet Genom= Yi chuanxue bao 43(8):529–532. https://doi.org/10.1016/j.jgg.2016.07.003

Zhang Y, Bai Y, Guangheng W, Zou S, Chen Y, Gao C, Tang D (2017) Simultaneous modification of three homoeologs of Ta EDR 1 by genome editing enhances powdery mildew resistance in wheat. Plant J 91(4):714–724. https://doi.org/10.1111/tpj.13599

Zhang F, Maeder ML, Unger-Wallace E, Hoshaw JP, Reyon D, Christian M, Li X, Pierick CJ, Dobbs D, Peterson T, Joung JK (2010) High frequency targeted mutagenesis in Arabidopsis thaliana using zinc finger nucleases. Proc Natl Acad Sci 107(26):12028–12033. https://doi.org/10.1073/pnas.0914991107

Zhu C, Bortesi L, Baysal C, Twyman RM, Fischer R, Capell T, Schillberg S, Christou P (2017) Characteristics of genome editing mutations in cereal crops. Trends Plant Sci 22(1):38–52. https://doi.org/10.1016/j.tplants.2016.08.009

Role of Site-Directed Mutagenesis in Genetic Improvement of Crop Plant

Nandakumar Vidya, Kumaresan Kowsalya, Jayachandran Halka, Packiaraj Gurusaravanan, Ravindran Jaganathan, Muthukrishnan Anand, Jesudass Joseph Sahayarayan, and Muthukrishnan Arun

Abstract

The quest for advanced agricultural technologies aims to combat hunger and unreliable food access, exacerbated by a rising global inhabitant, limited arable land, and climate change. These technologies, particularly in developing countries, have the capacity to reduce poverty and improve the access to sufficient, safe, and nutritious food. This chapter evaluates site-directed mutation methods such as vector-based mutation, PCR-based mutation and nucleases mediated mutation. Moreover, this chapter highlights their potential for the genetic improvements in crops. Site-directed mutagenesis has been successfully utilized for inducing targeted mutations in a wide range of crops, comprising staple food crops, oilseeds, and horticultural crops. This technique has successfully enhanced agricultural characteristics like productivity, nutritional value, and resilience to stress. Additionally, site-directed mutagenesis is highly utilized to eliminate selectable markers in transgenic crops. A majority of the research involving site-directed mutagenesis has concentrated on cereals, with horticultural crops following closely behind. Of the various techniques in inducing mutagenesis, CRISPR/Cas9 technology has been the most commonly utilized followed by TALENs. CRISPR/Cas9 genome editing technology is valued for its ability to create precise mutations with minimal off-target effects, making it the preferred choice in numerous studies. TALENs, on the other hand, are generally used to knockout genes associated with undesirable traits. The mutations generated through these techniques have been shown to be stable and heritable across generations. Therefore, the CRISPR/Cas9 system for site-directed mutagenesis could be a promising tool for advancing agricultural development, helping to ensure global food sustainability.

Keywords

Site-directed mutagenesis · Zing finger nucleases · TALENs · CRISPR/Cas9 · Crop improvement

1 Introduction

As global demographics trend toward a population of 10 thousand million by the year 2050, meeting these demands will require significant advancements in crop production along with enhanced nutritional quality (FAO 2017). Exploring potential technologies for crop improvement could play a substantial part in easing poverty and malnourishment particularly in developing countries and globally as a whole. Elimination or reduction of malnutrition among people can be achieved by providing them with high quality food products, and this could be achieved through the implementation of traditional breeding methods and modern technological approaches that can be very well employed to develop improved crop varieties. However, further advancements are needed to enhance nutrition, increase yields, and strengthen pliability against biological and environmental stress factors. Site-directed mutagenesis, a collection of molecular biology

N. Vidya · K. Kowsalya · J. Halka · M. Arun (✉)
Department of Biotechnology, Bharathiar University, Coimbatore, Tamil Nadu, India
e-mail: arun@buc.edu.in

P. Gurusaravanan
Department of Botany, Bharathiar University, Coimbatore, Tamil Nadu, India

R. Jaganathan
Preclinical Department, University Kula Lumpur Royal College of Medicine Perak, Ipoh, Perak, Malaysia

M. Anand
Ramco Institute of Technology, Rajapalayam, Tamil Nadu, India

J. J. Sahayarayan
Department of Bioinformatics, Alagappa University, Karaikudi, Tamil Nadu, India

J. Devi (ed.), *Mutation Breeding in Crop Improvement*, Sustainable Landscape Planning and Natural Resources Management,
https://doi.org/10.1007/978-3-032-17911-1_12

techniques, facilitates precise and targeted changes to an organism's DNA, offering a method to quickly introduce advantageous mutations into crops without incorporating any foreign DNA. Unlike the random mutations generated in conventional mutation breeding, site-directed mutagenesis enables specific and deliberate modifications to the target genes in plants. Moreover, these techniques pave a robust path toward further advancements in agriculture sector.

This approach of site-directed mutagenesis involves precise modifications to specific DNA sequences, allowing scientists to accurately edit genes. By focusing on key genetic loci, researchers can develop edible crops with superior characters namely improved nutritional value, crops with increased crop productivity, and greater tolerance to plant diseases and environmental stress. The potential applications of site-directed mutagenesis extend far beyond basic crop enhancement, encompassing a wide range of agricultural innovations. This technique not only aids in the study of gene functions but also promotes the expansion of novel plant varieties adapted to diverse ecological conditions. By harnessing these genetic advancements, several edible and economically valuable crops can be cultivated to thrive in diverse climates and soil types, contributing to global food resilience and ensuring enduring agricultural viability. Additionally, the precision of gene editing through this method provides opportunities to reduce reliance on synthetic fertilizers and pest control agents, fostering eco-friendly farming practices. The promising applications of site-directed mutagenesis underscore its vital contribution in transforming agricultural production, strengthening resilience, and advancing sustainable farming practices.

2 Site-Directed Mutagenesis and Types

Site-directed mutagenesis has widely employed for inducing mutations one or multiple locations within the genome. This technique involves introduction of specific (desired) alterations in the DNA sequence of any gene, while leaving the rest of the sequence unchanged. Site-directed mutagenesis (SDM) is crucial for exploring the connection between a gene's sequence and its function. By introducing specific modifications, scientists can analyze how these changes impact protein structure and activity, examine gene regulation processes, and design proteins with tailored characteristics. Currently, site-directed mutagenesis can be performed using three different approaches: PCR-based, vector-based, and nuclease-based methods. Methods like CRISPR/Cas9-based genome editing, and oligonucleotide-directed mutagenesis enable scientists to modify specific genes to either activate desirable traits or deactivate undesirable ones. Compared to earlier mutant breeding methods, site-directed mutagenesis offers a more accurate and effective approach for refining crop genotypes. The key benefit of this methodology is its precision in altering genes and regulatory elements, enabling the modification of protein functions or expression patterns. Unlike traditional plant breeding, which relies on random mutagenesis and screening, site-directed mutagenesis enables more targeted and effective advancements.

The core concept of site-directed mutagenesis (SDM) is to create precise mutations at a designated location within a gene, unlike random mutagenesis, which results in changes across the entire DNA sequence. The SDM process generally consists of the following steps (1) Design of mutagenic primer: A synthetic DNA primer is designed to include the intended mutation. This primer matches the target sequence except at the site of mutation, (2) Amplification of mutated DNA: The mutagenic primer is employed in polymerase chain reaction (PCR) to amplify the DNA segment containing the mutation. The DNA polymerase synthesizes the complementary strand, incorporating the mutation, (3) Transformation into host cells: The amplified mutated DNA is introduced into host cells, usually bacteria, allowing replication and expression of the modified gene, (4) Screening and Verification—The mutant DNA is sequenced to verify the successful incorporation of the intended mutation. The modified gene is expressed in the host, and the resulting protein is analyzed for functional changes.

2.1 PCR-Based Site-Directed Mutagenesis

This technique utilizes polymerase chain reaction to enable more effective site-directed mutagenesis without the need for sub-cloning. The megaprimer approach, inverse PCR, and overlap extension PCR are the methods by which site-directed mutagenesis could be achieved through PCR procedures. PCR-based site-directed mutagenesis is an accurate method that employs the polymerase chain reaction (PCR) to introduce targeted mutations into a gene. The process starts with designing a mutagenic primer, a short synthetic DNA strand that matches the target sequence except for the specific mutation introduced. This primer, combined with a high-fidelity DNA polymerase, facilitates the amplification of the DNA segment containing the mutation. During amplification, the polymerase synthesizes new DNA strands while incorporating the designed mutation. After PCR, the amplified mutated DNA is treated with a digestion enzyme to remove the parental (non-mutated) template, ensuring that only the newly synthesized mutant DNA remains. The mutated DNA is then introduced into host cells, such as bacteria, for replication and expression. Finally, sequencing and functional analysis are carried out to check the mutation and assess its effects on gene expression and protein function. This method allows researchers to investigate the functions of genes, protein interactions, and biological processes at the molecular level with

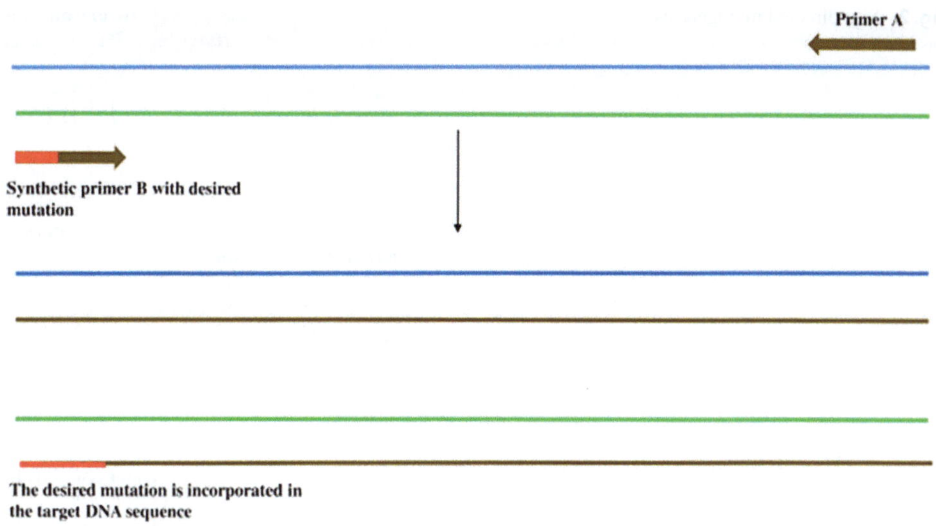

Fig. 1 Site-directed mutagenesis through conventional PCR

high precision. PCR-based methods are simple to use and may alter big constructs with a reasonably high efficiency. However, they call for a number of stages and a unique primer design for every mutation site. The introduction of mutation through conventional PCR is illustrated in Fig. 1.

2.2 Vector-Based Site-Directed Mutagenesis

One of the earliest strategies for site-directed mutagenesis involved the use of vectors. This method includes inserting the target gene into a plasmid vector through sub-cloning, mutating the target gene using techniques like cassette mutagenesis or single-primer mutagenesis, and then screening for mutants after transforming the bacteria. The advantages of vector-based techniques are their affordability and ease of use, but they are generally less effective, and the length of the sequence that can be modified is limited. The site-directed mutagenesis using plasmid cloning vector is presented in Fig. 2.

2.3 Nuclease-Based Site-Directed Mutagenesis

Targeted modifications to specific genes can result in the overexpression or inhibition of desired proteins or characteristics through site-directed mutagenesis. These techniques have been transformed more recently by nuclease-based mutagenesis like Clustered Regularly Interspaced Short Palindromic Repeats-CRISPR-associated proteins (CRISPR/Cas), Transcription activator like endonuclease (TALENs), and Zinc finger nucleases (ZFNs) (Fig. 3). In comparison to earlier methods, these nuclease-based techniques are more efficient and precise, as these tools create double-strand breaks at specific DNA target sites. Owing to its affordability, user-friendliness, and adaptability, CRISPR/Cas9 has had the greatest influence. Hence, this technology has been very well tested in various crops to enhance characteristics namely fruit quality, abiotic stress tolerance, and disease resistance. Even though they are used less frequently these days, TALENs and zinc finger nucleases may still be adapted for tough target areas and have a high specificity. CRISPR/Cas9 has become the leading genetic editing technology for site-directed mutagenesis in crops owing to its affordability, effectiveness, and ability to simultaneously alter multiple genetic targets. The progression of site-directed mutagenesis techniques, from vector-based methods to more advanced PCR and nuclease-driven approaches, has significantly improved the precision and efficiency of genome editing. Nucleases mediated mutations has both advantages and disadvantages. These techniques are highly precise, allowing for targeted modifications to DNA. They can be directed at any gene, providing researchers with significant flexibility in genetic alterations. Additionally, they are relatively user-friendly, making them a viable option for scientists with limited experience in gene editing. However, molecular biology-based technologies can be costly to develop and implement. Moreover, there is a risk of unintended mutations, potentially causing undesirable side effects.

2.3.1 Nucleases-Mediated Mutagenesis Using Zinc Finger Nucleases

Zinc finger nucleases (ZFNs) comprises of a zinc-finger DNA-binding domain connected to a FokI nuclease domain through a linker sequence. These domains are specifically engineered to recognize target gene sequences. A single zinc finger (ZF) binds to a distinctive 3-base-pair (bp) DNA sequence, and multiple zinc fingers (usually 4–6) are combined together to recognize longer sequences spanning

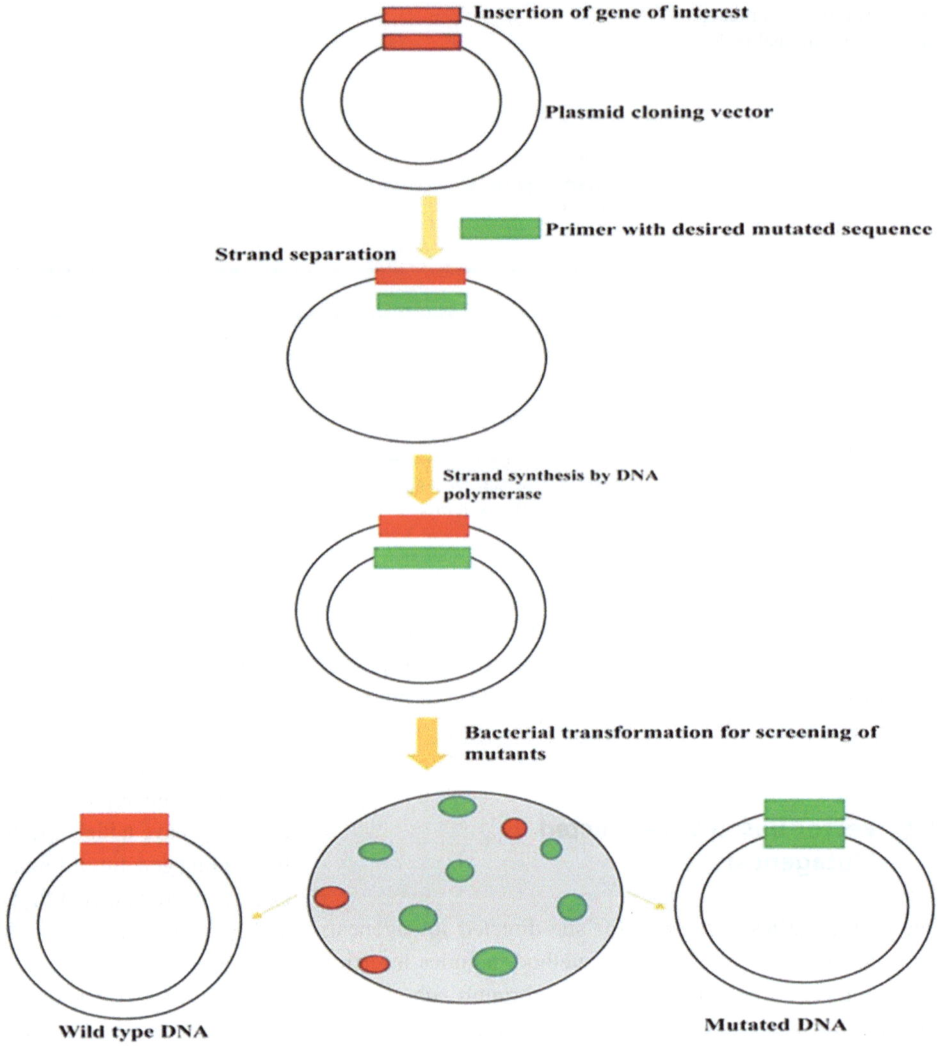

Fig. 2 Site-directed mutagenesis based on vectors

12–18 bp. The FokI endonuclease, a non-specific restriction enzyme, requires dimerization to cleave DNA. To minimize unintended cuts, one half of the FokI enzyme is attached to one DNA strand while the rest of the half is positioned on the complementary strand. This technique has been widely used to introduce precise genomic modifications, leading to mutant plants with gene insertions and deletions. The successful expression of ZFN genes and their precise cleavage activity at specific genomic sites have been demonstrated in several plants comprising of cereals like rice, maize, and in other plants like soybean, and *Arabidopsis*. Even though ZFNs has been successfully implemented in several crops, but it has certain limitations. The requirement to design and assemble a precisely engineered unit compatible with certain amino acids makes this process technically challenging and restricts its broader application. Additionally, ZFN technology is labor-intensive, expensive, and prone to off-target DNA cleavage, potentially causing unintended genetic modifications. Highlighting these drawbacks provides a balanced perspective on its advantages and constraints, especially when comparing it to newer gene-editing technologies like CRISPR/Cas9.

2.3.2 Nucleases-Mediated Mutagenesis Using Transcription Activator like Endonuclease

Other genetic technologies like the transcription activator-like effector nuclease (TALEN) system, induce DNA breaks in both strands similar to zinc finger nucleases (ZFNs). However, unlike ZFNs, which recognize three-base-pair sequences, TALENs target individual nucleotides. TAL effectors originate from *Xanthomonas* species, and their DNA-binding central domain consists of approximately 34 repeating amino acid sequences. The binding specificity of each repeat is dictated by repeat variable diresidues (RVDs) located at positions 12 and 13. Integrating the FokI nuclease with TALEs offers greater accuracy and effectiveness than zinc finger nucleases (ZFNs). This is because TALENs target single

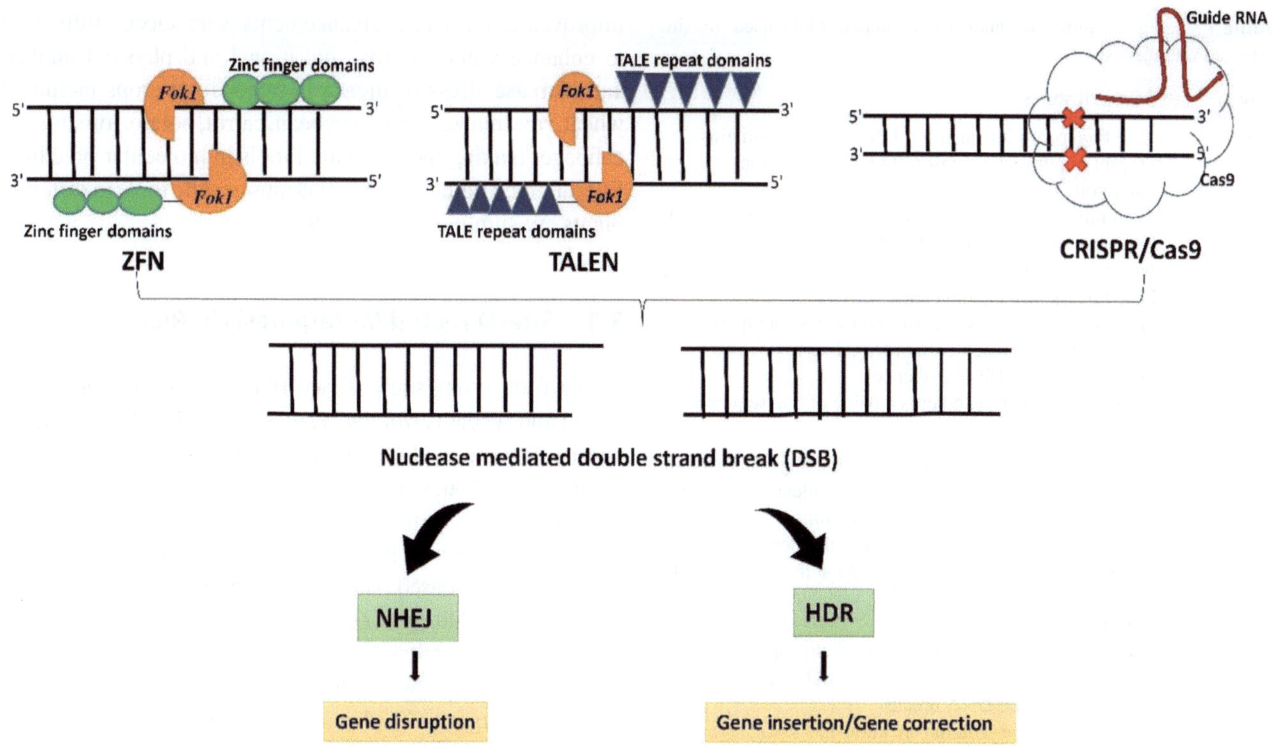

Fig. 3 Site-directed mutagenesis induced by nucleases such as zinc-finger nucleases (ZFNs), transcription activator-like effector nucleases (TALENs), and CRISPR/Cas9 systems can rapidly scan the genome, precisely target specific DNA sequences, and induce double-strand breaks (DSBs) with high efficiency. These breaks are primarily repaired through non-homologous end joining (NHEJ), which can lead to gene disruption, or homology-directed repair (HDR) when a donor DNA template is available, enabling gene correction or insertion

nucleotides rather than three-base-pair sequences, allowing for more specific and flexible genome editing.

2.3.3 Nucleases-Mediated Mutagenesis Using Clustered Regularly Interspaced Short Palindromic Repeats-CRISPR-Associated Proteins

Past decade has witnessed yet another potent technique for precisely altering genes, and has seen rapid advancements. Unlike zinc finger nucleases and transcription activator like endonuclease, which depend on protein-based DNA recognition, the CRISPR/Cas9 system utilizes small, easily customizable guide RNAs (gRNAs) in combination with Cas9 nucleases to create DNA beaks at both the strands. This simplicity in design and construction makes CRISPR/Cas9 a more effectual and accessible technology for genome editing. The most widely use CRISPR/Cas9 approach primarily depend on two key components: Cas9 endonuclease and a single guide RNA (sgRNA). The single guide RNA consists of two small RNA molecules: the CRISPR RNA (crRNA), a 20-nucleotide sequence that complements the target DNA, and the trans-activating CRISPR RNA (tracrRNA), that will offer a binding platform for the Cas9 enzyme. The Cas9 enzyme locates its DNA target through a three-base PAM (protospacer-adjacent motif) sequence, situated directly beside the target site. SpCas9, derived from *Streptococcus pyogenes*, preferentially binds to the PAM sequence 5'NGG3'. The Cas9 enzyme harbors two distinct catalytic domains: RuvC and HNH, that are responsible for cleaving DNA. Upon identifying the target sequence, Cas9 unwinds the DNA, facilitating the binding of the sgRNA to the complementary strand. It then makes a precise cut, or "nick," usually three bases before the PAM sequence, enabling targeted genome editing. After a double-strand break (DSB) occurs, the DNA can be repaired through one of two pathways. The first is non-homologous end joining (NHEJ), which frequently leads to insertions or deletions (INDELs) at the target site, often leading to disruptions in gene. The second is homology-directed repair (HDR), which enables precise modifications in genome, like gene knock-ins, by utilizing a homologous DNA template. The salient features of nucleases mediated mutagenesis is summarized in Table 1. The site-directed mediated mutation by nucleases is illustrated in Fig. 3.

Table 1 Salient features of nuclease mediated tools used in site-directed mutagenesis

Tools	Salient features
ZFN	Zinc fingers are fundamental DNA-binding domains present in numerous transcription factors within eukaryotic cells Each finger consists of an approximately 30-amino-acid ββα structure, which usually interacts directly with three base pairs of DNA Zinc fingers lack endonuclease activity Endonuclease activity is achieved by FokI, a bipartite restriction endonuclease that recognizes a non-palindromic DNA sequence Induces double-stranded cuts outside of that sequence, resulting in a 5' overhang
TALEN	TALENs are a group of naturally occurring transcription factors that were first discovered as virulence factors in the bacterial plant pathogen *Xanthomonas* TAL effectors recognize DNA through a series of 34-amino-acid repeats that vary only at the 12th and 13th amino acids, known as the repeat variable di-residues (RVDs) TAL effectors lack endonuclease activity and are typically used in pairs, fused to the FokI cleavage domain, to induce DNA breaks These engineered proteins are known as TAL effector-like nucleases (TALENs)
CRISPR/ Cas9	Derived from bacteria and archaea Requires CRISPR RNA (crRNA), and tracrRNA, and forms a complex with the CRISPR-associated endonuclease (Cas9).This complex directs Cas9 to a 20 bp complementary DNA sequence for targeted binding and cleavage DNA breaks occur at complementary target sites that include the correct protospacer adjacent motifs (PAMs) These PAM sequences ensure cleavage occurs only at the intended target site, preventing self-cleavage by the host Cas9 has inherent endonuclease activity, it does not require additional domainsto induce a DNA break

3 Application of Site-Directed Mutagenesis in Genetic Improvement of Crop

Meeting the demand of escalating global populace presents an increasing challenge, with a major portion of the solution lying in the establishment of novel crop varieties that offer greater crop productivity and improved stress tolerance. With the increasing demand to enhance worldwide agricultural production, there is an urgent requirement to expedite the incorporation of good qualities into crops. Among the diverse methods for improving crop, genome editing tools emerge as a highly effective strategy for developing new traits. In recent years, various techniques have been adopted to introduce targeted mutations, either to enhance beneficial traits or eliminate undesirable ones in crops. The swift progress in this field has made it possible to create effectual, accurate, and cheaper methods for producing crops with improved traits. These advancements were successfully used to enhance yield, improve nutritional and physical quality, and increase stress resilience in major food crops including wheat, rice, maize, barley, soybean, carrot, potato, tomato, and cabbage, making them essential tools in molecular breeding. Site-directed mutagenesis by various genetic tools in different plants are summarized in Table 2.

3.1 Site-Directed Mutagenesis in Rice

Staple food crops such as rice (*Oryza sativa*), maize (*Zea mays*) and wheat (*Triticum aestivum*) are globally consumed and its demand is always rising every year. There are many studies carried out in rice to improve the quality of grains and rice plants. Mutagenesis via TALENS has been reported in rice. Rice variety with high-tryptophan content was developed using a vector-based site-directed mutagenesis approach. Accurate gene editing of *OASA2*, which encodes the α-subunit of anthranilate synthase, a crucial enzyme in rice tryptophan (Trp) biosynthesis, was achieved. As a result, mature seeds from homozygous GT plants accumulated Trp at levels 230 times higher than those in control plants. However, there was no noticeable morphological changes (Saika et al. 2011). Another attractive quality of rice is its aroma, which is one of the major factors that consumers look for, and the aroma of rice increases its marketability. Highly fragrant rice varieties were successfully created using TALENs by editing the *OsBADH2* gene, thereby triggering the biosynthesis of 2-acetyl-1-pyrroline (2AP), the main compound contributing to rice aroma. TALENs were designed to specifically target and disrupt the *OsBADH2* gene, resulting in the recovery of six T_0 heterozygous *BADH2* mutant rice plants. The mutant plants exhibited an increase in 2-acetyl-1-pyrroline (2AP), from 0 to 0.35–0.75 mg/kg (Shan et al. 2015). TALEN-based editing to modify the dominant TFIIAγ5 allele in rice, created mutants that had reduced susceptibility to *Xanthomonas oryzae* (Xoo) (Han et al. 2020). ZFNs were used to create double-stranded break in SSIVa, a soluble starch synthase that plays a role in the starch biosynthesis pathway, thus enabling the regulation of SSIVa expression. Furthermore, the introduction of premature stop codons and point mutations in transgenic lines led to the loss of SSIVa transcript expression, significantly decreased starch content, and pronounced dwarfism (Jung et al. 2018). In addition to TALENs and ZFN, CRIPSR/Cas9 genome editing technique is largely used in rice to improve its morphological and physiological traits. Genome editing of *BADH2* gene in rice through CRISPR/Cas9 has enhanced the fragrance (Imran et al. 2023). Beyond enhancing the aroma of rice, CRISPR-Cas9-mediated editing has been done to increase the yield namely editing in genes such as GRAIN NUMBER 1A (Gn1a), DENSE AND ERECT PANICLE 1 (*DEP1*), GRAIN SIZE 3 (*GS3*), IDEAL PLANT ARCHI-

Table 2 Genetic improvement in crop through site-directed mutagenesis

Crop	Genetic tool	Target gene	Trait improved
Rice	ZFN	*SSIVa*	Low Starch content and dwarf phenotype
	TALEN	*OsCSA, OsPMS3, OsDERF1*	Detection of mutations in the
		OsGN1a, OsTAD1, OsMST7, and *OsMST8*	Target genes Trait improvement is not mentioned
	CRISPR/Cas9	*BADH2*	Fragrance
Rice	CRISPR/Cas9	*OsLCD*	Low cadmium content
Maize	ZFN	IPK1	Herbicide tolerance and alteration of the inositol phosphate profile
	CRISPR/Cas9	*PSY1*	White colored seeds
Wheat	ZFN	*AHAS*	Resistance to imidazolinone herbicides
	TALEN	*UidA and lr21Ψ*	Detection of mutation
	CRISPR/Cas9	*DREB2,ERF3,and TSalI*	Drought tolerance genes
Barley	CRISPR/Cas9	*Gbss1a and Pts1l*	Composition of starch
	CRISPR/Cas9	*HPT* and *HGGT*	Vitamin E
	CRISPR/Cas9	*COMT1*	Lignin content
Tomato	ZFN	*L1L4*	Phenotypic traits and improved nutritional content
	TALEN	*PROCERA*	Increased Gibberellin response
	TALEN CRISPR	*ANT1*	Detection of mutation
Carrot	CRISPR	*F3H*	Blockage of the anthocyanin biosynthesis
Potato	TALEN	*SSR2*	Reduced steroidal glycosides
	CRISPR	*StDMR6*	Increased resistance to early blight disease
Soybean	ZFN	*DCL*	Detection of mutation
	TALEN CRISPR	*GmPDS11, GmPDS18*	Albino phenotype
	TALEN	*FAD2-1A FAD2-1B*	Improved seed oil quality
	CRISPR	*GmTAP1*	High oleic acid
	CRISPR	*GmPDCT*	Resistance to blast disease

TECTURE 1 (*IPA1*) (Li et al., 2016), and GRAIN WEIGHT 2 (*GW2*) (Zhou et al. 2019). Thus, highlighting the potential of CRISPR/Cas9 as an effective tool for creating novel alleles that enhance crop productivity. Recently, CRISPR/Cas9 genome editing of NF-YC4 promoters resulted in 68% rise in leaf protein and a 17% increase in seed protein in rice (Wang et al. 2025b).

3.2 Site-Directed Mutagenesis in Maize

Maize (*Zea mays*), the world's most widely produced crop, with the highest production at the global level, serves as a staple for human consumption, a primary source of animal feed, and a key component in biofuel production. Maize is a globally consumed staple crop, and extensive research has focused on introducing targeted mutations in it through various site-directed mutagenesis techniques. Lys606, a highly conserved lysine residue in maize C4-form phosphoenolpyruvate carboxylase (PEPC), was mutated to Asn (Aspargine), Glu (Glutamine), or Arg (Arginine) using site-directed mutagenesis. However, replacing Lys606 with Asn or Glu significantly reduced the maximum velocities (Vmax) to 22% and 2% of the wild-type PEPC, respectively. Furthermore, mutant enzymes showed enhanced activation by glucose 6-phosphate and glycine (2–3 times higher than wild-type) (Dong et al. 1997). Substituting threonine with serine through double-stranded plasmid vector site-directed mutagenesis enhanced the regulatory capacity of the pyruvate orthophosphate dikinase (PPDK) enzyme in maize, while the catalytic efficiency remained unaffected (Chastain et al. 1997). TALENs were used to create stable, heritable mutations at the maize glossy2 (gl2) locus. Transgenic lines with mono- or di-allelic mutations were generated from the Hi-II maize genotype at a frequency of approximately 10% (nine mutated events out of 91 transgenic events) (Char et al. 2015). CRISPR/Cas9 genome editing of *ZmACO2* has the potential to greatly enhance traits such as ear length, kernel number per row, ear weight, and overall grain yield in hybrid varieties (Ning et al. 2021). As consumer interest in specialty corn types like sweet, waxy, and baby corn, continues to grow, CRISPR/Cas9 technology presents a powerful tool for developing these unique germplasms (Dong et al. 2019). For example, simultaneously targeting *ZmBADH2a* and *ZmBADH2b* can lead to the creation of aromatic corn through increased production of 2-acetyl-1-pyrroline (Wang et al. 2021). Additionally, super sweet and waxy maize has been developed through CRISPR-Cas9 editing of the *SH2* and *WX* genes (Dong et al. 2019). Male sterile lines have been created using CRISPR-Cas9 technology by targeting male fertility genes such as *ZmTMS5*, *ZmMs7*, *Ms8*, *Ms26*, and *Ms45* (Chen et al. 2018). In addition, CRISPR technique has been employed to impart disease resistance in maize. CRISPR-Cas9 targeting of *ZmGDIa* significantly enhances maize resistance to maize rough dwarf virus without any agronomic drawbacks (Liu et al. 2022). *ZmCOIa*, by interacting with *ZmJAZ15*, modulates maize immunity to Gibberella stalk rot (GSR), caused by *Fusarium graminearum*. Knocking out *ZmCOIa* boosts maize resistance to GSR (Ma et al. 2021). These reports highlight the prospective of gene editing tools for developing cultivars with improved tolerance to insects, fungus and bacteria.

3.3 Site-Directed Mutagenesis in Wheat

Several studies have explored the use of CRISPR/Cas9 gene-editing technology to enhance wheat (*Triticum aestivum*) grain yield-related traits. Targeted mutations were introduced in four genes known to negatively regulate grain development (*TaCKX2-1*, *TaGLW7*, *TaGW2*, and *TaGW8*), including a 1160 bp homozygous deletion in the *TaCKX2-D1* gene, which significantly increased the number of grains per spikelet (Zhang et al. 2019). Silencing the *TaGW7* gene and introducing mutations in the B or D genome (or both) led to an increase in both grain width and weight. Double-copy mutants exhibited greater yield improvements compared to single-copy mutants (Wang et al. 2019). Additionally, CRISPR/Cas9 was used for site-specific deletions targeting three genes, *TaABCC6*, *TaNFX1*, and *TansLTP9.4* in wheat. Among these, deletions were successfully induced in two genes, with TaNFXL1 showing the largest deletion, demonstrating its potential adaptability (Cui et al. 2019). In addition to morphological traits, physiological traits such as abiotic stress tolerance has also been improved through CRISPR/Cas9 technology in maize. In this regard, CRISPR/Cas9-mediated editing of *ZmPL1* gene improved tolerance to drought stress in maize (Wang et al. 2025a).

3.4 Site-Directed Mutagenesis in Barley

Only few studies have appraised the effectiveness of site-directed mutagenesis and its transfer to subsequent generations using TALEN and CRISPR/Cas9 systems in barley (*Hordeum vulgare*). The first transformation, utilizing TALEN, involved gene knockout in barley through pollen-regenerable cells, establishing true-breeding generations. A green fluorescent protein (GFP)-specific TALEN, delivered via *Agrobacterium*-mediated gene delivery, resulted in 22% homologous primary mutants and successfully knocked out the GFP gene through nucleotide deletions (4-36bp) (Gurushidze et al. 2014). Another study used TALENs to target the GFP gene, leading to a single amino acid change and also targeted for the production of yellow fluorescent protein (YFP) through site-specific mutations (Budhagatapalli et al. 2015). Barley was also modified using CRISPR/Cas9, achieving a high mutation rate of 88% in single-gene editing of the cytokinin oxidase/dehydrogenase *HvCKX1* gene in T_0 plants. Multiplex editing of *HvCKX1* and *HvCKX3* led to successful edits in 21% of the plants. Furthermore, knockout of the Nud gene produced hull-less barley grains with noticeable phenotypic alterations, simplifying post-harvest processing. These mutations were stably inherited by the T1 generation (Gasparis et al. 2018).

3.5 Site-Directed Mutagenesis in Soybean

Soybean (*Glycine max*) is one of the major crops used for human consumption, animal feed, and also used for industrial purpose. Site-directed mutagenesis is done to enhance several morphological and physiological traits in soybean to make it more marketable. Enhancing the quality of soybean oil was achieved through targeted mutagenesis of the *FAD2-1A* and *FAD2-1B* fatty acid desaturase gene families using TALEN technology (Haun et al. 2014). In homozygous mutant seeds of FAD2-1A and FAD2-1B, oleic acid levels (a monounsaturated omega-9 fatty acid, 18:1 cis-9) rose significantly from 20% to 80%. Meanwhile, linoleic acid, a polyunsaturated omega-6 fatty acid, dropped from 50% to 4%. Moreover, the mutation was confirmed to be heritable (Haun et al. 2014). These desaturase enzymes catalyze the removal of hydrogen from fatty acids, increasing polyunsaturation, which may pose risks to cardiovascular and neurological health. (Schattenberg and Bergheim 2019). Extensive studies have been carried out in soybean via CRISPR/Cas9 technology. Improving soybean oil quality using the CRISPR/Cas9 system to edit the FAD2-2 gene reported a mutation rate of 21%. The results showed a significant increase in oleic acid content, reaching up to 65.58%, while linoleic acid levels were reduced to 16.08%, indicating enhanced oil quality (Al Amin et al. 2019). Multiplex gene editing with the CRISPR/Cas9 system was used to target three genes in soybean like *GmF3H1*, *GmF3H2*, and *GmFNSII-1*, which act as negative regulators of isoflavone biosynthesis. These genes were successfully knocked out, achieving a triple mutation efficiency of 44.44%. In the T3 generation, plants with triple-gene homologous mutations demonstrated a two-times higher isoflavone accumulation in their leaves. This improved isoflavone production also contributed to increased resistance against soybean mosaic virus (Zhang et al. 2020). In Addition to nutritional improvement, morphological traits have been improved. For example, recent research on developing climate-resilient soybean has focused on modifying the time of flowering via targeted mutagenesis based on CRISPR/Cas9 system. In this study, the soybean cultivar Jack was genetically altered by inducing mutations at a specific site in the *GmFT2a* gene (positioned on chromosome 16 with four exons). In the T1 generation, homozygous plants carrying null alleles of *GmFT2a* were produced through frameshift mutations which showed 1 base pair insertion or a short deletion, which effectively led to knock out of *GmFT2a*. Consequence, these plants exhibited a delayed flowering period, enabling them to better withstand environmental stress. Furthermore, the mutation was confirmed to be heritable (Cai et al. 2018). Soybean β-conglycinin is a primary allergen that negatively impacts the nutritional quality of soybean, posing challenges for its consumption and use in food products. Lately, CRISPR/Cas9 genome editing technique was employed to knock out lincRNA gene which is associated with β-conglycinin (Song et al. 2024). Currently, as done in rice, a 25% boost in leaf protein and an 11% increase in seed protein in soybean was observed by CRISPR/Cas9 editing of *NF-YC4* promoters (Wang et al. 2025b).

3.6 Site-Directed Mutagenesis in Tomato

Tomato is consumed globally and is widely used in different cooking methods. Hence, the demand for high quality tomatoes are increasing in the global market. Site-directed mutation is a promising tool for enhancing the nutritional value and other desirable traits in tomatoes. One key marketable trait that can be improved through this technique is an extended shelf life. Tomato shelf life was extended through CRISPR/Cas9 genome editing, delivered via *Agrobacterium*-mediated transformation. Genetic modifications, including deletions and insertions, were introduced in the RIN gene, which encodes an MADS-box transcription factor involved in fruit ripening. Mutations in the RIN protein altered ripening physiology and ethylene production, enabling tomatoes to stay fresh for several months. Mutations induced by CRISPR/Cas9 editing through *Agrobacterium*-mediated gene delivery, has been reported for both single- and multi-site mutations. This technology holds promise for precise genetic modifications in important traits of tomatoes and other crops. Tomatoes are highly susceptible to powdery mildew; however, CRISPR/Cas9-mediated editing of the *DND1* gene has enhanced their resistance to this disease (Li et al. 2024).

3.7 Site-Directed Mutagenesis in Potato

Potato is a globally grown tuber crop that serves as a staple ingredient in diverse cuisines. It is in high demand for making potato chips, one of the most commercially popular products globally. Cold storage is usually employed to minimize sprouting and extend the shelf life during the post-harvest period of potato tubers. Though, low temperatures stimulate accumulation of sugar in the tubers. TALENs have been used to knock out the *VInv* gene, leading to a significant reduction in reducing sugars. The resulting chips exhibited reduced acrylamide content and a lighter coloration (Clasen et al. 2016). TALENs mediated targeted DNA insertion was successfully done in potato with high frequency and also resulted in considerable expression of the target gene (Forsyth et al. 2016). Targeted editing of the *stALS1* gene using the CRISPR/Cas9 system resulted in mutation rates between 3% and 60%, with successful transmission of mutations to the next generation (Butler et al. 2015). CRISPR/Cas9-mediated mutagenesis was used to modify starch quality by targeting the *GBSS* gene. CRISPR/Cas-mediated knockout of the *GBSS* gene in tetraploid potato protoplasts successfully induced mutations in all four alleles. Editing was directed at three loci within the granule-bound starch synthase gene, with mutation frequencies ranging from 2–12% in individual alleles and multiplex mutation rates reaching up to 67%. The complete loss of GBSS enzymatic function significantly altered starch composition, as evidenced by reduced amylose synthesis and a corresponding elevation in the amylopectin/amylose ratio (Andersson et al. 2017). Norouzi et al. (2024) has reported that targeted mutation of the *StNRL1* gene using CRISPR/Cas9 technology improved late blight resistance while simultaneously heightening susceptibility to early blight in potatoes.

3.8 Site-Directed Mutagenesis in Carrot

The genetic improvement through site-directed mutagenesis is very limited in carrot (*Daucus carota*). Klimek-Chodacka et al. (2019) successfully knocked out the *F3H* gene, which is responsible for anthocyanin pigment synthesis, leading to the development of white carrots. This mutation eliminated the purple pigmentation by deleting and suppressing anthocyanin expression. Similarly, Xu et al. (2019) reported that the targeted knockout of the *DcPDS* gene in orange carrot using CRISPR/Cas9 resulted in albino phenotypes, with a mutation efficiency of 35.3%. In parallel, CRISPR-mediated editing of the *DcMYB113-like* gene in purple carrot induced significant depigmentation, with a mutation rate of 36.4%.

3.9 Site-Directed Mutagenesis in Cabbage

Cabbage is widely cultivated across the world, with its leaves being the primary edible part of the vegetable. Very few studies have been currently reported about site-directed mutagenesis in cabbage. A customized TALEN-based nuclease was constructed through the "unit assembly" method, to specifically target the endogenous *FRIGIDA* gene in cabbage. This approach successfully introduced a deletion at the target site, demonstrating the effectiveness of site-directed mutagenesis for genetic modification. Furthermore, Ma et al. (2019) also reported that knocking out the *BoPDS* gene in cabbage produced albino shoots, with a mutation frequency of 1.14%.

4 Limitations and Challenges

Site-directed mutagenesis is essential for the production of elite variety of crops as it allows precise modifications to target DNA sequences, leading to changes in expression of genes and improved crop characteristics. In contrast to recombinant DNA technology, which randomly integrates foreign genes, site-specific mutagenesis alters genes at predetermined sites. Random gene insertion can sometimes silence important genes or cause unintended gene expression. However, not all existing technologies are equally effective in achieving genetic improvements through site-directed mutagenesis. Over time, advancements in targeted mutation methods have led to the development of more efficient techniques. Traditional approaches, such as vector-based and PCR-based site-directed mutagenesis, have become less common in plant transformation due to less efficiency and limited impact on agricultural traits. As a result, newer and more effective technologies have been introduced for crop transformation through site-directed mutagenesis. Despite these advancements, newly developed mutagenic tools still face challenges in achieving highly efficient and precise genetic alterations.

Zinc Finger Nucleases (ZFNs) have been utilized for site-directed mutagenesis but pose several challenges, including off-target effects, difficulties in delivery owing to their large size, dimerization issues, higher costs, and labor-intensive processes. As an alternative, Transcription Activator-Like Effector Nucleases (TALENs), is more efficient, simpler to construct, and less expensive than ZFNs for creating site-specific deletions or insertions. However, TALENs exhibit dimerization and off-target effects, although at lower levels than ZFNs and CRISPR/Cas9. Among these technologies, the CRISPR/Cas9 gene editing tool offers numerous benefits over both ZFNs and TALENs. Its short single-guide RNA sequence (sgRNA) facilitates easy delivery, lowers costs, reduces labor requirements, and enhances efficiency. However, despite these benefits, CRISPR/Cas9 is not entirely free from limitations, as it can still result in off-target effects.

5 Conclusion

The use of site-directed mutagenesis for improving crops have immense potential to enhance the agricultural sector, which can directly boost the economies of any country. The CRISPR-based system. From the studies, it is evident that the CRISPR-based system is the most frequently utilized method (66.67%) for improving crop traits through targeted mutagenesis. TALENs, on the other hand, were employed to knock out genes associated with undesirable traits. Owing to its efficiency and fewer off-target effects, CRISPR/Cas9 technology is widely utilized to improve crops and promote food sustainability. Numerous agronomic, physiological, and stress-resistant traits in crops have been enhanced using this technology. Site-directed mutagenesis has proven to be especially beneficial for cereal crops, which are difficult to transform with recombinant DNA technology. Maize and rice are among the most frequently transformed cereal crops using site-directed mutagenesis. The overall trend shows that CRISPR/Cas9 is the dominant technology used globally for efficient crop transformation and improvement. Additionally, TALENs are considered ideal for knocking out genes responsible for undesirable traits. By targeting and improving key staple crops, gene editing technologies play a vital role in strengthening worldwide food stability.

Acknowledgements The authors are thankful to the Department of Science and Technology (DST), Ministry of Science and Technology, Government of India (Reference no: SR/WOS-A/LS-405/2018), for the financial support provided to Vidya Nandakumar to carry out the present work.

References

Al Amin N, Ahmad N, Wu N, et al (2019) CRISPR-Cas9 mediated targeted disruption of FAD2– 2 microsomal omega-6 desaturase in soybean (*Glycine max*.L). BMC Biotechnol 19:9. https://doi.org/10.1186/s12896-019-0501-2

Andersson M, Turesson H, Nicolia A et al (2017) Efficient targeted multiallelic mutagenesis in tetraploid potato (*Solanum tuberosum*) by transient CRISPR-Cas9 expression in protoplasts. Plant Cell Rep 36:117–128. https://doi.org/10.1007/s00299-016-2062-3

Budhagatapalli N, Rutten T, Gurushidze M et al (2015) Targeted modification of gene function exploiting homology-directed repair of TALEN-mediated double-strand breaks in barley. G3 Genes|Genomes|Genetics 5:1857–1863. https://doi.org/10.1534/g3.115.018762

Butler NM, Atkins PA, Voytas DF, Douches DS (2015) Generation and inheritance of targeted mutations in potato (*Solanum tuberosum* L.) Using the CRISPR/Cas System. PLoS One 10:e0144591. https://doi.org/10.1371/journal.pone.0144591

Cai Y, Chen L, Liu X et al (2018) CRISPR/Cas9-mediated targeted mutagenesis of GmFT2a delays flowering time in soya bean. Plant Biotechnol J 16:176–185. https://doi.org/10.1111/pbi.12758

Char SN, Unger-Wallace E, Frame B et al (2015) Heritable site-specific mutagenesis using TALEN s in maize. Plant Biotechnol J 13:1002–1010. https://doi.org/10.1111/pbi.12344

Chastain CJ, Lee ME, Moorman MA et al (1997) Site-directed mutagenesis of maize recombinant C 4-pyruvate,orthophosphate dikinase at the phosphorylatable target threonine residue. FEBS Lett 413:169–173. https://doi.org/10.1016/S0014-5793(97)00884-3

Chen R, Xu Q, Liu Y et al (2018) Generation of transgene-free maize male sterile lines using the CRISPR/Cas9 system. Front Plant Sci 9. https://doi.org/10.3389/fpls.2018.01180

Clasen BM, Stoddard TJ, Luo S et al (2016) Improving cold storage and processing traits in potato through targeted gene knockout. Plant Biotechnol J 14:169–176. https://doi.org/10.1111/pbi.12370

Cui X, Balcerzak M, Schernthaner J et al (2019) An optimised CRISPR/Cas9 protocol to create targeted mutations in homoeologous genes and an efficient genotyping protocol to identify edited events in wheat. Plant Methods 15:119. https://doi.org/10.1186/s13007-019-0500-2

Dong L-Y, Ueno Y, Hata S, Izui K (1997) Effects of Site-directed Mutagenesis of Conserved Lys606 Residue on Catalytic and Regulatory Functions of Maize C4-form Phosphoenolpyruvate Carboxylase. Plant Cell Physiol 38:1340–1345. https://doi.org/10.1093/oxfordjournals.pcp.a029127

Dong L, Qi X, Zhu J et al (2019) Supersweet and waxy: meeting the diverse demands for specialty maize by genome editing. Plant Biotechnol J 17:1853–1855. https://doi.org/10.1111/pbi.13144

Forsyth A, Weeks T, Richael C, Duan H (2016) transcription activator-like effector nucleases (TALEN)-mediated targeted DNA insertion in potato plants. Front Plant Sci 7. https://doi.org/10.3389/fpls.2016.01572

Gasparis S, Kała M, Przyborowski M et al (2018) A simple and efficient CRISPR/Cas9 platform for induction of single and multiple, heritable mutations in barley (Hordeum vulgare L.). Plant Methods 14:111. https://doi.org/10.1186/s13007-018-0382-8

Gurushidze M, Hensel G, Hiekel S et al (2014) True-breeding targeted gene knock-out in barley using designer TALE-nuclease in haploid cells. PLoS ONE 9:e92046. https://doi.org/10.1371/journal.pone.0092046

Han J, Xia Z, Liu P et al (2020) TALEN-based editing of TFIIAy5 changes rice response to Xanthomonas oryzae pv. Oryzae. Sci Rep 10:2036. https://doi.org/10.1038/s41598-020-59052-w

Haun W, Coffman A, Clasen BM et al (2014) Improved soybean oil quality by targeted mutagenesis of the fatty acid desaturase 2 gene family. Plant Biotechnol J 12:934–940. https://doi.org/10.1111/pbi.12201

Imran M, Shafiq S, Tang X (2023) CRISPR-Cas9 -mediated editing of *BADH2* gene triggered fragrance revolution in rice. Physiol Plant 175. https://doi.org/10.1111/ppl.13871

Jung Y-J, Nogoy FM, Lee S-K et al (2018) Application of ZFN for site directed mutagenesis of rice SSIVa gene. Biotechnol Bioprocess Eng 23:108–115. https://doi.org/10.1007/s12257-017-0420-9

Klimek-Chodacka M, Oleszkiewicz T, Baranski R (2019) Visual assay for gene editing using a CRISPR/Cas9 system in carrot cells, pp 203–215. https://doi.org/10.1007/978-1-4939-8991-1_15

Li J, Zhang H, Si X et al (2017) Generation of thermosensitive male-sterile maize by targeted knockout of the ZmTMS5 gene. J Genet Genomics 44:465–468. https://doi.org/10.1016/j.jgg.2017.02.002

Li M, Li X, Zhou Z et al (2016) Reassessment of the four yield-related genes Gn1a, DEP1, GS3, and IPA1 in rice using a CRISPR/Cas9 system. Front Plant Sci 7. https://doi.org/10.3389/fpls.2016.00377

Li R, Cui L, Martina M et al (2024) Less is more: CRISPR/Cas9-based mutations in DND1 gene enhance tomato resistance to powdery mildew with low fitness costs. BMC Plant Biol 24:763. https://doi.org/10.1186/s12870-024-05428-3

Liu C, Kong M, Yang F et al (2022) Targeted generation of Null Mutants in ZmGDIα confers resistance against maize rough dwarf disease without agronomic penalty. Plant Biotechnol J 20:803–805. https://doi.org/10.1111/pbi.13793

Ma C, Liu M, Li Q et al (2019) Efficient BoPDS gene editing in cabbage by the CRISPR/Cas9 system. Hortic Plant J 5:164–169. https://doi.org/10.1016/j.hpj.2019.04.001

Ma L, Sun Y, Ruan X et al (2021) Genome-wide characterization of jasmonates signaling components reveals the essential role of ZmCOI1a-ZmJAZ15 action module in regulating maize immunity to Gibberella stalk rot. Int J Mol Sci 22:870. https://doi.org/10.3390/ijms22020870

Ning Q, Jian Y, Du Y et al (2021) An ethylene biosynthesis enzyme controls quantitative variation in maize ear length and kernel yield. Nat Commun 12:5832. https://doi.org/10.1038/s41467-021-26123-z

Norouzi M, Nazarain-Firouzabadi F, Ismaili A et al (2024) CRISPR/Cas StNRL1 gene knockout increases resistance to late blight and susceptibility to early blight in potato. Front Plant Sci 14. https://doi.org/10.3389/fpls.2023.1278127

Saika H, Oikawa A, Matsuda F et al (2011) Application of gene targeting to designed mutation breeding of high-tryptophan rice. Plant Physiol 156:1269–1277. https://doi.org/10.1104/pp.111.175778

Schattenberg JM, Bergheim I (2019) Nutritional intake and the risk for non-alcoholic fatty liver disease (NAFLD). Nutrients 11:588. https://doi.org/10.3390/nu11030588

Shan Q, Zhang Y, Chen K et al (2015) Creation of fragrant rice by targeted knockout of the Os BADH 2 gene using TALEN technology. Plant Biotechnol J 13:791–800. https://doi.org/10.1111/pbi.12312

Song B, Luo T, Fan Y et al (2024) Generation of new β-Conglycinin-Deficient Soybean lines by editing the lincRNA lincCG1 using the CRISPR/Cas9 system. J Agric Food Chem 72:15013–15026. https://doi.org/10.1021/acs.jafc.4c02269

Wang C, Zhou Y, Wang Y et al (2025a) CRISPR-Cas9-mediated editing of ZmPL1 gene improves tolerance to drought stress in maize. GM Crops Food 16:1–16. https://doi.org/10.1080/21645698.2024.2448869

Wang L, O'Conner S, Tanvir R et al (2025b) CRISPR /Cas9-based editing of *NF-YC4* promoters yields high-protein rice and soybean. New Phytol 245:2103–2116. https://doi.org/10.1111/nph.20141

Wang W, Pan Q, Tian B et al (2019) Gene editing of the wheat homologs of TONNEAU 1- recruiting motif encoding gene affects grain shape and weight in wheat. Plant J 100:251–264. https://doi.org/10.1111/tpj.14440

Wang Y, Liu X, Zheng X et al (2021) Creation of aromatic maize by CRISPR/Cas. J Integr Plant Biol 63:1664–1670. https://doi.org/10.1111/jipb.13105

Xu Z-S, Feng K, Xiong A-S (2019) CRISPR/Cas9-mediated multiply targeted mutagenesis in orange and purple carrot plants. Mol Biotechnol 61:191–199. https://doi.org/10.1007/s12033-018-00150-6

Zhang P, Du H, Wang J et al (2020) Multiplex CRISPR/Cas9-mediated metabolic engineering increases soya bean isoflavone content and resistance to soya bean mosaic virus. Plant Biotechnol J 18:1384–1395. https://doi.org/10.1111/pbi.13302

Zhang Z, Hua L, Gupta A et al (2019) Development of an *Agrobacterium* -delivered CRISPR /Cas9 system for wheat genome editing. Plant Biotechnol J 17:1623–1635. https://doi.org/10.1111/pbi.13088

Zhou J, Xin X, He Y et al (2019) Multiplex QTL editing of grain-related genes improves yield in elite rice varieties. Plant Cell Rep 38:475–485. https://doi.org/10.1007/s00299-018-2340-3

Mutation Breeding for Yield and Quality Improvement: A Case Study of Rice

Shedbale Namrata, Kamble Sadhana, Londhe Ranjit, Dhole Vinod, Bapat Vishwas, and Gaikwad Nikhil

Abstract

Rice (*Oryza sativa* L.) is a major staple food crop which serve as important food source for more than half of the global population. Improving grain quality, along with production, remains a primary concern to meet the food demands of the world's growing population. Methods for crop improvement were needed now more than ever due to the continuous push to enhance the quality of grains as well as yield in rice. Mutation breeding, which involves inducing genetic changes in plants using different mutagens to improve plant varieties, has been proven to be an effective tool in plant improvement. Unlike traditional breeding, mutation breeding can add variety to a crop without changing the genetic background of the successful variety. To understand the process of mutation breeding, this case study focuses on exploring the relationship between mutation breeding and yield and quality improvement in rice. The findings supported the utilization of mutation breeding as a viable and efficient approach to manage both concerns with stable rice food supply and the need for qualitative improvement of rice.

Keywords

Crop · Mutagens · Mutation breeding · Rice · Quality · Yield

S. Namrata · K. Sadhana · L. Ranjit · G. Nikhil (✉)
Department of Botany, Shivaji University, Kolhapur, India
e-mail: nbgaikwadsu@gmail.com

D. Vinod
Nuclear Agriculture and Biotechnology Division, Bhabha Atomic Research Centre, Mumbai, India

B. Vishwas
Department of Biotechnology, Shivaji University, Kolhapur, India

1 Introduction

Rice is a monocot plant species classified in the genus *Oryza* of the order Poaceae. The global rice production is nearly 95% reported from Asian countries (Bhattacharjee et al. 2002). Rice is a primary food source for two-thirds of the population, in addition to being a significant caloric provider (Monks et al. 2013). There are over 40,000 rice varieties, of which most of us know and are familiar with only a few (Kushwaha 2016). With an increase in food demands on global food systems, there is an effort to enhance rice yield and grain quality characteristics in rice breeding programs. The fast-growing global population, urbanization, changing food habits and other emerging demographic changes are all stimulating higher international demand for food from minimal and potentially diverse sources; therefore, it is essential to step-up rice production with greater attention on grain quality characteristics and specific traits, i.e., aroma, texture, cooking quality and nutritional quality.

Conventional breeding has limitations in generating the amount of genetic variability required for significant rice improvement. Mutagenesis is a powerful breeding tool that utilizes physical or chemical agents to induce heritable changes in the genetic background, thereby introducing new genes into the rice genome. In the twenty-first century, it is crucial to adopt low-input agriculture to address this issue. The development of improved crop cultivars that enhance genetic diversity, resilient to environmental change, with yield improvement along with nutritional value and adaptability to different agro-ecosystems and management practices, can be part of that solution. Overcoming the above challenges requires promoting low-cost agricultural improvements by utilizing improved crops with high-yielding varieties for better quality (FAO 2011).

In recent times, mutation breeding of crops has been proven as a reliable method for enhancing crop production, nutritional value and resistance to abiotic and biotic factors.

J. Devi (ed.), *Mutation Breeding in Crop Improvement*, Sustainable Landscape Planning and Natural Resources Management,
https://doi.org/10.1007/978-3-032-17911-1_13

This process has significantly advanced to improve cultivars with attractive traits and enhanced crop productivity, thereby meeting the growing demand for food. Traditional breeding measures have increased yield for many agronomic categories, and these measures rely on the genetic variability present in breeding populations, which can take time. In mutation breeding, mutations are induced in a plant's genetic material using either physical or chemical mutagens. These mutagens induce genetic variability by affecting point mutations or chromosomal changes, as a point of origin for new traits. Mutation breeding also enables the improvement of traits of interest in elite cultivars with minimal genetic change. This case study provides an overview of the applications of mutation breeding in rice and evaluates those findings in terms of increased yield and quality. The study demonstrates a successful application of mutation breeding, which can be utilized in rice improvement programs, as it addresses both food and quality related concerns.

2 Historical Background of Released Mutant Varieties Across the Globe

It was generally accepted that induced mutation can generate genetic diversity in varieties of rice that were otherwise genetically uniform. With early successes in mutation breeding in barley and wheat, rice was one of the first cereal crops improved through mutation breeding. Since then, researchers and breeders have used physical and chemical mutagens to produce and release numerous rice varieties, which have been developed for various agronomic traits, such as reduction in height, short maturity period, enhanced grain quality, enhanced biotic as well as abiotic stress resistance, and increased yield in Asia, Africa, Europe, and the Americas. Several international organizations, particularly the International Atomic Energy Agency (IAEA) and the Food and Agriculture Organization (FAO), have played a crucial role in promoting improvement in mutation breeding strategy worldwide, resulting in a diverse range of rice varieties.

2.1 Rice

Rice (*Oryza sativa* L.) is a staple food along with cereal crops, especially for Asian peoples (Orthoefer 2005). In developing countries, particularly in Asia, rice consumption is exceptionally high. After sugarcane and maize cultivation, rice ranks as the third most important agricultural commodity. It serves as a primary energy source in Asia and the Pacific (17 countries), South and North America (nine countries), and Africa (eight countries). And provide nutritional value and different bioactive compounds. The rice grain is a good source of fibre, minerals, vitamins, and various biomolecules. The nutritional property which contains 80% carbohydrates, 7–8% protein, 3% fat, 3% fibre and 7–6% minerals (iron, calcium, phosphorus), glutelin, crude as well as soluble protein, reducing and nonreducing sugar, and different other bioactive compounds (Burlando and Cornara 2014). Rice grain contains some minute components like vitamins, minerals, and various phytochemicals like carotenoids, flavone, flavanols, flavonoids, phenols, polyphenols, anthocyanins, saponins, saturated, and unsaturated fatty acids (Verma and Srivastav 2017). In rice production, some significant constraints like late maturity, lodging, low yield due to biotic and abiotic stresses are reported. With the help of induced mutation, several traits have been improved, including dwarf and semi-dwarf varieties, high yield, lodging resistance, early maturity, and resistance to both biotic and abiotic stresses in improved mutant varieties.

2.2 Rice Mutant Varieties Released Worldwide

Mutation is an excellent source for creating genetic variation and is used for analyzing gene structure and function. Asia has a remarkable diversity of rice mutant varieties (706), boasting the highest number among all the continents. Countries such as China, Japan, India, Vietnam, and Indonesia have made significant contributions to this achievement. North America has developed 45 mutant varieties, demonstrating its efforts to enhance crop yields and quality. Africa has released 36 mutant varieties, with Nigeria, Mali, Senegal, and Cote d'Ivoire being major contributors. South America has developed 25 mutant varieties. Europe has introduced 14 mutant varieties from France, Italy, Hungary, Portugal, and Romania, which are the major contributors (Fig. 1).

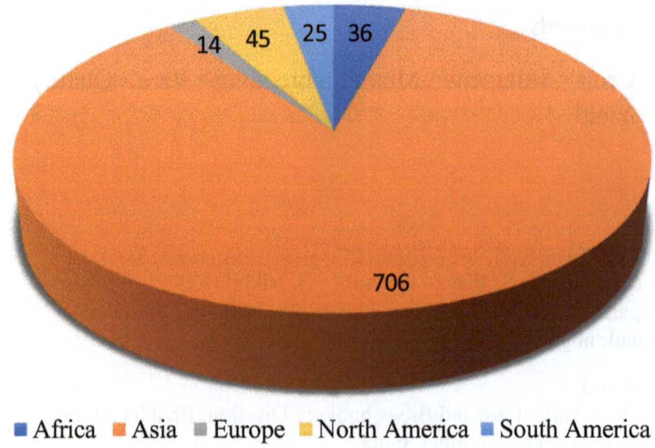

Fig. 1 Rice mutants varieties developed worldwide

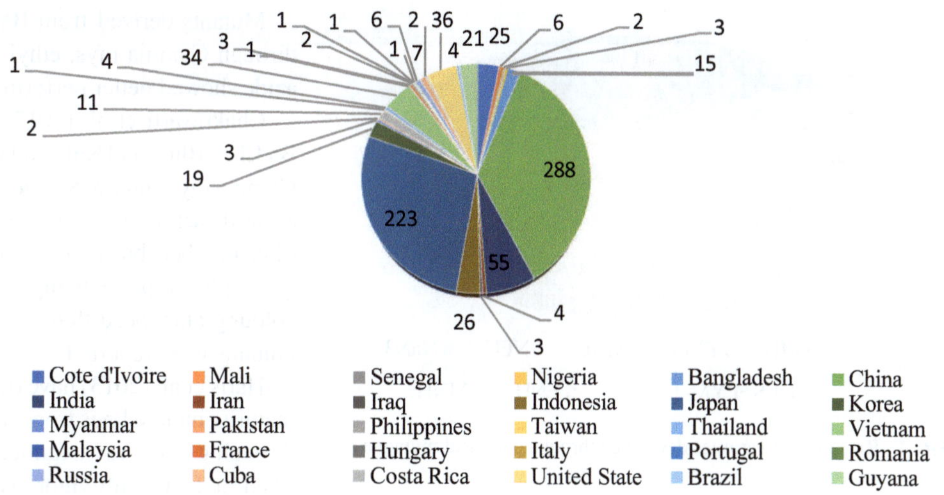

Fig. 2 Rice mutant varieties released from different countries

2.3 Rice Mutant Varieties Released from Different Countries

China leads in rice mutant variety development, accounting for 36% of the total mutant varieties. Japan follows with 28%, while India accounts for 7%. Other notable contributors include Vietnam (4%), Indonesia (3%), and Bangladesh (2%). Thailand also plays a crucial role in the improvement of rice mutant cultivars. In contrast, European countries such as France, Italy, Hungary, Portugal, and Romania have significantly fewer rice mutant varieties, suggesting that mutation breeding efforts in these countries are less extensive. African countries, including Nigeria, Mali, Senegal, and Côte d'Ivoire, also have a minor share. Similarly, Brazil, Cuba, the United States, and Costa Rica contribute only a small number of rice mutant varieties (Fig. 2).

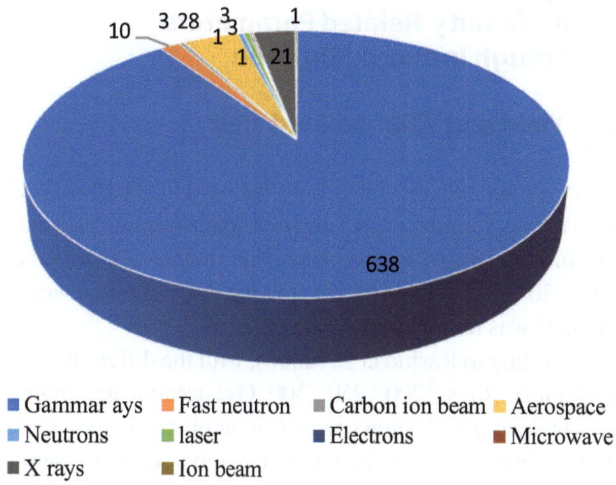

Fig. 3 Rice mutant varieties improved through physical

2.4 Rice Mutant Varieties Improved Through Physical Mutagens

Gamma rays are the most extensively used as a radiation source for rice improvement. Other types of radiation include neutrons, X-rays, electrons, and fast neutrons, but their share is significantly smaller compared to gamma rays. Carbon ion beam, ion beam, laser, microwave, and aerospace methods have the least contribution. This suggests that gamma rays are the most commonly used method, while other types of radiation contribute minimally. By using such mutagenic treatment, mutant varieties were developed with improved characters like early maturity, improved in grain and cooking quality, dwarf, lodging resistance, improved maturity period, high grain yield, disease resistance (blast, blight, viruses and fungus) and tolerant to drought and low temperature (Fig. 3).

2.5 Rice Mutant Varieties Improved Through Chemical Mutagens

The chart suggests that chemical mutagens, such as N-Nitroso-N-methylurea (NMU) and Ethyl methane sulfonate (EMS), are the most commonly used mutagens, while others, including N-Ethyl-N-Nitrosourea (NEU), and Sodium azide (NaN3), play secondary roles. Treatments such as diethyl sulfonate, Pingyangmycin (PYM), Ethyleneimine (EI), ENH, and Nitrosoguanidine (MNNG) represent additional but less prevalent methods. By using such mutagenic treatment, mutant varieties were developed with improved characters like agronomical characters, aroma enhancement, brewing quality, improved maturity period, high yield, abiotic and biotic stress tolerance, biochemical character, dwarf or semi-dwarf parameter, and cooking quality in different varieties with different mutagenic treatment (Fig. 4).

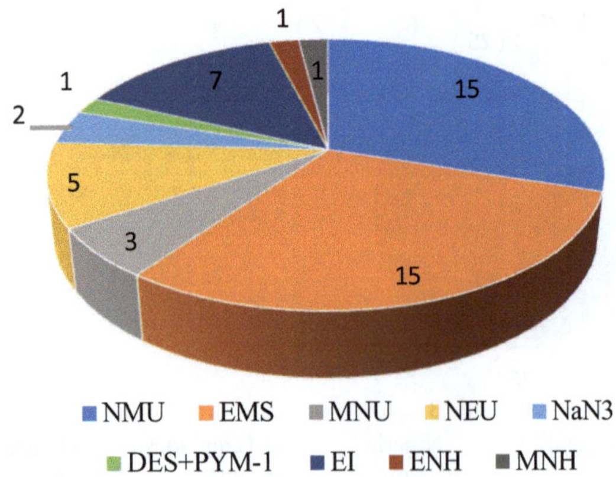

Fig. 4 Rice mutant varieties developed through chemical mutagen

3 Improvement in Yield and Quality-Related Parameters Through Induced Mutation

3.1 Yield and Yield-Related Trait

A semi-dwarf mutant, CI 11048, was produced from WC-1403, a long, lodging prone, and rice water weevil-resistance line. In comparison to the parent, the mutant was approximately 20 cm shorter in height, and the level of water-weevil tolerance was retained.

According to Rashid et al. (2003), with the different doses of Gamma Rays (200, 250, 300 Gy) treatments, mutant was obtained which shows reduction in plants height (18.3–33.6%), dwarf with maximum tiller number, early flowering (7–21 days), and increase in grain length (10 mm –10.76 mm).

Singh (2006) used gamma rays (50–300 Gy) for induction of mutation in Taraori Basmati and Pusa Basmati-1, where only 100 Gy doses was found to be most effective, 15 type of morphological mutants were reported such as short-statured, semi-dwarf, long, early maturing, profuse tillering, large panicle, partial or complete sterile, bushy, colored pericarp, red kernel, large grain, awned, awn less, pigmented awn and high yielding.

Baloch et al. (2006) reported that a high-yielding mutant, IR6-25A, was derived from gamma-ray (150 Gy) treatment, which is significantly better than the control IR6 and check variety Shadab. Based on a three-year average performance, it shows a 16% increase in yield over its parent, IR6, and the check variety, Shadab.

According to Bughio et al. (2007), rice mutant variety Mehak with high-yielding, has been improved from Basmati-370, with the help of gamma rays (150 Gy). Mehak, the mutant variety, was found to be significantly superior to its parent variety, Basmati-370, in terms of yield and yield-related parameters, including excellent aroma and other physicochemical properties.

Mutants derived from Basmati 370 and Sanwal Basmati through Gamma rays, ethyl methane sulfonate, and sodium azide showed better performance over generations.

Chakravarti et al. (2012) reported induced variability in Badshah Bhog and Kalanamak aromatic rice cultivars through Gamma rays and EMS alone, as well as combined treatments. Semi-dwarf, early maturing, plant types with an increase in tiller number, bushy and stiff stems, broad and narrow leaf types, plant types with pigmented nodes, small, slight, yellow colour grain type, different grain size with yield enhancement mutants were reported.

Haris et al. (2013) have developed high-yielding and early maturity mutant lines from Ase Lapang and Mandoti cultivars through gamma-ray treatment. Mohapatra et al. (2014) identified mutants with vigorous growth, flowering, early maturity, number of grains, shape, size, yield enhancement, and efficient phosphorus use from the Nagina 22 cultivar through EMS (0.6, 0.8, 1.0, and 1.5%) treatments. Among all the treatments, 0.8% EMS was the most effective, showing the most significant response, including increases in yield, biomass, and harvest index.

Dwarf, long maturity period, average blast tolerance, mutants with improved yield, slender grains, and enhanced cooking quality mutants were derived from SCS118 Marques cultivar through Gamma rays (250 Gy) treatment (Schiocchet et al. 2014). Photoperiod-insensitive, early-flowering dwarf mutants were derived from gamma ray treatment by Aung and Khai (2015).

Gowthami et al. (2016) induced mutations in ADT 37 and ADT 45 rice cultivars through Gamma ray and electron beam treatments and reported various functional mutants for early maturity, high yield, a higher number of tillers per plant, a higher number of seeds per panicle, and leaf length and breadth, among other traits. The electron beam was more effective than gamma rays in inducing variability.

TCDM-1(Trombay Chhattisgarh Dubraj Mutant-1) and TKR Kolam (Trombay Karjat Rice Kolam) mutants, derived from gamma rays, show dwarf stature, high yielding, average maturity, and good aroma traits, and were released for cultivation in Chhattisgarh and some regions of Konkan in Maharashtra (Kumar et al. 2017).

Improved mutants from Jawaphool and Dubraj cultivars, which showed dwarf, semi-dwarf, semi-tall, early maturity, mid-late maturity, increased tillers with high yield, bushy and wide-grassy leaves, cytoplasmic male sterility, and aggregated grain with red kernel color traits are available. Among the treatments, 300 Gy was found to be the most effective treatment.

From the Tulpanji cultivar, eight mutants were derived through gamma-ray treatment, retaining photoperiod-insensitive, dwarf, semi-dwarf, and yield-enhancing traits, as well as pleasant aroma, as reported by Roy et al. (2018).

Through Sodium azide (NaN3) (1.5 and 2.5 mM) and cytokinin (25 and 50 ppm) combined treatments in Indonesian traditional rice varieties, genetic variability was induced. The number of leaves were prolonged, delaying senescence and increase in chlorophyll as well as the number of stomata and the thickness of leaves and stems, while reducing the length of the stem and leaf epidermal cells was reported along with increase in growth and slight improvement in yield compared to the parent plants (Dewi et al. 2019).

Purwanto et al. (2019) have reported the induction of mutation in Cempo Ireng, Cempo Melik and Melik rice cultivars that were treated with 100 to 300 Gy Gamma rays, where high-yielding, early-flowering, early-maturity plant types were derived.

Through N-methyl-N-nitrosourea (MNU) treatment, Xuan et al. (2019) developed high-yielding mutant plant types. Phanchaisri et al. (2019) have reported the induction of high-yielding mutants through low-energy heavy ion beam treatment with improved grain quality, and this was confirmed with cDNA fingerprinting from Thai Jasmine rice cultivar.

Hanifah et al. (2020) reported the induction of lodging resistance, high yield, early flowering, and early maturity mutants in Mentik Susu rice through gamma ray treatment from Indonesia.

Two high-yielding mutants, Fukuhibiki H6 and Fukuhibiki H8, were obtained through gamma-ray treatment, as reported by Kato et al. (2020) from Japan.

Heavy ion beam radiation was used to induce mutations in Pare Ambo black rice, resulting in improved mutants with higher numbers of grains per panicle, panicle density, grain weight per panicle, and grain weight per line (Putra et al. 2020).

Gautam et al. (2021) developed improved mutants from famous rice variety Bapatla 2231 (BPT 2231) with the help of EMS and gamma ray alone and combined treatments and reported early flowering, early maturing mutants (40–50 days than control), grain quality, biochemical traits, high yielding with fine as well as bold grain type.

Okasa et al. (2021) used an Argon-ion dose of 10 Gy and a carbon-ion dose to induce mutations and identified improved plant types in terms of tall plant, tiller numbers, number of panicles, and grain weight per panicle.

Pujiasmanto et al. (2021) irradiated a black rice cultivar with gamma rays and reported mutants with high yield performance.

Desai et al. (2021) developed high-yielding, early-maturity, dwarf, and lodging-resistant mutants from Ajara Ghansal and Kala Jirga non-basmati aromatic rice cultivars from Kolhapur, Maharashtra, through gamma ray, EMS, and SA treatments.

Andrew-Peter-Leon et al. (2021) identified useful mutants with early-flowering, early-maturity, high yield, and dwarf plant types through gamma ray treatment, and mutations were confirmed by using SSR markers.

Black rice (Cempo Ireng) was treated with gamma rays and reported mutants with higher yield, higher anthocyanin, amylose, fat, and protein content by Nandariyah et al. (2021). Sao et al. (2022) developed improved mutants through gamma ray treatment with reduction in plant height, early maturity, and higher yield performance as compared with the mother plant. Ren et al. (2023) have developed high-yielding mutants through ion beam treatment. Riysanto et al. (2024) reported broad genetic variation in tiller number, number of grains/panicle and grain yield, and height of the plant through gamma ray treatment.

3.2 Biotic and Abiotic Stress Tolerance

By using different types of mutagen treatment in rice disease resistance, plant types were identified for bacterial leaf blight (*Xanthomonas oryzae*) and blast (*Pyricularia oryzae*). Mutagens like fast neutrons, X-rays, γ-rays, EMS, ethidium bromide, nitroso methyl urea (NMU), Sodium azide, and a combination of both chemical (EMS) and physical (γ-rays) agents were used to induce mutations. The improved traits show dwarf, early maturity, high tiller number, tolerance to blast and leaf blight, resistance to lodging and cold, grain quality improvement, and yield enhancement (Sharma and Singh 2013). Nagina 22 variety was improved for resistance to blast, bacterial leaf blight diseases, and tolerance to drought, salinity, and herbicides through various treatments of EMS, among all the treatments, 0.8% EMS treatment was found to be more effective (Mohapatra et al. 2014).

By using gamma rays (300 Gy), disease and pest-resistant lines were developed from three aromatic rice cultivars, including Dubraj, Jawaphool, and Pusa Basmati 1 (Kumar et al. 2017). A Glyphosate (Glycel) herbicide-tolerant mutant in the CSR27 rice variety has been developed using proton beam radiation, which designates that a proton beam dose of 170-180 Gy was the most effective and optimum dose.

Through N-methyl-N-nitrosourea (MNU) treatment, Xuan et al. (2019) developed disease and pest-resistant mutant plant types. Mutant MM1 was identified through gamma-ray treatment, which shows higher yield potential under the reproductive stage of drought stress. After a panicle transcriptome analysis, it was revealed that biological reactions and molecular regulators are included in the development of panicle, yield, and the integration of stress, such as drought (Eragam et al. 2023).

3.3 Nutritional Quality

In rice varieties such as indica and japonica, a considerable number of mutants have been developed for various aspects, including nutritional quality, specifically raw protein content and protein patterns distribution in the embryo. With the help of different mutagens, mutants in four rice varieties (T (N)-I, IR-8, Taichung-65, and Tainan-3) that exhibit variations in grain weight, protein content, and seed sterility grade have been reported (Vilawan and Siddiq 1973).

Kumamaru et al. (1988) investigated the development of improved variants for nutritional quality (protein content) and those with waxy endosperm through MNU treatment. Aung and Khai (2015) reported that mutants with reduced amylose content were developed through gamma-ray treatment.

Gamma rays were used to develop an improved plant type of Kashmiri Basmati, which exhibits resistance to cold stress, early maturity, and retains cooking quality and characteristics alike to those of the mother plants (Raina et al. 2016). Thai black rice irradiated with 20–250 Gy shows the most effective results at 150 Gy. The results suggest that the lower dose of gamma rays shows a positive trend in rice growth, whereas rice growth is inhibited at higher gamma treatment. Likewise, the aroma content in rice shoots treated with high gamma dose seemed higher than that of irradiated rice shoots treated with low gamma dose (Sansenya et al. 2017). According to Purwanto et al. (2019), gamma-rays (100–250 Gy) was applied to improve biochemical, nutritional properties, resistance to abiotic stress, and industrial quality of rice. In varieties Cempo Ireng, Cempo Melik, KDML105, and Ayarmin biochemical properties were enhanced, plant development, and nutritional quality were improved in the Ciherang variety, and in varieties like RD6 and Sangyod Phatthalung, biochemical and nutritional properties were enhanced. According to Viana et al. (2019), the use of different types of mutagenic treatment resulted in an enhancement of qualities. With the MNU (0.25–1.00 mM), AS (1–10 mM), and colchicine (0.04–0.3%) treatments, there was an improvement in nutritional quality and yield. With the help of γ rays (50–350 Gy) and IBR (Carbon 20–50 Gy (up to 220 MeV), Neon 10 Gy (60 80 keV/μm)), there was an enhancement in nutritional quality. Through gamma irradiation, several mutants were obtained in white Ponni; these mutants exhibit improvements in cooking quality characteristics, such as grain elongation ratio, gel consistency, and gelatinization temperature (Andrew-Peter-Leon et al. 2021).

4 Conclusion

Mutation breeding is a crucial tool in modern crop improvement programs, enabling the generation of new genetic variability that can be utilized to enhance key agronomic traits. In rice, it is beneficial for generating increased yield potential and grain quality and can overcome limitations of conventional breeding methods. It enables the rapid development of new varieties or cultivars with desired characteristics, providing a more scientific foundation by introducing changes through mutation. As breeding technologies continue to advance, the combination of mutation breeding with molecular and genomic tools has exciting opportunities for improvement of the resilient and high-performing rice varieties. To address future food security issues and ensure sustainable rice production systems, continued support for this tool is necessary.

References

Andrew-Peter L, Ramchander S, Kumar K, Mehanathan M, ArumugamPillai M (2021) Assessment of efficacy of mutagenesis of gamma-irradiation in plant height and days to maturity through expression analysis in rice. Plos One 16(1):1–20. https://doi.org/10.1371/journal.pone.0245603

Aung N, Khai A (2015) Developing early-flowering or photoperiod-insensitive mutants from rice var. Ayarmin (Machando) by using gamma radiation from 60Co source. IJSR 6(5):600–604. https://doi.org/10.21275/ART20173272

Baloch A, Soomro A, Bughio H, Bughio M, Odhano I, Asad M, Abbasi S (2006) Breeding for high grain yield through mutagenesis in rice (*Oryza sativa* L.). PJBT 3(1–2):11–16. https://pjbt.org/index.php/pjbt/article/view/174/172

Bhattacharjee P, Singhal R, Kulkarni P (2002) Basmati rice: a review. Int J Food Sci Technol 37(1):1–12. https://doi.org/10.1046/j.1365-2621.2002.00541.x

Bughio HR, Asad MA, Odhano IA, Bughio MS, Khan MA, Mastoi NN (2007) Sustainable rice production through the use of mutation breeding. Pak J Bot. 39(7):2457–61. https://www.researchgate.net/profile/Athar-Khan-4/publication/298285507_SUSTAINABLE_RICE_PRODUCTION_THROUGH_THE_USE_OF_MUTATION_BREEDING/links/56f6a44d08ae81582bf2f849/SUSTAINABLE-RICE-PRODUCTION-THROUGH-THE-USE-OF-MUTATION-BREEDING.pdf

Burlando B, Cornara L (2014) Therapeutic properties of rice constituents and derivatives (*Oryza sativa* L.): a review update. Trend Food Sci Technol 40:82–98. https://doi.org/10.1016/j.tifs.2014.08.002

Chakravarti S, Singh S, Kumar H, Lal J, Vishwakarma H (2012) Study of induced polygenic variability in M_1 and chlorophyll mutations in M_2 generations in aromatic rice. The Bioscan 8(1):49–53. file:///C:/Users/Admin/Downloads/CHAKRAVARTI STUDYOFINDUCED-POLYGENICVARIABILITYINM.pdf

Desai S, Jadhav A, Ramteke A, Dhole V, Bapat V, Gaikwad N (2021) Genetic improvement of two Indian non-basmati aromatic rice landraces through physical and chemical mutagenesis. Int J Radiat Biol 98:82–89. https://doi.org/10.1080/09553002.2021.1987567

Dewi A, Dwimahyani I, Sobrizal (2019) Application of induced mutation technique to improve genetic variability of Indonesian traditional rice varieties. Environ Earth Sci 482:012016. https://doi.org/10.1088/1755-1315/482/1/012016

Eragam A, Mohapatra A, Shukla V, Kadumuri R, George A, Putta L, Akkareddy, Chavali S, Vemireddy L, Ramireddy E (2023) Panicle transcriptome of high-yield mutant indica rice reveals physiological mechanisms and novel candidate regulatory genes for yield under reproductive stage drought stress. BMC Plant Bio 2–18. https://doi.org/10.1186/s12870-023-04507-1

FAO (2011) State of Food and Agriculture. ISSN-0081-4539

Gautam V, Swaminathan M, Akilan M, Gurusamy A, Suresh M, Kaithamalai B, Joel A (2021) Early flowering, good grain quality mutants through gamma rays and EMS for enhancing per day productivity in rice. (*Oryza sativa* L.) Int J Radiat Bio 97(12):1716–1730. https://doi.org/10.1080/09553002.2021.1987563

Gowthami R, Vanniarajan C, Souframanien J, Arumugam P (2016) Effect of gamma rays and electron beam on various quantitative traits of Rice (*Oryza sativa* L.) in M1 Generation. Adv Life Sci 5(5):1876–1882. https://doi.org/10.1080/09553002.2021.1889702

Hanifah W, Parjanto P, Hartati S, Yunus A (2020) The performances of M4 generation of Mentik Susu rice mutants irradiated with gamma-ray. Biodiversitas 21:4041–4046. https://doi.org/10.1088/1755-1315/142/1/012063

Haris A, Abdullah Bakhtiar, Subaedah Aminah, Jusoff K (2013) Gamma ray radiation mutant rice on local aged dwarf. Middle East J Sci Res 15(8):1160–1164. https://doi.org/10.5829/idosi.mejsr.2013.15.8.11541

Kato H, Li F, Shimizu A (2020) The selection of gamma-ray irradiated higher yield rice mutants by directed evolution method. Plants 9(8):2–16. https://doi.org/10.3390/plants9081004

Kumamaru T, Satoh H, Iwata N, Omura T, Ogawa M, Tanaka K (1988) Mutants for rice storage proteins: 1. Screening of mutants for rice storage proteins of protein bodies in the starchy endosperm. Theor Appl Genet 76(1):11–6. https://doi.org/10.1007/BF00288825

Kumar V, Sharma D, Das B, Tiwari A, Sahu P, Singh S, Baghel S (2017) Identification of semi-dwarf and high-yielding mutants in Dubraj rice variety of Chhattisgarh through gamma rays based induced mutation. Int J Genet Sci 9(9): 298–303. https://doi.org/10.13140/RG.2.2.16073.03682

Kushwaha U (2016) Black Rice. Springer International Publishing Switzerland 5(1):1–199. https://doi.org/10.1007/978-3-319-30153-2

Mohapatra T, ROBIN S, Sarla N, Sheshashayee M, Singh A, Singh K, Singh N, Vamitha S, Sharma R (2014) EMS Induced Mutants of Upland Rice Variety Nagina 22: generation and characterization. Proc Indian Natl Sci Acad 80:163–172. https://doi.org/10.16943/ptinsa/2014/v80i1/55094

Monks J, Vanier N, Casaril J, Berto R, Oliveira M, Gomes C, Carvalho M, Dias A, Elias M (2013) Effects of milling on proximate composition, folic acid, fatty acids and technology. J Food Compos Analy 30(2):156–164. https://doi.org/10.1016/j.jfca.2013.01.009

Nandariyah Rahayu M, Pujiasmanto B, Luxrensa V (2021) Yield quality and performance of three stains black rice from gamma ray irradiation. Earth Environ Sci 905:012027. https://doi.org/10.1088/1755-1315/905/1/012027

Okasa A, Sjahril R, Riadi M, Abe T (2021) Evaluation of Toraja (Indonesia) local aromatic rice mutants developed using heavy-ion beam irradiated. Biodiversitas 22(8):3474–3481. https://doi.org/10.13057/biodiv/d220846

Orthoefer F (2005) Rice Brain Oil. In Bailey's Industrial Oil and Fat Products A book published by Wiley. https://doi.org/10.1002/047167849X.bio015

Phanchaisri B, Khitka B, Techarang J (2019) Low-energy heavy-ion-beam-induced mutation of novel high-yielding drought-tolerant Thai Jasmine rice. Nucl Instrum Methods Phys Res 492:34–42. https://doi.org/10.1016/j.nimb.2021.02.003

Pujiasmanto B, WicaksonA Supriyono, Nandariyah Sutarno, Riyatun Suharyana (2021) Yield quality and performance of line 13, 46 and 52 black rice from gamma rays irradiation. Earth Environ Sci 905:012033. https://doi.org/10.1088/17551315/905/1/012033

Purwanto E, Suhadi S, Mercy B (2019) Induced mutation for genetic improvement in black rice using gamma-ray. J Agric Sci 41(2): 213–220. https://doi.org/10.17503/agrivita.v41i2.876

Putra M, Riadi M, Sjahril R (2020) Black rice mutant strain selection results of M3 generation mutation breeding. Earth Environ Sci 486:1–5. https://doi.org/10.1088/1755-1315/486/1/012094

Raina A, Laskar R, Khursheed S, Amin R, Tantray Y, Parveen K, Khan S (2016) Role of mutation breeding in crop improvement past, present, and future. Asian Res J Agric 2(2):1–13. https://doi.org/10.37446/volbook102024/46-61

Rashid M, Cheema A, Ashraf M, Qamar Z, Mahmood Z (2003) Development of basmati rice varieties through the use of induced mutation and related technique. Pak J Bot 35(5):811–818. https://www.pakbs.org/pjbot/PDFs/35(5)/PJB35(5)16.pdf

Ren W, Wang H, Du Y, Li Y, Feng Z, Zhou X, Kang G, Shu Q, Guo T, Guo H, Yu L, Jin W, Yang F, Li J, Ma J, Li W, Xu C, Chen X, Liu X, Yang C, Liu L, Zhou L (2023) Multi-generation study of heavy ion beam-induced mutations and agronomic trait variations to accelerate rice breeding. Plant Sci 14:1–16. https://doi.org/10.3389/fpls.2023.1213807

Riysanto A, Oktaviani E, Wulansari N, Haryanto T (2024) Genetic parameters of yield component and yield in M1 rice (*Oryza sativa* L.) generation irradiated with gamma-ray. J Kultivasi 23(2):187–200. https://doi.org/10.24198/kultivasi.v23i2.55129

Roy B, Kumar V, Tulsiram S, Das B (2018) Development of high yielding aromatic mutants of rice (*Oryza sativa* L.) from a local aromatic cultivar, Tulaipanji by using gamma radiation. Ind J Genet 78(4): 409–416. https://doi.org/10.31742/IJGPB.78.4.2

Sansenya S, Hua Y, Chumanee S, Winyakul C (2017) Effect of gamma irradiation on the 2-acetyl-1-pyrroline content during growth of Thai black glutinous rice (Upland rice). Aust J Crop Sci 11(05):631–637. https://doi.org/10.21475/ajcs.17.11.05.p286

Sao R, Sahu P, Patel R, Das B, Jankuloski L, Sharma D (2022) Genetic improvement in plant architecture, maturity duration and agronomic traits of three traditional rice landraces through gamma ray-based induced mutagenesis. Plants 11(24):3448–3454. https://doi.org/10.3390/plants11243448

Schiocchet M, Noldin J, Raimondi J, Neto A, Marschalek R, Wickert E, Martins1, Eduardo Hickel G, Knoblauch R, Scheuermann K, Eberhardt D, Andrade A (2014) SCS118 Marques-New rice cultivar obtained through induced mutation. Crop Breed Appl Biotechnol 14:68-70. https://doi.org/10.1590/s1984-70332014000100012

Sharma A, Singh S (2013) Induced mutation a tool for creation of genetic variability in rice. (*Oryza sativa* L.) J Crop Weed 9(1):132–138. https://www.cropandweed.com/vol9issue1/pdf2005/23.pdf

Singh S (2006) Gamma rays induced mutation in Basmati rice. (*Oryza sativa* L.) Indian J Genet Pl Br 66(2):143–144. https://www.isgpb.org/journal/index.php/IJGPB/article/view/1517/1846

Verma D, Srivastav P (2017) Proximate composition, mineral content and fatty acids analyses of aromatic and non-aromatic Indian rice. Rice Sci 24(1):21–31. https://doi.org/10.1016/j.rsci.2016.05.005

Viana V, Pegoraro C, Busanello C, Oliveira A (2019) Mutagensis in rice: the basis for breeding a new super plant. Front Plant Sci 10:1326. https://doi.org/10.3389/fpls.2019.01326

Vilawan S, Siddiq E (1973) Study on mutational manipulation of protein characteristics in rice. Theor Appl Genet 43(6):276–280. https://doi.org/10.1007/BF00277788

Xuan T, Anh T, Tran H, Khanh T, Dat T (2019) Mutation Breeding of a N-methyl-N- nitrosourea (MNU)-Induced Rice (*Oryza sativa* L. ssp. Indica) population for the yield attributing traits. Sustainability 11:1–15. https://doi.org/10.1155/2012/949038

Mutation Breeding for Yield and Quality Improvement: A Case Study for Millets

Kamble Sadhana, Shedbale Namrata, Londhe Ranjit, Dhole Vinod, Bapat Vishwas, and Gaikwad Nikhil

Abstract

Genetic diversity is the foundation for plant breeders to create improved crop varieties with improved traits suited to different environments. Variation in plants originates from factors like hybridization, recombination, and mutations. However, natural mutations occur at a very low rate, making them less useful for crop improvement. Hence, induced mutation is considered to be a more effective way to create variation and develop improved crop varieties. But problems like limited agricultural land, less water, and a changing climate make it difficult to increase yield using traditional methods. To produce enough food for a growing population, we need new ways to improve crop production, make crops stronger, and boost their quality. Mutation breeding, which involves using physical or chemical agents to create genetic variation, and played an essential role in crop improvement. Cereals play the most significant role as a main part of humans' diets, providing vital nutrients and energy for daily life. This article provides the use of mutation breeding for enhancing yield and quality traits in major cereal crops: sorghum, oats, bajra, foxtail millet, proso millet, little millet, finger millet, rye, kodo millet, etc.

Keywords

Crop · Cereals · Mutation breeding · Mutagens · Quality · Yield

K. Sadhana · S. Namrata · L. Ranjit · G. Nikhil (✉)
Department of Botany, Shivaji University, Kolhapur, India
e-mail: nbgaikwadsu@gmail.com

D. Vinod
Nuclear Agriculture and Biotechnology Division, Bhabha Atomic Research Centre, Mumbai, India

B. Vishwas
Department of Biotechnology, Shivaji University, Kolhapur, India

1 Introduction

Ensuring food security has become one of the main challenges of the developed world all over the globe due to population growth, which is estimated to be up to 9.1 billion by 2050, hence, an increase in food production by a striking percentage of 70% must be made for the demand to be fulfilled (FAO 2009). Such an increase in food production are relatively complex to realize, especially since weather and environmental conditions have become pretty erratic (Charles et al. 2014). Historically, agri-economic parameters have been genetically diversified using induced mutagenesis to fulfill the requirements of the developing world (Mba et al., 2012). The mutation breeding for crop improvement was initiated by Huge de Vries in 1902, firstly on barley. Now, induced mutation breeding is a well-recognized method for crop improvement, which involves exposing plant seeds or other propagules to physical or chemical mutagens, which induce random changes in the genetic material (DNA). Such mutations may affect gene expression and phenotypic variations. Mutation breeding has been regarded as a more rational breeding technique for the induction of genetic variability in various crops. It has become a most valuable and convenient tool for crop improvement, particularly in increasing yield and quality. Mutagens act by altering the DNA sequence of the plant and may result in the development of beneficial mutations like improved grain quality, disease resistance, stress tolerance, and high yield. Secondly, the selection and further breeding are carried out to stabilize those desirable mutations. Mutation breeding is a useful technique to create new varieties without introducing foreign genes. Since the 1920s, when Lewis John Stadler established X-rays effects in the induction of mutations, he identified induced mutation breeding is a promising method for crop improvement (Shu et al. 2012). Through mutation breeding, more than 3400 mutant varieties were released worldwide FAO/IAEA (2022).

J. Devi (ed.), *Mutation Breeding in Crop Improvement*, Sustainable Landscape Planning and Natural Resources Management,
https://doi.org/10.1007/978-3-032-17911-1_14

Mutations occurs in two ways: spontaneous mutations, which occur naturally without using any mutagens; and induced mutations, which happen when plant parts (like seeds, cuttings, stems, pollen, or ovules) are exposed to mutagens. The process of improving crops by using induced mutations is called mutation breeding (Pathirana 2011). Mutagenesis, or induced mutation, refers to genetic changes that can be passed on and are caused by external agents like radiation, chemicals, or viruses, rather than by normal genetic processes. Spontaneous mutations are naturally occurring, genetic changes that arise without any external influence or mutagenic agent. Common causes of spontaneous mutation are DNA replication errors, tautomeric shifts, spontaneous lesions, transposable elements, etc. Induced mutations in plants or their parts, such as stems, seeds, seedlings, etc., using physical and chemical mutagens can help to create new plants with better traits. The mutagens are of two types such as chemical and physical agents that are responsible for changing the genome of plants. The selection of mutagens and their concentration is vital in the mutation breeding program. Chemical and physical mutagens are commonly used in mutation breeding to improve crops and their traits. Many researchers worked with chemical mutagens to improve crops over several decades. Thom and Steinberger initially used nitrous acid to induce mutations in *Aspergillus* in 1939. The chemical mutagens are equally effective for mutation like physical mutagens. Now, more than a hundred chemical mutagens have been identified, such as sodium azide, ethyl methanesulfonate, nitroso compounds, base analogues, acridine dyes, etc. (Kharkwal 2012).

Physical mutagens were first used in crop improvement before the 1920s after researchers discovered that radium could cause mutations while testing it on fruit flies (Kharkwal et al. 2004). Physical mutagens are preferred over chemical ones, because they do not require washing after treatment (Shah et al. 2000). X-rays, gamma rays, fast and slow neutrons, and UV rays are commonly applied physical mutagens in plant mutation studies. Over the last few decades, mutation breeding is now an extensively used method to improve cereal crops by increasing yield, enhancing nutrition, and strengthening tolerance to both biotic and abiotic stresses. This approach has contributed significantly for creating new varieties with improved traits and rising productivity in cereals, thereby meeting the growing food demand.

Among agricultural crops, cereals are essential for securing world food supply, being primary calorie and nutrition provision for billions of people. Cereals play a vital role in worlds food requirements, being the essential food. Cereals belonging to the family *Poaceae* and cultivated for food which are rich in carbohydrates, proteins, vitamins, and important minerals. Common cereals such as wheat (*Triticum aestivum*), rice (*Oryza sativa*), maize (*Zea mays*), sorghum (*Sorghum bicolor*), barley (*Hordeum vulgare*), and oats (*Avena sativa*) are required for both humans and animals. Cereals have adapted to diverse agro-environmental climatic conditions, making them a critical component of farming systems worldwide. Mutation breeding for cereal crops has been a powerful tool to overcome some significant challenges faced by crop production. The first disease-resistant mutant for barley was discovered in 1942 (Usharani and Kumar 2015). The modern varieties, therefore, need to be based on wide genetic diversity in these subspecies, coupled with a large diversity within the crop species. Such management of the unpredictable nature of ecological and climatic conditions aims to boost food production (Charles et al. 2014). Use of low-input farming methods is essential for sustainable agriculture in today's world, so that this challenge can be addressed. Creating improved crop varieties, with genetic diversity, climate-resilient, high-yielding, nutritionally rich, and adaptation to different agro-ecosystems and farming practices, can be a significant part of that solution (FAO 2011). To fulfill this food scarcity demand, there is an urgent need to improve crop plants, especially cereal crops. To overcome these challenges, the FAO (2011) recommended low-cost agricultural practices by developing and utilizing improved crop varieties with higher yield, enhanced quality, and superior nutritional value.

In addition to agricultural practices, some regulatory measures will support food security for the global population by 2050. The use of growth hormones, fertilizers, pesticides, and antibiotics in agriculture has raised many health concerns. Efforts are being made to cut down their use. Once their use diminishes, research should also be directed at eco-friendly, sustainable farming techniques that lower production costs, such as better water use methods like those developed in Israel and regulation of groundwater use. Along with the use of cheap renewable energy, the application of a carbon tax can be a feasible alternative (Ehrlich and Harte 2015). An ever-evolving global environment poses a formidable threat to crop yields and also a difficult challenge to plant breeders. Thus, crop improvement projects predominantly need to aim at developing high-yield cultivars that can adapt efficiently to the new climate scenario (Raina et al. 2016; Laskar et al. 2018). This study aims to examine the potential outcomes of mutation breeding for enhancing the yield and quality of cereal crops, with a particular focus on its application and results. This study aims to lay a foundation for future opportunities and breeding programs to develop climate-resilient, high-quality cereal varieties that support global population growth. Considering all these points, the study will highlight the importance of mutation breeding techniques and their contribution to the development of improved cereal varieties.

2 Historical Background of Released Mutant Varieties Across the Globe

The concept of improving crops through induced mutation began in the early twentieth century. Over the last five decades, induced mutation has significantly contributed to the successful cultivation of many crops by helping to create varieties with improved qualities. Worldwide application of induced mutation in crop breeding has resulted in the release of 3,362 improved mutant varieties in 240 plant species from 75 countries FAO/IAEA (2022).

According to data collected from the FAO/IAEA Mutant Variety Database 2022, the 1602 of mutant types were released through induced mutation across different cereal crops. There have been 851 mutant varieties of rice (*Oryza sativa* L.) released, followed by barley (*Hordeum vulgare* L.) with 311 and bread wheat (*Triticum aestivum* L.) with 265 (Table 1). Significant advancements in mutation breeding are also reported in other cereals, such as oats (*Avena sativa* L.) with 23 mutant types and maize (*Zea mays* L.) with 96. Crops such as rye (*Secale cereale* L.), emmer wheat (*Triticum dicoccum* L.), and finger millet (*Eleusine coracana*) also provide minor contributions. The extensive use and effectiveness of induced mutation in cereal crop improvement, with a cumulative total of 1,602 mutant varieties, was reported in cereal crops FAO/IAEA (2022).

2.1 Sorghum

Sorghum (*Sorghum bicolor*) is a major cereal crop cultivated for food, animal feed, and biofuel. It is highly valued for its adaptability to harsh environments, especially in drought regions. Globally, sorghum is grown on approximately 41 million hectares, yielding around 58 million tonnes. The main countries contributing to this production are the USA, Nigeria, Mexico, and India (Surashe et al. 2022). According to **ICRISAT/FAO 1996**, Sorghum is the world's most widely grown cereal crop after rice, wheat, barley, and maize. The major regions for sorghum cultivation include Africa and Asia, with India being one of the major contributors. Sorghum grains are utilized for human consumption, livestock feed, and as a base for alcohol and biofuel production. With its high content of carbohydrates, fiber, protein, vitamins, and minerals, sorghum offers considerable health benefits. Its increasing use in food and biofuel shows its value for sustainable farming. Sorghum grows in harsh conditions like drought and high heat, so creating new genetic variations through mutation breeding can develop better varieties that survive in such environments.

Table 1 Number of mutant varieties released through induced mutation from cereals

Latin name	Number of mutant varieties
Avena sativa L. (Oat)	23
Eleusine coracana (Finger millet)	2
Hordeum vulgare L. (Barley)	311
Oryza sativa L. (Rice)	851
Secale cereal L. (Rye)	4
Sorghum bicolar L. (Sorghum)	17
Triticum aestivum L. (Bread wheat)	265
Triticum dicoccum L. (Emmer wheat)	2
Triticum turgidum ssp. durum Desf	31
Total	**1506**

Source **FAO/IEAE Mutant Variety Database** (2022)

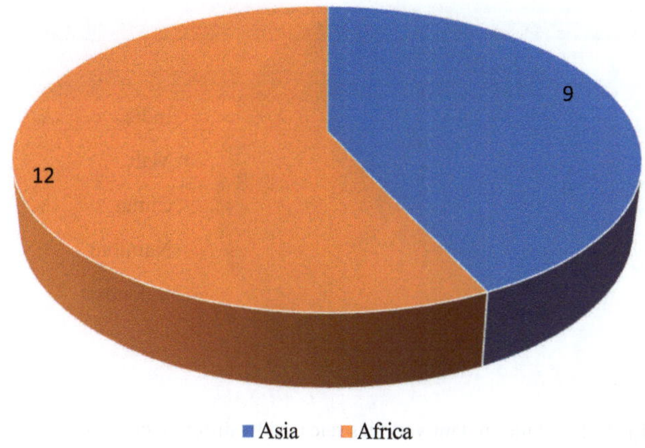

Fig. 1 Sorghum mutant varieties released by different continents

2.1.1 Sorghum Mutant Varieties Released Worldwide

Induced mutation breeding has been successfully used to create improved varieties of sorghum. Through induced mutation, 21 sorghum varieties have been officially released. Among these, Africa has released 12 mutant varieties, ranking first, while Asia has introduced nine varieties (Fig. 1).

2.1.2 Sorghum Mutant Varieties Released by Different Countries

Through induced mutation, 21 sorghum varieties have been officially released. Among the contributing countries, Mali ranks first with the release of eight sorghum varieties, followed by China with five varieties. Namibia has released four varieties, while Indonesia has contributed three. India has released one sorghum variety through mutation breeding (Fig. 2).

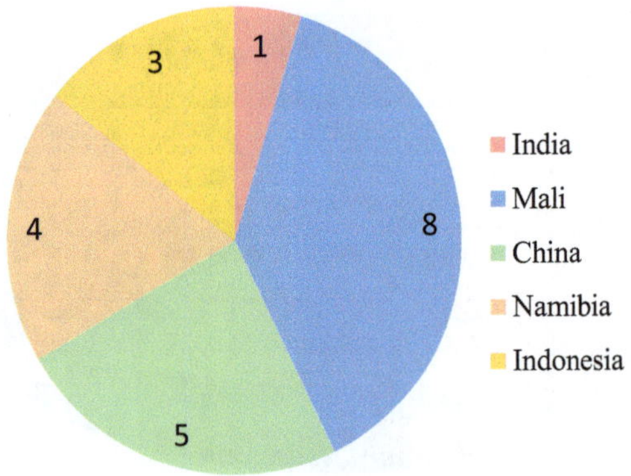

Fig. 2 Sorghum mutant varieties released by different countries

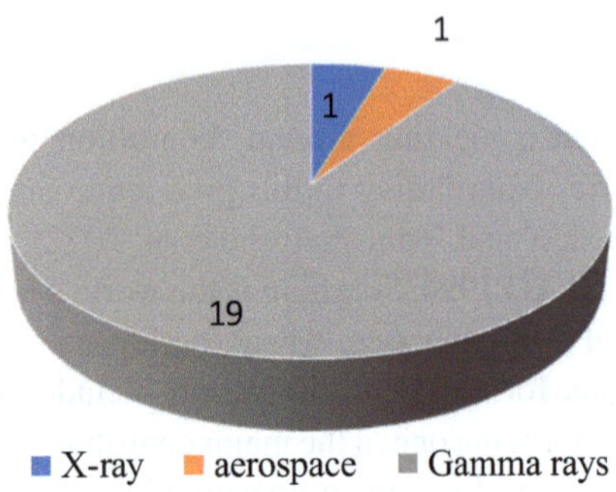

Fig. 3 Sorghum mutant varieties released through physical mutagens

2.1.3 Varieties Released Through Physical Mutagens

There are 21 mutant varieties of sorghum has been released using physical mutagens. Among them, the majority of varieties were released through gamma rays, which contributed 19 varieties. One variety was developed using X-rays, and one variety resulted from aerospace-induced mutation (Fig. 3).

2.2 Oat

Oat (*Avena sativa*) is a major cereal crop cultivated mainly for food and animal feed. It is well known for its excellent nutritional profile, particularly its high fiber, protein, vitamin, and mineral content. Oats are commonly used to produce oatmeal, porridge, and various baked goods. It contains a good amount of beta-glucan, which helps to reduce cholesterol and improve heart health. Oats are a hardy crop that can grow in cool climates and poor soils, making them suitable for many regions worldwide. It is an important export product for Chile. Due to its health benefits and adaptability, oat continues to be an important crop in both human and animal diets.

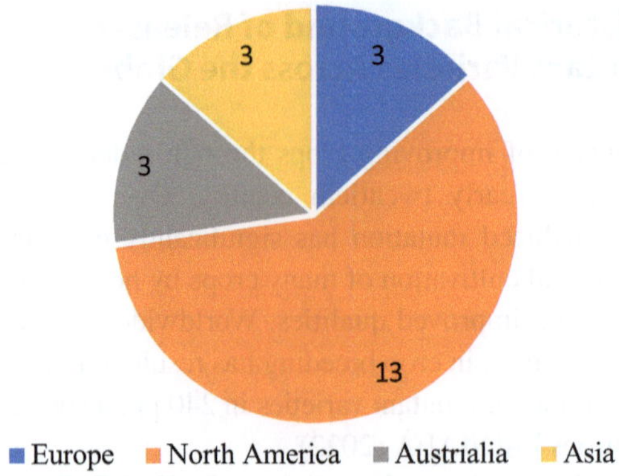

Fig. 4 Oat mutant varieties released across the continents

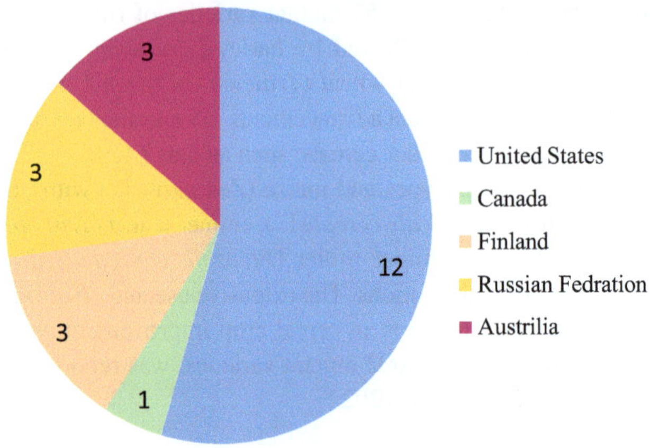

Fig. 5 Oat mutant varieties released by different countries

2.2.1 Oat Mutant Varieties Released Worldwide

There are 22 Oat mutant varieties that have been released through mutation breeding. Among different continents, North America leads with 13 varieties, followed by Europe, Australia, and Asia, each contributing three varieties (Fig. 4).

2.2.2 Oat Mutant Varieties Released by Different Countries

United States has released the maximum number (12) of mutant varieties of Oats. It is followed by Australia, Finland, and the Russian Federation, each of which contributed three varieties, while Canada has released one variety (Fig. 5).

2.2.3 Oates Mutant Varieties Released Using Physical and Chemical Mutagens

There are 15 mutant varieties that were developed using physical mutagens. Among them, thN contributed 10 varieties, X-rays were used for four varieties, and FN was responsible for one variety development (Fig. 6). Through chemical mutagen treatment two mutant varieties were released.

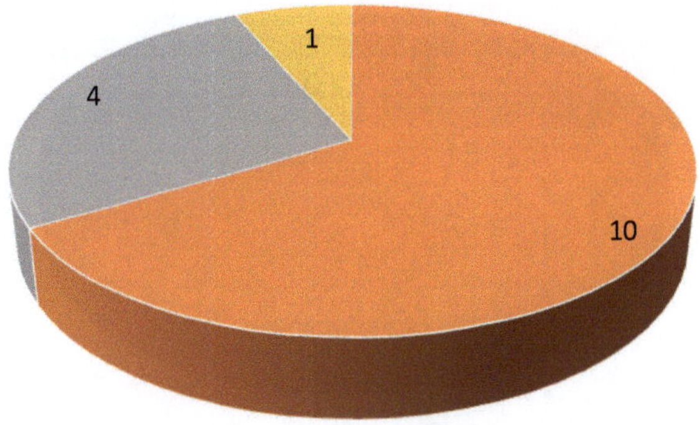

Fig. 6 Oat mutant varieties released through physical mutagens

2.2.4 Pearl Millet

Pearl millet ranks third among food crops grown in India due to its ability to thrive in dry areas, low-nutrient soils, and hot climates. With the help of gamma rays, five varieties were released with resistance to downy mildew, *Sclerospora graminicola* (ICMH-451, New hybrid Bajra-5, NHB-36, NHB-4, Pusa 46) (Fig. 7).

2.2.5 Foxtail Millet

In warm and temperate climates, the genus *Setaria* is extensively cultivated. Among the genus, foxtail millet has the highest ecological value. One of the earliest crops ever grown is foxtail millet. In foxtail millet, only China developed 17 varieties while using physical mutagens like gamma rays and fast neutrons with improved characters like resistance to *Sclerospore graminicola,* and resistance to other fungal diseases (Anghu-221, Changwei-74, Changwei-75, Chigu-4, Fugu-3,4,6, etc.) (Fig. 8).

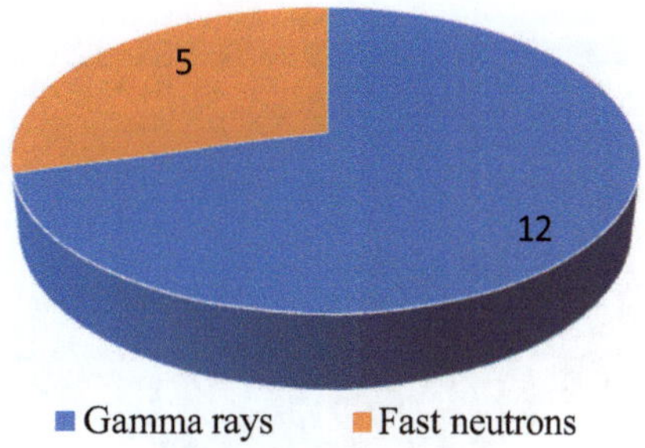

Fig. 8 Foxtail millet mutants varieties developed through radiations

2.2.6 Proso Millet

Proso millet has a short growth season, high nutritional content, and minimal water and fertilizer requirements. However, very little research has been conducted on proso millet, and it has been primarily done in Russia. Proso millet holds great promise for addressing the world's problems of malnutrition, water scarcity, and increasing food production needs. Additionally, it can help to achieve the objective of a varied and sustainable production system (Hoogeveen et al. 2015). With the help of NMU and DMS, mutations were induced in proso millet, and two varieties were released with improved characters like resistance to drought and smut and enhanced cooking quality (Fig. 9).

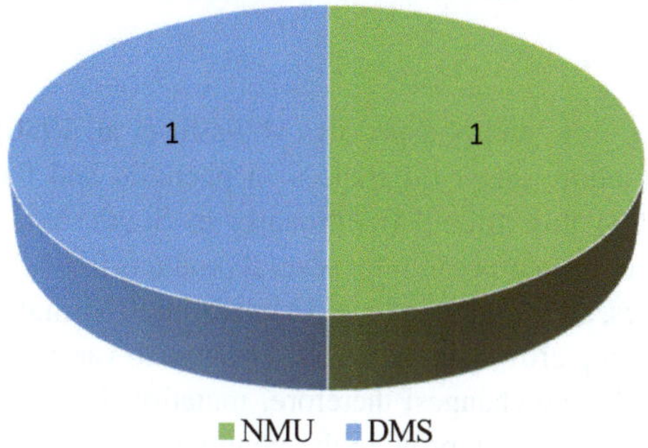

Fig. 9 Proso millet varieties developed through chemical mutagens

2.2.7 Finger Millet

Finger millet is the most crucial millet in some parts of India. It is a very essential source of carbohydrates, dietary fiber, starch, and lipids, with some mineral content, including

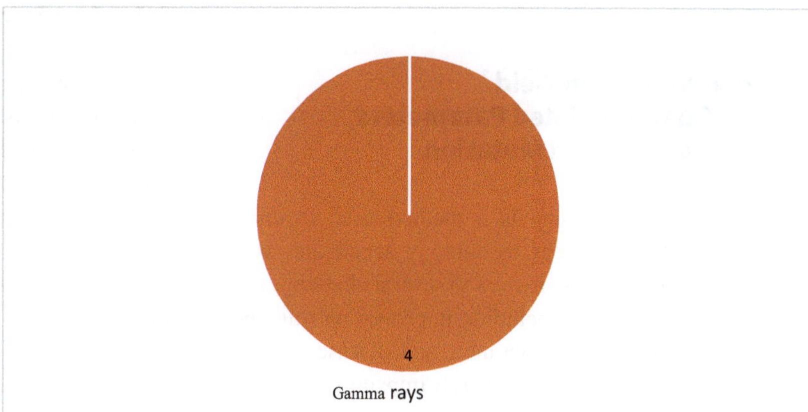

Fig. 7 Pearl millet mutant varieties released through physical mutagen

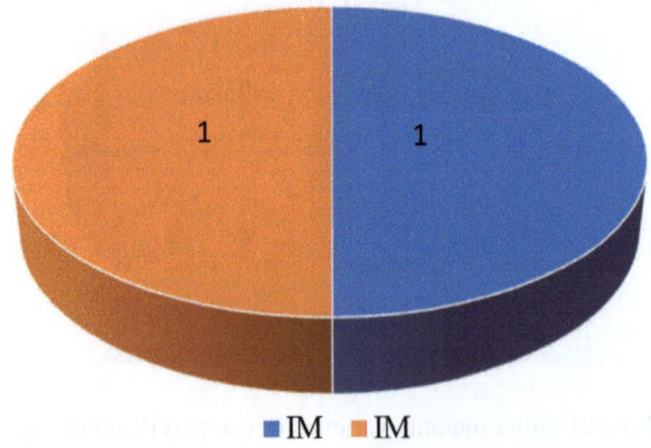

Fig. 10 Finger millet mutants varieties developed through induced mutation

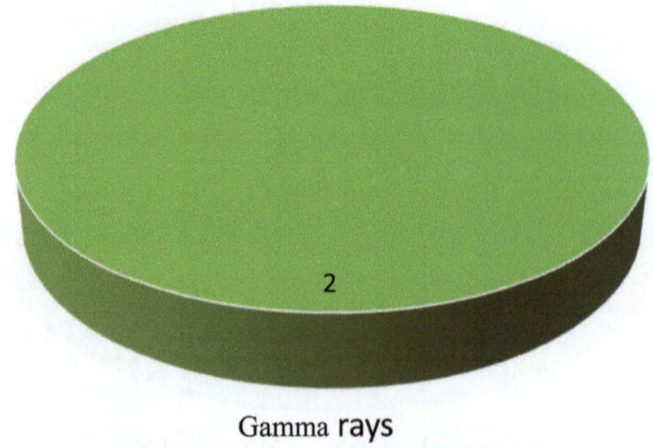

Fig. 11 Rye mutant varieties developed through radiation

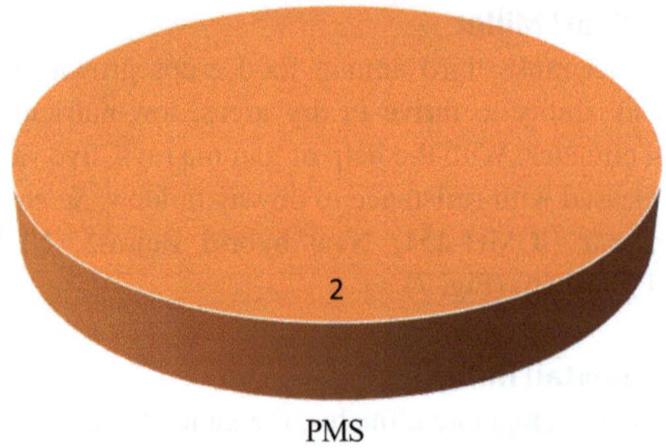

Fig. 12 Rye mutants varieties developed through chemical mutagen

calcium, iron, and phosphorus (Malleshi et al. 1986). Varietal variations show differences in phenolic and flavonoid content in this millet. Traditionally in South Asia, finger millet has been processed for several products like beverages, porridges, idli, dosa, and roti (Malathi and Nirmalakumari 2007). Crop growth and yield are sensitive to various stresses due to climate changes; therefore, mutational breeding is a valuable tool for improving the genotypic traits with desirable results. In finger millet, by using induced mutation, two varieties (FMM165 and FMM 175) were developed in Zambia (Fig. 10).

2.2.8 Rye

Rye is the second most significant crop after wheat used for bakery products, and it is a rich source of dietary fibres and functional bioactive constituent. Despite the complex chemical composition, it has a wide range of nutritional potentials. Rye is more unique for its source of fibre, nutritional content and bioactive substances (Ikram et al. 2023).

In Germany and Finland two mutant varieties (Donar, Hankkija's Jussi (Hja 6900), HJA 6902, Pollux, Donar) each were released by using chemical mutagen (PMS) and physical mutagen (gamma rays) respectively with improved qualities like shorter culm, dwarf and sprouting resistance, tolerant to low temperature, short stiff straw and better quality (Figs. 11 and 12).

3 Improvement in Yield and Quality-Related Parameters Through Induced Mutation

Induced mutation breeding is a method used to enhance crop yield and quality. The technology accelerates genetic progress as well as offers answers to several challenges faced by the agriculture sector, including improved nutritional and processing quality, stress tolerance, and disease resistance. Particularly, climate change is severely impacting the capacity to produce higher yields and nutritional quality. Along with yield enhancement, induced mutation breeding significantly enhances crop quality. Increasing the levels of vitamins, minerals, and essential amino acids in plants can improve their nutritional value and reduce malnutrition, especially in staple food crops such as rice and wheat. Induced mutations could also enhance properties affecting taste, texture, and appearance, including the sweetness of fruits and increasing shelf-life period. Induced mutation significantly increases food security through cost-effective improvements in crop quality and yield.

3.1 Sorghum

Over the years, many mutant sorghum varieties with enhanced agronomic traits like higher yields and improved grain quality have been developed and released for farming. Combining mutation breeding with modern biotechnology has made it faster and easier to create superior sorghum varieties. This approach plays a significant role in enhancing food availability and promoting eco-friendly agriculture, particularly in areas affected by climate change.

3.1.1 Yield and Yield-Related Traits

Various types of mutants, including early maturity, dwarf, altered head morphology, and seed size, as well as other morphological traits, have been reported. With the help of chemical mutagens, Malinovaskii et al. (1975) identified CMS lines through ethyleneimine and NMU treatment. Through gamma-ray-induced mutagenesis, Human et al. (2011) developed mutant lines with enhanced yield and improved quality traits. Three new mutant varieties (Pahat, Samurai-1, and Samurai-2) of sorghum were developed by Soeranto et al. (2020) using different gamma-ray doses. The sorghum varieties Samurai-1, Samurai-2, and Pahat, developed through mutation breeding, were released for their resilience to drought and superior yield characteristics. Wijaya et al. (2022) identified the mutant lines of sorghum which was improved for yield as compared to Pahat, Samurai-1, and Samurai-2.

3.1.2 Nutritional Quality

Offei et al. (2014) reported higher iron, carotenoid, and protein content in mutants compared with controls derived through gamma-ray treatment. Emendack et al. (2022) reported early flowering and a 14 to 24% protein-containing mutant through EMS treatment recommended for animal feed.

3.2 Oats

Improving yield and quality in oats is crucial to meet the growing global for high-quality grain. Although traditional breeding methods have helped to boost oat production, they still face challenges like limited genetic diversity and long breeding cycles. To overcome these challenges, induced mutation breeding has become an effective method for developing new genetic changes and improving important traits in oats.

3.2.1 Yield and Yield-Related Traits

Mutants with higher yield were identified using thermal neutrons (Headquarters et al. 1990; Mitchell et al. 2003) and through X-ray treatment, lodging-resistant mutants of oat were developed with n-nitroso-n-methylurea (FAO/IAEA 2022). Dwarf mutants with higher yield were identified by Mitchells et al. (2003) from thermal neutron treatment.

3.2.2 Abiotic and Biotic Traits

Crown rust-resistant and stem rust resistance mutants in oat were identified by Headquarters et al. (1990) with the help of thermal neutron treatment. Similarly, resistance mutants against Victoria blight and crown rust disease were reported with the help of X-rays by Headquarters et al. (1990).

3.3 Pearl Millet

Pearl millet (*Pennisetum glaucum*) is a prominent grain and fodder source in the dry areas of Africa and Asia. Pearl millet is known for its tolerance to drought and high temperatures. Pearl millet is a significant food security crop, providing a decent livelihood to millions of farmers. Still, improved pearl millet varieties would need to meet the needs of higher yield, disease resistance, and increased nutritional value. Pearl millet forms an important source of food for the people of Africa and South Africa. Mutation breeding is a strong tool for crop improvement, from which genetic diversity can be exploited for the selection of superior traits. Unlike conventional breeding, which relies on the natural diversity in the gene pool, mutation breeding creates new phenotypic traits by inducing such mutations in the genetic constitution of an individual through physical or chemical mutagens.

3.3.1 Yield and Yield-Related Traits

According to Burton and Hanna (1982), stable cytoplasmic male sterility was reported following treatment with Streptomycin at 200 and 500 ppm, and Mitomycin at 50 ppm. Addai and Salifu (2016) treated the Naara variety with 100, 200, and 300 Gy doses and reported that enhancements in agronomic traits were recorded as a result of gamma-ray treatments. The increased number of effective tillers, early maturity, and higher yield were induced under 100 and 200 Gy doses. A moderate dose of gamma rays (100Gy and 200Gy) had a beneficial effect on some yield-related traits of Naara as evidenced by these results.

Ambli and Mullainathan (2018) studied mutagenic effect of gamma and EMS in Bajra where all the mutagenic treatments were effective in improving most of the quantitative traits. Compared with the control and other treatment levels, the average plant height, 1000-grain weight, yield per plant, chlorophyll content, and carbohydrate content were significantly increased by the 20 kR gamma ray and 30 mmol EMS treatments. It was noted that at the maximum concentration of EMS (40 mmol) and gamma rays (30 kR), plant performance was poorer than that of the control as well as from lower concentration treatments, which could reveal that excess mutagenic levels may exert detrimental effects on plant growth and yield. Ati and Shehu (2023) reported a mutant with higher yield through sodium azide treatments.

3.3.2 Resistance to Biotic Disease

Murty (1980) developed downy mildew-resistant mutant plants through gamma-ray treatment. Ati and Shehu (2023) reported downy mildew-resistant plants through sodium azide treatments.

3.3.3 Nutritional Quality

Ambli and Mullainathan (2018) observed the impact of gamma rays and ethyl methanesulfonate on carbohydrate contents, which were significantly increased by the 20 kR gamma ray and the 30 mmol EMS treatments. The inhibition level of PPO and POX, the enzymes that affect the shelf life of Bajra, was studied. During this study, the impact of gamma radiation on Bajra was assessed. The data reveal that grains irradiated with gamma radiation possessed 13–15% less PPO activity, while POX activity decreased to a greater extent, ranging from 53 to 55%. This data is evidence that gamma irradiation inhibits POX more than PPO. Gamma irradiation improves the taste, appearance, and flavor of Bajra by inhibiting these enzymes.

Singh et al. (2023) studied the effect of gamma rays on seed storage quality, reported enhanced shelf life and higher protein solubility, and reduced antinutrients like phytic acid and tannins under storage conditions.

3.4 Foxtail Millet

3.4.1 Yield Yield-Related Traits and Nutritional Quality

The effect of EMS on foxtail millet at various concentrations (0.15%, 0.30%, 0.45%, 0.60%, and 0.75%) yielded significant findings regarding growth, yield, and nutritional content. At 0.45% EMS treatment, there was an increase in the number of effective tillers, while 0.30% concentration led to a rise in total grain weight. In comparison to other doses, riboflavin levels were highest at a 0.30% EMS concentration, and proline content increased at 0.75%. Overall, the findings suggest that EMS treatment at specific concentrations can enhance the growth characteristics and nutritional value of Foxtail Millet (Gopinath et al. 2017).

The impact of nitrous acid on plant growth attributes was studied by Esson et al. (2017), who observed moderate stimulation in plant growth parameters at a 0.1% concentration. This implies that 0.1% nitrous acid is beneficial for growth, although not optimally efficient in terms of yield. The nitrous acid treatment improved various growth parameters, and the treatments were considered ideal for optimizing plant growth and yield.

3.5 Proso Millet

Proso millet grows in various soil types and climates, requiring a short growing period and minimal water. Additionally, it adapts well to high elevations and plateau environments Matz (1986). Evaluating the diversity of proso millet, which has various agronomic traits and grain quality characteristics, can significantly aid in breeding high-yielding cultivars, diversifying crops, and reducing micronutrient and protein malnutrition. This assessment is essential for enhancing crop improvement efforts (Vetriventhana and Upadhyaya 2018).

3.5.1 Yield and Yield-Related Traits

Gamma radiation treatment was used for the induction of variability, where the highest variability in different morphological traits was reported at a lower dose, as this lower dose of radiation encouraged such a response Kate et al. (2018). The mutagenic effect of gamma rays (100to 1000 Gy) on the ATL1 variety was evaluated in terms of yield-attributing traits, including panicle length, number of tillers, number of panicles, and grain yield per plant. From the study, it was observed that a 400 Gy dose of gamma rays was most efficient for yield improvement (Francis et al. 2021).

Effect of sodium azide on three millet varieties (Pavlodarskoe 4, K-10275-Kvartet, and PI289324) was studied, and yield and yield-related parameters were evaluated. From these results, it was observed that both the treatment time and the concentration of NaN3 affect germination and yield (Zeinullina et al. 2023).

3.6 Little Millet

Little millet is a homozygous self-pollinating crop. The strains released have primarily been products of either mass selection or pure-line selection. Due to the problem with the achenes on fragile pedicels, difficulty in hybridization has been reported; therefore, breeding has not played an active role in improvement. The induced mutagenesis technique is utilized to create variability and to select lodging-resistant and high-yield little millet varieties.

3.6.1 Yield and Yield-Related Traits

CO3 and CO(Samai)4 little millet varieties treated with gamma radiation, 300 to 700 Gy doses. In all mutagenic treatments, the genetic advance and heritability were greater for the number of total tillers and grain yield in both varieties (Nirmalakumari et al. 2007). Effects of different doses of gamma radiation, 5 to 50 kR, on various growth parameters of Samai CO-4(little millet) was studied. The study showed that germination and seedling growth declined with increasing doses of gamma irradiation. Under 20 kR treatment, significant seed germination and overall growth were observed (Ramkumar and Dhanavel 2020).

3.7 Finger Millet

Intake of finger millet based food reduces the chances of diabetes, cardiac problems, and gastrointestinal cancers and is mainly suggested as a substitute cereal for patients of celiac disease (Chandrashekara and Shahidi, 2010).

3.7.1 Yield and Yield-Related Traits

Nayar et al. (1979) developed an early maturity variety with increased yield through gamma rays. Induced mutation breeding was studied in varieties of finger millet, i.e., GPU 26 and VR 708, using ethyl methanesulfonate, nitroso guanidine, and gamma rays alone and in combination. Among all the treatments, EMS has produced more mutants compared to other treatments. The frequency of high yielding mutants with number of tillers, number and length of finger and productivity were found higher in 0.30% and 0.45% EMS, along with 0.030% NG treatment as well as the combined treatment of 300 Gy gamma rays with 0.30% EMS shows best results (Muduli and Misra 2007). Finger millet cultivars Dapoli and Dapoli-1 were exposed to gamma-rays. From the Dapoli-1 cultivar, two early maturing and three high-performing variants were identified from 500 and 600 Gy gamma-ray treatments, respectively (Ambavane et al. 2014). Mutations were induced in the TRY1 variety by using ethyl methanesulfonate (EMS), where 0.15% EMS treatment was found to be most effective treatment, which produces the desirable traits like difference in height, tiller number, finger length, grain weight, and grain yield compared with other treatments (Eswari et al. 2014).

In the finger millet, Co-13 variety was treated with EMS and diethyl sulphate. In M2 and M3 generations, improved plant types were identified with plant height, length of leaf, number and finger length, time of first bloom, yield enhancement, and weight of grains, where EMS was found to be slightly effective than DES (Aviya and Mullainathan 2018). In CO-13 variety, nine mutants were identified with desirable traits at 0.03% methyl glyoxal treatment, like ear head, fingers per ear head, high yield, and early maturity (Nivetha et al. 2018).

Finger millet cultivars MR 6, GPU 28, and GPU 67 were exposed to gamma rays, four desirable mutants were identified, each from the GPU28 and GPU67, and one from MR-6 variety with higher grain yield and other related characters like tillers per plant and showing resistance to blasts (Ganapathy et al. 2021). With the help of gamma rays (100 Gy-1000 Gy), mutants were induced in Paiyur-2 variety, and it was concluded that lower doses could be necessary for crop improvement (Sellapillaibanumathi et al. 2021). Behera et al. (2022) have induced mutation in GPU-28 and KMR 204 varieties with gamma rays (200–500 Gy) and high yielding mutants were reported from 300 and 400 Gy treatment.

3.8 Kodo Millet

Kodo millet is a member of the family Poaceae and also commonly referred to as cow grass and ditch millet. It is native to India, where it is also known as the land of origin of kodo millet. The domestication of kodo millet (*Paspalum scrobiculatum* L.) took place around 3000 years ago. The tropical and sub-tropical zones are most suitable for its production (Saxena et al. 2018). In India, its cultivation is limited to Chhattisgarh, eastern Madhya Pradesh, Gujarat, Karnataka as well as several regions of Tamil Nadu. This type of species that reaches up to 90 cm in stature with different grain colors.

3.9 Yield and Yield-Related Traits

Two kodo millet varieties ATL 1 and CKMV 1 were treated with gamma rays (300 Gy-700 Gy). Two early maturing mutants were isolated from ATL 1 at a 400 Gy dose, and three high-yielding mutants from CKMV 1 variety were reported at 500 Gy dose (Vaithiyalingan et al. 2024).

4 Conclusion

Mutation breeding has proven to be an effective tool for improving quality and yield traits in large cereal crops, such as barley, wheat, sorghum, and oats. Through the use of chemical and physical mutagens, significant genetic variability has been generated, leading to the development of high-yielding and nutritionally enhanced mutants. Mutagenesis has helped to overcome genetic limitations and introduced desirable traits such as early maturity, improved grain size, higher protein content, and stress tolerance. This demonstrates how mutation breeding can help to meet the increasing global need for better nutrition and food. In the future, combining mutation breeding with modern genetic tools can speed up crop improvement. This will lead to the development of better cereal varieties with higher yields, improved quality, and the ability to adapt to changing environments.

References

Addai I, Salifu B (2016) Selection of mutants with improved growth and total grain yield in the M2 generation of Pearl millet (*Pennisetum glaceum* L.) in the northern region of ghana. J Agron 15(2): 88–93.https://doi.org/10.3923/ja.2016.88.93.

Ambavane A, Sawardekar S, Sawantdesai S, Gokhale N (2014) Studies on mutagenic effectiveness and efficiency of gamma rays and its effect on quantitative traits in finger millet (*Eleusine coracana* L. Gaertn). J Radiat Res Appl Sci 8:120–125. https://doi.org/10.1016/j.jrras.2014.12.004

Ambli K, Mullainathan L (2018) Effect of mutagens on quantitative characters in M3 generation of Pearl millet (*Pennisetum typhoides* (Burn. F) Stapf. And C. E. Hubb.) J Phytol 10:1–5. https://doi.org/10.25081/jp.2018.v10.3404.

Ati M, Shehu H (2023) Inducing genetic variability in Pearl millet using sodium azide and nitrous acid. Trends Agric Sci 2(2):106–111. https://doi.org/10.17311/tas.2023.106.111

Aviya K, Mullainathan L (2018) Studies on effects of induced mutagenesis on quantitative characteristics of finger millet (*Eleusine coracana* (l.) gaertn.) var-co 13 in m2 and m3 generations. Int J Contemp Stud 3(4):1–6. https://doi.org/10.25081/hbr.2018.v4.3485

Behera K, Rangaiah S, Vyshnavi R, Vaidya N, Sindhu K, Barela A, Prajapati S (2022) Exploring the relationship between yield and its attributing traits in the induced (M2) and recurrent irradiated (RM2) population of finger millet (*Eleusine coracana* L. Gaertn) using path coefficient analysis. Int J of Environ Clim Chang 12(12):1871–1879. https://doi.org/10.9734/ijecc/2022/v12i121676

Burton G, Hanna W (1982) Stable cytoplasmic male sterile mutants induced in Tift 23DB, pearl millet with mitomycin and streptomycin. Crop Sci 22(3):651–652. https://doi.org/10.2135/cropsci1982

Chandrasekara A, Shahidi F (2010) Content of insoluble bound phenolics in millets and their contribution to antioxidant capacity. J Agric Food Chem 58:6706–6714. https://doi.org/10.1021/jf100868b

Charles H, Godfray H, Garnett T (2014) Food security and sustainable intensification. Philos Trans 369(1639):20120273. https://doi.org/10.1098/rstb.2012.0273

Ehrlich P, Harte J (2015) Opinion: To feed the world in 2050 will require a global revolution PNAS 112(48):14743–14744. https://doi.org/10.1073/pnas.1519841112

Emendack Y, Xin Z, Hayes C, Burrow G, Sattler S, Bean S, Smolensky D (2022) Registration of three new bmr12 sorghum mutant from an ethyl methanesulfonate induced BTx623 mutant population. J Plant Regist 1–6. https://doi.org/10.1002/plr2.20219

Esson E, Adamu A, Adelanwa M, Adebola M (2017) Nitrous acid-induced variations in some quantitative traits of foxtail millet. (*Setaria italica* (L.) P. Beauv.) J Sci Agric 1:303–306. https://doi.org/10.25081/jsa.2017.v1.844.

Eswari K, Gogulan G, Hari P, Sakila M (2014) Development of early maturing mutants in finger millet. JAFS 2(6):1–9. https://www.isca.in/AGRI_FORESTRY/Archive/v2/i6/1.ISCA-RJAFS-2014-026.pdf

FAO (2009) How to feed the World in 2050. https://doi.org/10.1111/j.1728-4457.2009.00312.x

FAO (2011) State of Food and Agriculture. ISSN 0081–4539

FAO/IAEA (2022) Mutant variety database. https://mvd.iaea.org/#!Search.

Francis N, Rajasekaran R, Krishnamoorthy I, Muthurajanc R, Thiyagarajand C, Alagarswamye S (2021) Gamma irradiation to induce beneficial mutants in Proso millet (*Panicum miliaceum* L.): an underutilized food crop. Int J Radiat Biol 98(1):1–33. https://doi.org/10.1080/09553002.2022.2024292

Ganapathy K, Patro T, Palanna K, Das I, Elangovan M, Prashant B, Venkatesha S, Anuradha N, Nagaraja T, Tonapi V (2021) Development of improved mutants for grain yield and related traits in Finger Millet (*Eleusine coracana* L. Gaertn.) Through Gamma Mutagenesis. Int J Plant Sci 33(18):225–235. https://doi.org/10.9734/ijpss/2021/v33i1830592

Gopinath L, Saranaya M, Archaya S, Bhuvaneswari R (2017) Genetic improvement of Foxtail millet from Kolli hills landraces through chemical mutation. Indian J Appl Res 7(11):304–306.https://www.worldwidejournals.com/indian-journal-of-appliedresearch%28IJAR%29/recent_issues_pdf/2017/November/November_2017_1509970153__181.pdf

Headquarters A, FRG AM, Maluszynski M, Amano E, Ashri A, Halgand L (1990) Symposia Reports. https://www.iaea.org/publications/3710/plant-mutation-breeding-for-crop-improvement

Hoogeveen J, Faures J, Peiser L, Burke J, Giesen N (2015) Globwat-A global water balance model to assess water use in irrigated agriculture. Hydrol Earth Syst Sci 12(1):3839–3844. https://doi.org/10.5194/hess-19-3829-2015

Human, S., Andreani, S., Sihono and Indriatama, W.M. (2011) Stability test for sorghum mutant lines derived from induced mutations with gamma ray irradiation. At Indones 37(3):102–106. https://doi.org/10.17146/aij.2011.76

Ikram A, Saeed F, Noorb R, Imranb A, Rasheeda M, Islam F, Iqbala A, Zahoorc T, Nazc S, Waheedc W, Shahida M, Khand A, Kinki A (2023) A comprehensive review on biochemical and technological properties of rye (*Secale cereale* L.). Int J Food Prop 26(1):2212–2228. https://doi.org/10.1080/10942912.2023.2244697.

Kate S, Desai S, Bhave S, Thorat B, Bal C (2018) Mutagene induced variability in Proso millet. (*Panicum miliaceum* L.) Int J Chem Stud 6(5):13–16. https://www.chemijournal.com/archives/2018/vol6issue5/PartA/6-4-806-804.pdf

Kharkwal M, Pandey R, Pawar S (2004) Mutation breeding for crop improvement. A book published by Springer, pp 601–645. https://doi.org/10.1007/978-94-007-1040-5

Kharkwal M (2012) A brief history of plant mutagenesis. In: Plant mutation breeding and biotechnology. A book published by CABI, pp 21–30. https://doi.org/10.1079/9781780640853.0021.

Laskar R, Laskar A, Raina A, Khan S, Younus H (2018) Induced mutation analysis with biochemical and molecular characterization of high yielding lentil mutant lines. Int J Biol Macromol 109:167–179. https://doi.org/10.1016/j.ijbiomac.2017.12.067

Malathi D, Nirmalakumari A (2007). Cooking of small millets in Tamil Nadu. In: Krishne Gowda KT, Seetharam A (eds) Food uses of small millets and avenues for further processing and value addition: Indian Council of Agricultural Research. Project coordination cell, All India Co-ordinated small millets improvement project. Indian Council of Agricultural Research, UAS, GKVK, Bangalore, India, pp 57–63. https://www.ijfans.org/uploads/paper/78007891b134293f93d3853c7f7b0949.pdf

Malinovskii B, Zoz N, Kitaev A (1975) Induction of cytoplasmic male sterility in sorghum by chemical mutagens. Genetika 9(6):682. PMID: 4798243

Matz S (1986) Millet, wild rice, adlay, and rice grass. In: Cereal science. Avi Press, Westport, CT. 225–229

Malleshi N, Desikachar H, Tharanathan R (1986) Free sugars and non-starchy polysaccharides of finger millet (Eleusine coracana), pearl millet (*Pennisetum typhoideum*), foxtail millet (*Setaria italica*) and their malts. Food Chem 20:253–261. https://doi.org/10.1016/0308-8146(86)90095-6

Mba C, Guimaraes E, Ghosh K (2012) Re-orienting crop improvement for the changing climatic conditions of the 21st century. Agric Food Secur 1:7. https://doi.org/10.1186/2048-7010-1-7

Mitchell Fetch JW, Brown PD, Duguid SD, Chong J, Haber SM, Menzies JG (2003) Ronald oat. Can J Plant Sci 83(1):101–104. https://doi.org/10.4141/P02-019

Muduli K, Misra R (2007) Efficacy of mutagenic treatments in producing useful mutants in finger millet (*Eleusine coracana* Gaertn.). Indian J Genet 67(3):232–237

Murty B (1980) Breakthrough in breeding for resistance to downy mildew in pearl millet. Bull Organ Eur Mediterr Prot Plant JO 311.https://doi.org/10.1111/j.1365-2338.1980.tb02703.x

Nayar K, Pushpa G, Reddy B (1979) "Open ear headHamsa" true breeding early mutants of finger millet Current Research 8:212–213. https://api.semanticscholar.org/CorpusID:82378925

Nirmalakumari A, Arulselvi S, Ganapathy S, Souframanian J, Senthil N, Devan P (2007) Gamma ray induced variation for lodging resistance and its associated characters in little millet. (*Panicum sumatrense*

Roth Ex-roem and schult) Madras Agric J 94(7–12):151–155. https://doi.org/10.29321/MAJ.10.100656
Nivetha C, Nivetha S, Palani M, Sakila M (2018) Effect of mutagenic treatments of methyl glyoxal in producing high yielding mutants in finger millet (*Eleusine coracana*, gaertn). Int J Agric Sci Res 8(1):137–146. https://ssrn.com/abstract=3195205
Offei K, Danquah E, Owusu-Darko R, Eleblu J, Adjei E (2014) Improving food and nutritional security in Ghana through mutation breeding of Sorghum. In: book: mutagenesis: exploring novel genes and pathways, pp 125–142. https://doi.org/10.3920/978-90-8686-787-5_5
Pathirana R (2011) Plant mutation breeding in agriculture. CABI Reviews 6(032):1–20. https://doi.org/10.1079/PAVSNNR20116032
Raina A, Laskar R, Khursheed S, Amin R, Tantray Y, Parveen K, Khan S (2016) Role of mutation breeding in crop improvement past, present, and future. Asian Res J Agric 2(2):1–13. https://doi.org/10.9734/ARJA/2016/29334
Ramkumar R, Dhanavel D (2020) Studies on effect of induced physical mutagenesis on seed germination, seedling survival and lethal dosage of little millet. (*Panicum sumatrense* roth ex roemer & schults) Int J Botany Stud 5(2):152–156. https://www.botanyjournals.com/assets/archives/2020/vol5issue2/5-2-48-421.pdf
Saxena R, Vanga S, Wang J, Orsat V, Raghavan V (2018) Millets for food security in the context of climate change: a review. Sustainability 10:1–31. https://doi.org/10.3390/su10072228
Sellapillaibanumathi L, Dhanarajan A, Raina A, Ganesan A (2021) Effects of gamma radiation on physio-morphological traits of finger millet (*Eleusine coracana* (L.) Gaertn.). Plant Sci Today 1–7. https://doi.org/10.14719/pst.1142
Shah J, Hamid U, Rahim D, Malik M, Yusuf Z (2000) Development of sugarcane mutants through *in vitro* mutagenesis. Pak J Biol Sci 3(7):1123–1125. https://doi.org/10.3923/pjbs.2000.1123.1125
Shu Q, Forster B, Nakagawa H (2012) Plant mutation breeding and biotechnology. A book published by CABI. https://doi.org/10.1079/9781780640853.0000
Singh T, Ansheef A, Ranjeet R, Vinutha T, Suneha G (2023) Effect of gamma irradiation and variability in polyphenol oxidase and peroxidase in determining flour quality of pearl millet during storage. J Pharm Innov 12(1):1130–1134. https://www.thepharmajournal.com/archives/2023/vol12issue1/PartL/12-1-153-217.pdf
Soeranto H, Sihono WM (2020) Sorghum improvement program by using mutation breeding in Indonesia. Earth Environ Sci 484:1–5. https://doi.org/10.1088/1755-1315/484/1/012003
Surashe S, Kalpande H, Ashok B, Ambika M, Ganapathi T (2022) Mutation-induced phenotypic variation for yield traits in sorghum [*Sorghum bicolor* (L.) Moench]. Electron J Plant Breed 13(2):312–318. https://doi.org/10.37992/2022.1302.053
Usharani K, Kumar C (2015) Mutagenic effects of gamma rays and EMS on frequency and spectrum of chlorophyll mutations in urdbean (*Vigna mungo* (L.) Hepper). Ind J Sci Technol 8(10):927–933. https://doi.org/10.17485/ijst/2015/v8i10/54201
Vaithiyalingan M, Nirmalakumari A, Manimozhi V (2024) Mutation effectiveness and efficiency in kodo millet (*Paspalum scrobiculatum* L.). Electron J Plant Breed 15(2): 454–458. https://doi.org/10.37992/2024.1502.052
Vetriventhana M, Upadhyaya H (2018) Diversity and trait-specific sources for productivity and nutritional traits in the global Proso millet (*Panicum miliaceum* L.) germplasm collection. Crop J 6(5):451–463. https://doi.org/10.1016/j.cj.2018.04.002
Wijaya I, Galuh G, Teguh S, Sihono SH (2022) Yield productivity test and morphological characterization of 19 sorghum lines resulted from mutation breeding. Conf Ser: Earth Environ Sci 484:1–9. https://doi.org/10.1088/1755-1315/484/1/012022
Zeinullina A, Rysbekova A, Dyussibayeva E, Zhirnova I, Zhanbyrshina N, Zhunusbayeva Z, Yancheva S (2023) Application of sodium azide for chemical induced mutagenesis of Proso millet culture. (*Panicum miliaceum* L.) Bulg J Agric Sci 29(4):623–631. https://doi.org/10.5555/20230358717

Advancing Sugarcane Breeding Through Induced Mutagenesis and Somaclonal Variation—Prospects, Challenges, and Opportunities

Madhavi V. Purankar, Rachayya M. Devarumath, Dipak K. Sarode, and Penna Suprasanna

Abstract

Sugarcane is a critical crop for global sugar and bioenergy production. The genetic improvement of sugarcane is challenging due to complex polyploidy, long breeding cycles, and susceptibility to biotic and abiotic stresses. This chapter explores the transformative potential of induced mutagenesis and somaclonal variation as tools for advancing sugarcane breeding. Induced mutagenesis, employing chemical or physical agents, offers a viable solution by harnessing ability to create novel genetic variation, while somaclonal variation utilizes natural diversity arising from tissue culture processes. Together, these techniques have successfully contributed to improvements in traits such as disease resistance, drought tolerance, and agronomical parameters. This chapter explores the methodologies, advantages, and applications of induced mutagenesis and somaclonal variation in sugarcane breeding. Furthermore, it highlights the prospects of integrating these methods with modern genomic tools, such as CRISPR-based editing for the development of superior sugarcane varieties. Key challenges, including the stability of variations and advantages of these techniques, are critically analyzed to provide a balanced perspective. This chapter illustrates the pivotal role of induced mutagenesis and somaclonal variation in shaping the future of sugarcane breeding for sustainable agriculture. It aims to provide researchers and breeders with insights and actionable strategies for exploiting these tools in the pursuit of resilient and high-performing sugarcane varieties.

Keywords

Sugarcane · Mutagenesis · Crop improvement · In vitro tissue culture · Somaclonal variation

1 Introduction

Somaclonal variation term is coined by Larkin and Scowcroft to describe the variation that arises in plants by tissue culture. Their experiments suggested that cultured plant cells can exhibit genetic changes sometimes in exceptionally higher frequencies. These observations changed the thinking that plant cultured cells behave similar to microbes, with a similar lower rate of expected mutation frequencies (10^5–10^6) (Larkin and Scowcroft 1983). Induced mutagenesis in crops was initiated in the late 1920s after a pioneering experiment by Lewis Stadler proved that X-rays can induce heritable mutations in crop plants such as maize and barley. In sugarcane, induced mutagenesis and somaclonal variation have provided effective means to broaden the narrow genetic basis of this complex and vegetatively propagated crop. In situations where conventional breeding faces multiple challenges like polyploidy, limited recombination, and heterozygosity, these techniques have opened avenues to create novel variations to unlock the true genetic potential of this crop.

The modern sugarcane cultivars are arisen by crossing between two species *S. spontaneum* (disease resistant) and *S. officinarum* (high sugar content) and further backcrossing of progeny with *S. officinarum* termed as nobilization and intercrossing between generated the hybrids. This led to allopolyploidy in sugarcane for containing a high number of chromosome copies (2n = 99–130) and aneuploidy (2n + n) (Piperidis and D'Hont 2001). In nature, the combination of

M. V. Purankar · R. M. Devarumath (✉) · D. K. Sarode
Molecular Biology and Genetic Engineering Section, Vasantdada Sugar Institute, Pune, Maharashtra, India
e-mail: rdevarumath@gmail.com

P. Suprasanna
Nuclear Agriculture and Biotechnology Division, Bhabha Atomic Research Centre, Mumbai, India

Amity Centre for Nuclear Biotechnology, Amity Institute of Biotechnology, Amity University, Mumbai, Maharashtra, India

J. Devi (ed.), *Mutation Breeding in Crop Improvement*, Sustainable Landscape Planning and Natural Resources Management,
https://doi.org/10.1007/978-3-032-17911-1_15

the divergent genome within hybrids containing the bispecific origin of chromosomes, its structural differences, and interspecific chromosome recombination could cause "genomic shock." Genomic shock profoundly causes genome modifications such as transposon activation, deletions, insertions, duplications, or chromosome alterations. It may also cause epigenic changes such as a change in DNA methylations. Most importantly, it also leads to genetic redundancy which further may result in diversification of gene function, adaptation to different environmental conditions, and phenotypic variation (Rutherford et al. 2014). This forms the basic platform of genomic diversification in any species. However, in sugarcane, it seems that major reshaping of genomic structure caused by genome shock has not taken place due to the allopolyploid nature of *S. spontaneum* and *S. officinarum*. Modern cultivars generated through the breeding program have already acquired the majority of genetic diversity present in *S. officinarum* (Shabbir et al. 2021). This basis of low genetic diversity has formed a bottleneck in the sugarcane improvement program. The long-term effects of reduced genetic diversity on the sustainability of sugarcane production can be quite significant. One major concern is the increased vulnerability of sugarcane crops to diseases and pests (Rutherford et al. 2014). With lower genetic variability, populations may lack the necessary adaptations to cope with emerging threats, potentially leading to widespread crop failures. Additionally, reduced genetic diversity can limit the overall adaptability of sugarcane to changing environmental conditions, such as shifts in climate, soil health, and water availability, which further hinder the crop's resilience and productivity over time. Moreover, diminished genetic diversity may affect the long-term viability of breeding programs, making it challenging to develop new cultivars that can thrive under diverse conditions or with enhanced traits, such as improved sugar content or stress resistance. Ultimately, these factors can result in decreased yields and economic viability for sugarcane farmers, causing challenges for the sugar industry and food security as a whole. Additionally, the sugarcane breeding program can take 10–15 years for the development, selection, and release of the variety. Thus biotechnological interventions such as somaclonal variation, i.e., induced genetic variations by tissue culture techniques and/or mutagenesis can facilitate diversification of the germplasm, reduce the time required for development of germplasm, and can also introduce novel traits.

2 Somaclonal Variation and Induced Mutagenesis

Somaclonal variation refers to the changes caused by tissue culture in the genome. Somaclonal variations are likely to occur due to many reasons such as a change in methylation pattern, chromosomal aberrations, and activation of TEs. The use of a mutagenic agent increases the frequency of these events. Today, somaclonal variation and induced mutagenesis remain a significant tool in plant breeding and genetic improvement. It is utilized to develop new cultivars with desirable traits, such as disease resistance, improved yield, and abiotic stress tolerance (Rutherford et al. 2014). Modern breeding programs often integrate somaclonal variation with other techniques, including mutagenesis and advanced molecular breeding, to accelerate the development of improved plants. However, challenges persist, such as maintaining genetic diversity and managing the potential for unintended genetic changes. Innovations in molecular techniques, coupled with traditional breeding strategies, continue to enhance the understanding and application of somaclonal variation in crop improvement, particularly in dynamic agricultural contexts where adaptability is crucial.

3 Molecular Basis of Somaclonal Variation and Induced Mutagenesis

Several research activities have been reported to check mechanism underlying mutagenesis and somaclonal variation in sugarcane. Wherein, some strategies observed common in other crops and sugarcane are discussed in below sections. Please also refer Fig. 1.

3.1 Change in Methylation Pattern

The addition of methyl group (methylation) is carried out by the group of enzymes called "methyl transferases." DNA methyl transferase adds a methyl group to the C-5 position of cytosine. Methylation is an essential control system for basic cellular functions such as genome imprinting replication, transcription, suppression of replication of repetitive elements, and translation. *Methyltransferase* also carries out histone protein methylation. Change in methylation can increase the hydrophobicity of a particular region of a chromosome which can aid in the attachment of other proteins required for replication, transcription, and translation. The addition of a methyl group to a particular gene is considered as silencing of the gene (Grand-Downton and Dickinson 2005). Mutagenic agents particularly chemical mutagen such as 5-azacytidine reduce the activity of methyl transferases by acting as cytosine analog and form a covalent bond with the enzyme. This hypomethylation/demethylation can induce silenced TEs and transgenes which may further accelerate mutational changes in nucleotide sequences (Rutherford et al. 2014). Use of 5-azacytidine for induced Al tolerance in sugarcane has reported to increase epigenetic variation by nearly 66.6% (Koetle et al. 2023).

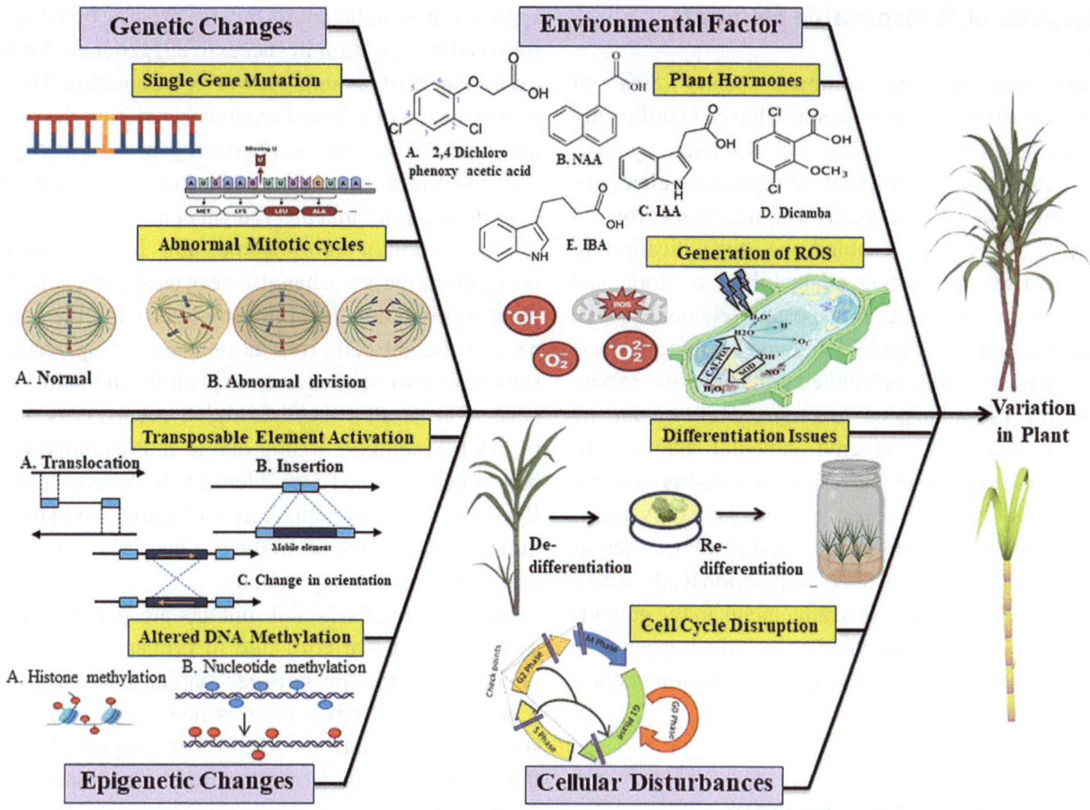

Fig. 1 Molecular basis of mutagenesis and somaclonal variation in plants

3.2 Stress and Generation of Reactive Oxygen Species

Stress can be generated due to rapid differentiation in tissue culture procedures. The cells in tissue culture undergo a rapid division phase. Particularly at the callus phase, many cellular rearrangements are involved when an unorganized group of cells to turn it into a more organized and structured form, such as shoots. This can lead to the formation of stress and the generation of reactive oxygen species (ROS) such as superoxide, hydrogen peroxide, peroxide, and alkoxy radicals.

The ionizing radiations such as gamma radiation can also cause lysis of water present in cell cytoplasm leading to the formation of reactive oxygen species such as hydroxyl ion (·OH), $\cdot^{-}O_2$ superoxide, H· hydrogen radical, and H_2O_2 hydrogen peroxide. The hydroxyl radical (·OH) can react with purines and pyrimidine and form hydroxypurine or hydroxypyrimidine of which 8-hydroxy guanine is a known biomarker for DNA oxidative damage. This hydroxynucleotide can act as a nucleotide analog and can induce point mutations. The ROS can remove hydrogen from the C4 position of deoxyribose sugar of DNA and lead to producing deoxyribose radical which further leads to DNA breaks. DNA breaks are repaired by the DNA repair system of the cell. The higher number of breaks can lead to errors in the reading of the opposite strand and causes base substitution.

3.3 Cytological Changes

The differences in ploidy (aneuploidy, polyploid, diploid, and mixoploid) observed in plant types obtained from colchicine-treated callus indicated that the asynchronous division of multinucleated cells formed during tissue culture might be a reason for chromosomal variation in the sugarcane somaclonal variants.

Chromosomal breaks, deletions, and change in methylation pattern can also cause chromosomal aberrations due to alteration of timing for replication. A normal cell cycle control system keeps a check on the synthesis of required proteins and other biological molecules to prevent cell division before DNA replication. It gets disrupted in tissue culture procedures leading to chromosomal breaks (Phillips et al. 1994). The process of repair of double-stranded breaks in DNA is error-prone. Breaks or deletions can lead to the removal of a specific gene and thus possibly alter the phenotype.

3.4 Activation of Transposable Elements

Transposable elements can comprise nearly 80% of the genome in grasses. Transposons have significantly contributed to the evolution of sugarcane, a C4 plant, by facilitating gene duplication, recruitment of regulatory elements, and modification of genetic pathways necessary for C4 photosynthesis. They have enabled the diversification of regulatory elements, promoting adaptability to biotic and abiotic stresses. In sugarcane, transposable elements (TEs) have shaped its complex polyploid genome, supported stress response mechanisms, and contributed to its genomic expansion, preserving essential gene functions while enhancing photosynthetic efficiency and environmental adaptability. About 2.3% of the sugarcane transcriptome consists of transposable elements (TEs); however, TEs are generally maintained in a silent state by epigenetic modifications such as methylation. Change in control over methylation leads to activation of TEs which can cause rearrangements in the genome or change in gene regulation mechanism by inducing genomic insertions, excisions, or non-homolog recombination of chromosomes.

3.5 Change in Nucleotide Sequence

All the above-mentioned mechanism induces change in nucleotide sequence either by insertions or excisions which may further lead to frame shift mutations, indels, or base substitutions. There are several possibilities that this may lead to either truncated or non-functional protein. Change in catalytic amino acid can hamper activity of the enzymes. Changes in promoter/controlling regions can lead to improvement in turnover time period. Sharma et al. (2023) reported changes in indels in COMT gene in sugarcane mutant produced by gamma radiation.

4 Integration of Mutagenesis and Somaclonal Variation in Sugarcane Breeding

In a somaclonal variation, the process of callus induction followed by the formation of embryogenic callus is generally carried out; however, apical meristem has also been used for obtaining variation through tissue culture. Application of somaclonal variation or induced mutagenesis preliminarily was thought to be an aid to a breeding program for broadening genome of the germplasm.

Two approaches for the use of mutagenesis are considered in sugarcane: first was to apply mutagenic treatment to the sugarcane cuttings followed by a selection of desired clones from the vegetatively generated population; secondly application of mutagenesis into the existing breeding program followed by selection in successive regeneration where there exist chances of obtaining novel recombination. Thus, in addition to the use of adapted sugarcane varieties, the research has also been carried out with hybrids generated through crosses between *Saccharum officinarum*, *Saccharum robustum*, and *Erianthus arundinaceous* (Nagai et al. 1991).

Desiccation of embryogenic callus for 4–6 h in desiccator post gamma radiation has also been used and resulted in higher plant regeneration capability by improving the competitiveness of mutant cells over normal calls. Sugarcane eye bud setts have also been used as explant for the induced mutagenesis (Rao et al. 1966) but higher occurrences of chimeras in MV1 generation arising due to multicellular structure of bud structure posed a problem in the selection of mutants. Recently, successful application of gamma radiation-induced mutagenesis in vivo on single eye bud setts of variety CoJ 85 to obtain high tillering mutant.

Many somaclones and mutants are not directly released as commercial varieties due to primary reason of lack of phenotypic variation between parent and mutant which is required for its registration as new variety and thus many somaclonal variants are released as "improved" versions of the original variety. However, these somaclones/mutants are used as germplasm for breeding programs. One successful example has been observed in our institute where TC-2513 a high sugared and early maturing somaclone of variety VSI434 (which itself is commercially released somaclone of variety CoC 671) has been used under sugarcane breeding program. A variety VSI 21121 formed thorough the cross has been found superior in sugar and yield characteristics under AICRP (All India Co-ordinated research Program) field trials and is under release for Maharashtra region (Data not published—as per communication with Dr. Repale, Sr. Scientist, Breeding section, VSI).

4.1 Methodology for Somaclonal Variation and Induced Mutagenesis

The overall methodology of induced mutagenesis is mentioned in Fig. 2. In somaclonal variation, nearly similar methodology is followed except only callus and embryogenic callus are used as explant and no mutagenic exposure is given.

4.1.1 Explant Selection

The process of development of somaclonal variation involves explant preparation from the intended variety. Usually, locally adapted variety with good agronomic characters susceptible for the specific trait is chosen. Various aspects of the experimentation are considered before the experiment such as identification of a problem, availability of source variety, whether to go for somaclonal variation or for induced mutations, type

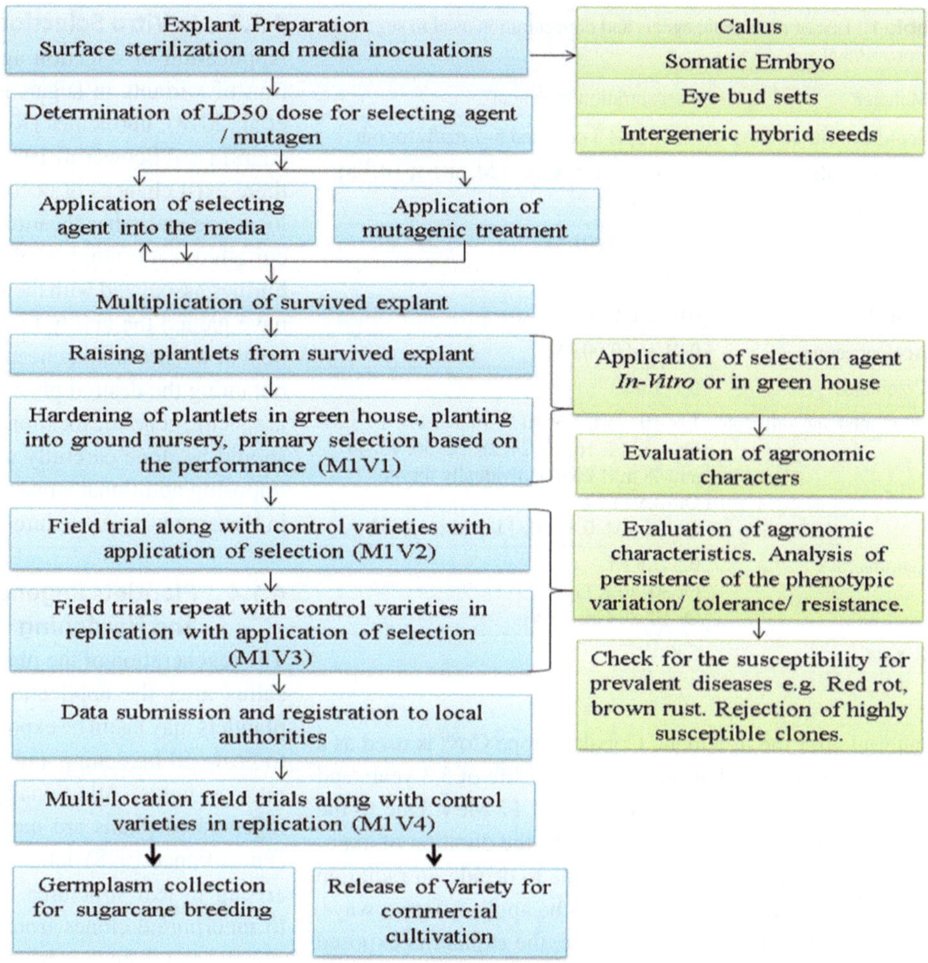

Fig. 2 Process of development of sugarcane variety through induced mutagenesis

of mutagen to be used, safety requirements of the procedure, and ease of handling the exposed material and ability of exposed explant to regenerate into the plant. Figure 2 represents various steps involved in sugarcane variety development through somaclonal variation or induced mutagenesis.

4.1.2 Decision of Mutagenic Dose

In physical mutagens, such as gamma rays, X-Rays, and UV-rays, LD_{50} is first evaluated by subjecting tissue to the different doses of such radiations. LD_{50} is most commonly applied dose as it is considered to be high as to induce genomic variations and not so high to cause lethality in plant cells. In case of chemical mutagens, along with LD_{50} value of mutagen, several properties of mutagen, such as half-life, toxicity, solubility, reactivity, interaction of tissue with the mutagen, stability of mutagen in culture medium, and pH of the medium, are carefully checked before considering it for application on any crop.

In somaclonal mutation, the presence of phytohormones such as 2, 4-D and kinetin is known to induce variations. The concentration of 2, 4-D is used between 3 and 4 mg/L whereas BAP is used between 0.5 and 3 mg/L. The addition of physical and chemical mutagen can capitalize the somaclonal variation by further accelerating the rate of mutation through various mechanisms which have been discussed in the chapter separately.

The application of both physical mutagen, such as X-rays and gamma rays, and chemical mutagens, such as EMS, MMS (methyl methane sulfonate), and colchicine, has been carried out in sugarcane. The choice of mutagen largely depends on the availability of infrastructure and the desired outcome. Chemical mutagen EMS or sodium azide induces point mutations whereas physical mutagens X-rays and gamma rays induce random mutations. When a change in phenotype is controlled by a multigene family a more randomized type of mutation technique followed by selection is used. Gamma radiation is the most commonly used physical mutagen in sugarcane and most other vegetatively propagated species because of their convenience in use, deep penetration ability in biological tissues, and no leftover/added radioactivity in a

Table 1 List of mutagenic agents and concentration used in sugarcane improvement

Mutagen	Concentration/treatment
Sodium nitrite (NaN_3)	0.01%, pH 3 or 5, and 5–7 mg/L for 6 h
Sodium azide	1×10^{-3} M, 5×10^{-3} M, or 1×10^{-2} M
5-Azacytidine	100 μM
Pyrogallol	0.01 and 0.001 M
Pyrocatechol	0.01 and 0.001 M
Quinol	0.01 and 0.001 M
Acetophenone	0.01 and 0.001 M
Colchicine	0.1–1.0%
Ethyl methane sulfonate	1×10^{-3} M, 5×10^{-3} M, and 1×10^{-2} M Or 4, 8, 12, 16, 20, 24, 28, 32, 36, 40, 44 and 48 mM EMS individually for 4 h Or (0.5%) 2 h Or 0.6 μM, 0.8 μM, 1.0 μM for 2–4 h
Gamma rays	0.5–7.5 Kr Or 30–50 K Gy Or 10–80 Gray (Gy)
X-rays	5 Kr

material after the treatment. Cobalt isotope Co^{60} is used as a source of gamma radiation. It has a half-life of 5.3 years and emits radiations of energies 1.33 and 1.17 MeV. Due to the fixed rate of emission from the source, the duration of exposure is varied from sample to sample to decide on radiation dose. The mutagenic treatment can be applied in two ways, acute and chronic. In the acute type, the explant is exposed to a higher concentration of mutagenic agent for a very short period whereas in chronic type the explant is exposed to a lower dose of a mutagenic agent for a relatively higher time exposure.

The success of mutagenesis depends on two basic factors: (1) the method to prevent detrimental effect of mutagenic treatments; (2) the stability of the regenerated population to maintain the desired phenotypic change in the adult stage, which are grown under natural conditions; and (3) ability to transfer the character in next generation with similar potency/vigor. In sugarcane, which is propagated vegetatively through eye setts/buds the chances of loss of characters by chromosomal segregation are less. Chronic type is preferred when the size of explant is larger, e.g., eye buds or setts. A sugarcane variety "Nanei" was developed by application of 420 Gy gamma rays on sets of "Ni 1" variety which showed a higher number of tillers, longer and thicker stocks, and improvement in yield and sugar. Similarly, variety Co 8153 was developed from 150 Gy gamma radiation treatments to hybrid seeds obtained by crosses between two varieties Co 6304 × Co 6406 which showed improvement in yield and better juice quality (Mutant variety database FAO-IAEA). The range of previously used mutagenic agent concentration is enlisted in Table 1.

4.1.3 In Vitro Selection

Application of selection agent has been used for the selection of variants in sugarcane, for example, PEG (Masoabi et al. 2018), herbicides (Koch et al. 2012), and fungal toxin (Larkin and Scowcroft 1983). Through somaclonal variation, there exist chances of occurrence of changes in both desired locations and other locations. Application of selection agent curtails on the number of undesired selections and further hurdles associated with their evaluation. Principally, cells that have gained the ability to survive against particular component at its threshold concentration enhance the likelyhood of obtaining the desired phenotype with required modifications at specific genomic locations. The addition of a selection agent should be done carefully after determining LD50 values as imposing additional stress also increases the rate of lethality in tissue culture procedures.

4.1.4 Plantlets Regeneration, Multiplication, and Hardening

The regeneration of the plantlets is carried out from survived callus after the application of the stress. The regenerated plantlets may again be exposed to the stress in vitro or allowed to grow till hardening and then transplanted into the ground nursery fields. Additionally, it was observed that in vitro regenerated plants are more prone to diseases such as rust (Perez Ponce 1998) due to temporary morphological changes arising in leaf structures. Thus, the protocol should follow to incorporate clones from a resistant variety, a susceptible variety, and from the test variety to check the degree of resistance. Usually, no stress is applied in-ground nursery trials and plantlets are allowed to attain full growth under normal conditions. However, selection based on phenotypic characters can be done to reduce the number of canes in further selection cycles. If buds are used for the mutagenesis, the exposed buds are allowed to germinate in vitro or in the greenhouse. Each tiller is again separately germinated and checked for the desired characteristics to avoid the selection of chimeric tissues.

4.1.5 Agronomic Field Trials

The agronomic field trials are conducted along with control varieties with known resistance or tolerance to the stress imposed and parental varieties. The first field trials are conducted generally in augmented block design/rod row trials due to the non-availability of sufficient material for the propagation but care must be taken to plant sufficient replications of the parent and control varieties for the comparative statistical analysis of the results. Phenotypic, agronomic characters and occurrences of prevalent diseases are recorded in the trial. Selective varieties are forwarded for further agronomic trails with replications. In vitro regenerated population shows an increase in tillering (15–50%) than vegetative bud proliferation, and this effect may persist even after two vegetative generations. Thus, the number of stalks/stool should not be only taken as the measure of selection in agronomic trials of somaclones. In many of the available research articles, either data of field performance has not been provided or data of only first-year trials are provided. For concluding the actual performance of the somaclones/mutant clones, this data needs to be reevaluated in larger field trials. Field data of the superior clones is submitted with local authorities (normally Agriculture Universities of the region) and further field evaluations are carried out at different locations. Based on overall performance and persistence the variety is registered and released for cultivation by farmers.

4.2 Screening of Somaclonal Variation/Mutant Through Physiological, Biochemical, Cytological, and Molecular Markers

Various methods have been used for the analysis of variation in the generated population such as isoenzymes pattern, molecular markers, phenotypic variation, and cytological examinations for the chromosome numbers and the methods applied for the analysis of variation are discussed below.

4.2.1 Physiological and Biochemical Variation

Analysis of phenotypic variation has the basic advantage of simplicity in performance. It is routinely carried out with other examinations such as field analysis. It is especially important in cases where the phenotypic change is the sole purpose. For example, formation of glabrous leaf sheath: spines on the sugarcane leaf sheath are not desired in countries where manual harvesting is practiced, and sugarcane tops are used as cattle feed. Early maturing variety VSI 434 produced through somaclonal variation of CoC 671 callus which has distinct color differences of the stem. The number of chlorophyll variants is often used to discriminate between healthy and off-types. The number of tillers, internode arrangement, flowering, stalk thickness, and stalk height are some of the common phenotypic characters which have been used for the selection of variants.

Nagai et al. (1991) observed changed isoenzymes pattern of three enzymes *peroxidase*, *phosphoglucomutase*, and *phosphoglucoisomerase* in somaclones generated from intergeneric hybrid H62-4671. Limited availability of standardized protocols for the isoenzymes analysis and difficulty in performance and reproducibility limits its application. Further, many enzymes are tissue specific and environmentally influenced thus employed at limited extent in in vitro variation analysis. Koetle et al. (2023) reported that one of the Al tolerant mutants showed nearly 8% higher stress tolerance index with 29% higher relative water content, about 35.8% higher green leaf area and about 23% longer roots over stressed parent control.

4.2.2 Cytological Analysis

The cytoplasmic analysis provides direct evidence of the effect of treatment on change in several chromosomes, additions deletions, and mitotic aberrations. As compared to the morphological markers, this process needs very less time. In sugarcane, occurrences of aneuploidy cells in callus cultures and the presence of mixaploid, diploid, and polyploidy in plants raised from colchicine-induced callus have been reported earlier. The presence of bi-, tri-, and pentanucleate cells in root tip cells and also variation in the number of chromosomes in raised plantlets have been found in somaclonal variants. The absence of a reference set of chromosomes for comparison and the time-consuming nature of the technique limit its application for the analysis of variation.

4.2.3 Molecular Markers

Some phenotypic changes observed in somaclones/mutants may also arise due to epigenic variations. Thus various molecular markers, such as ISSR and RAPD, have been frequently employed to understand the extent of genomic variation produced by the treatment. Several researchers used EST-SSR and SSR-SSCP primers along with ISSR and RAPD markers. RAPD markers have the advantage of non-specificity for a particular genomic location and hence can span the entire genome with an average of about 10 scorable markers per oligonucleotide primer. With the availability of many such markers, several hundred genetic loci can be compared and used RAPD markers and observed low variability within the regenerated plantlets. It was emphasized that a higher number of markers spanning different genomic locations should be used. Borras-Hidalgo et al. (2005) used AFLP primers on cDNA. Ahmed et al. (2019) used exon region primers of four genes: catalase, sucrose phosphate synthase (SPS), gibberellin 2-oxidase 4 (GA 2-oxidase 4), and tillering branched 1 (TB1) for detection of SNP but no variation was observed in SCMV-resistant somaclones.

Apart from the above methods, direct gene amplification method of COMT gene has also been reported by Sharma et al. (2023). Wherein, the length of the gene amplified was found similar both in parent and mutants but several SNP and indels were observed in COMT gene of mutated plants which authors suspected for reported change in lignin content of the mutant.

4.2.4 Transcriptomic Analysis

Somaclones (plants regenerated from tissue culture) and mutants (plants with induced genetic variations) often exhibit phenotypic and genetic variability. Transcriptomics can play a vital role in their characterization. Wherein, differentially expressed genes (DEGs) can reveal the genetic basis of traits like stress tolerance, yield improvement, disease resistance, or other characters such as floral development. Transcriptomic data might help to annotate novel genes and assign functional roles, aiding in linking specific traits to genetic changes in mutants. The study can also provide snapshot of pathway activations or suppressions, guiding targeted breeding or biotechnological interventions. Somaclones and mutants generated for stress tolerance can be assessed under stress conditions to determine their transcriptional responses, identifying potential regulatory genes for breeding programs. With the availability of a full sequence of sugarcane genome, the analysis of differences in genomic expression in situ between parent and improved variety has become possible. And with advancements in sequencing technologies like next-generation sequencing (NGS), the techniques have now become economical and time saving.

In sugarcane, however, very few transcriptomic studies are attempted for somaclonal variants and mutants. Borras-Hidalgo et al. (2005) used smut-resistant and eyespot-resistant somaclones to check the differential expression of these somaclones upon infection with the pathogen. AFLP analysis of C-DNA showed the presence of many polymorphic bands which upon sequencing showed homology with ACC oxidase, auxin binding proteins, lipid transfer protein precursor, fimbrin-like protein, and Papain-like proteins on infection with pathogenic spore suspension under field conditions. Negi et al. (2020) have compared the transcriptomic expression analysis of mutant clone M4209 and parent variety Co 86032 by gamma radiation-induced mutagenesis and NaCl. Three genes, phenylalanine ammonia-lyase (PAL), acyl-transferase-like (ATL), and salt-activated transcriptional activator (SATA), were identified as most associated with salt tolerance of M4209. Khan et al. (2004) carried out differential transcriptome analysis of high sugar mutant Guixuan B9 (GXB9) of low sugar variety (B9) in sugarcane. The transcriptomic data was also searched for SSR and SNP markers. Wherein, DEGs related to *sucrose phosphate synthase* and *sucrose synthase* were found to be up-regulated in mature internodes and DEGs related to invertase found to be up-regulated in immature internodes. Also DEG related to *trehalose synthase* was found to be down-regulated in mutant.

5 Application of Somaclonal Variation/ Induced Mutagenesis in Sugarcane Improvement

We have grouped the available research publications applying somaclonal variation or induced mutagenesis for sugarcane according to their objectives in areas of crop improvement. We have sincerely attempted to refer nearly all the available publications manually. Although some articles published other than the English language could not be accessed.

The pie diagram in Fig. 3 gives an overview that improvement in disease resistance was the main area of interest in with red rot resistance being the major area of focus followed by smut resistance and eyespot disease resistance. The second area of research focus is improvement in agronomical characters. Although the varieties developed for disease resistance and abiotic stress resistance have also been accessed for their agronomic performance, the instance, only publications with improvement in agronomical practices are considered here. The next area of interest was improvement in abiotic stress tolerance in which tolerance to salt stress was an area of focus. Below is the table that provides experimental details in brief relevant to the application of somaclonal variation and induced mutation in sugarcane (Table 2).

5.1 Abiotic Stress Tolerance

The somaclonal variation and induced mutation have been employed for inducing tolerance against several abiotic stress factors such as salt tolerance, drought tolerance, waterlogging stress, and aluminum tolerance.

Sugarcane being glycophyte is considered to be sensitive toward salt stress. The presence of high salt concentration in irrigation water, wrong irrigation practices in sugarcane can aggravate the problem by leaching and accumulation of dissolved salts on topsoils. Induced mutagenesis by EMS has been used for inducing salt tolerance in sugarcane. Many researchers have reported improvement in agronomical characteristics such as higher number of tillers, stem height, and cane girth in salt-tolerant somaclones or in mutant clones.

PEG has been frequently used as the primary selecting agent for obtaining drought-tolerant putative callus/plantlets. Both physical mutagens: gamma radiation (Majid et al. 2001) and chemical mutagen, viz., sodium nitrite and EMS (Masoabi et al. 2018) have been employed for the mutagenesis. Masoabi et al. (2018) used 16 mM EMS on the callus of sugarcane variety NCo 310. Application of 20% PEG in vitro led to the survival of only 18 plantlets.

Majid et al. (2001) used gamma radiation on callus of sugarcane for developing red rot resistance and tolerance to waterlogging. Three variants tolerant to waterlogging were reported to be obtained.

5.2 Biotic Stress Resistance

Sugarcane is affected by numerous fungal diseases such as red rot caused by *Colletotrichum falcatum*, smut caused by *Sporisorium scitamineum*, wilt caused by *Fusarium sacchari*, pineapple disease caused by *Ceratocystis paradoxa*, brown

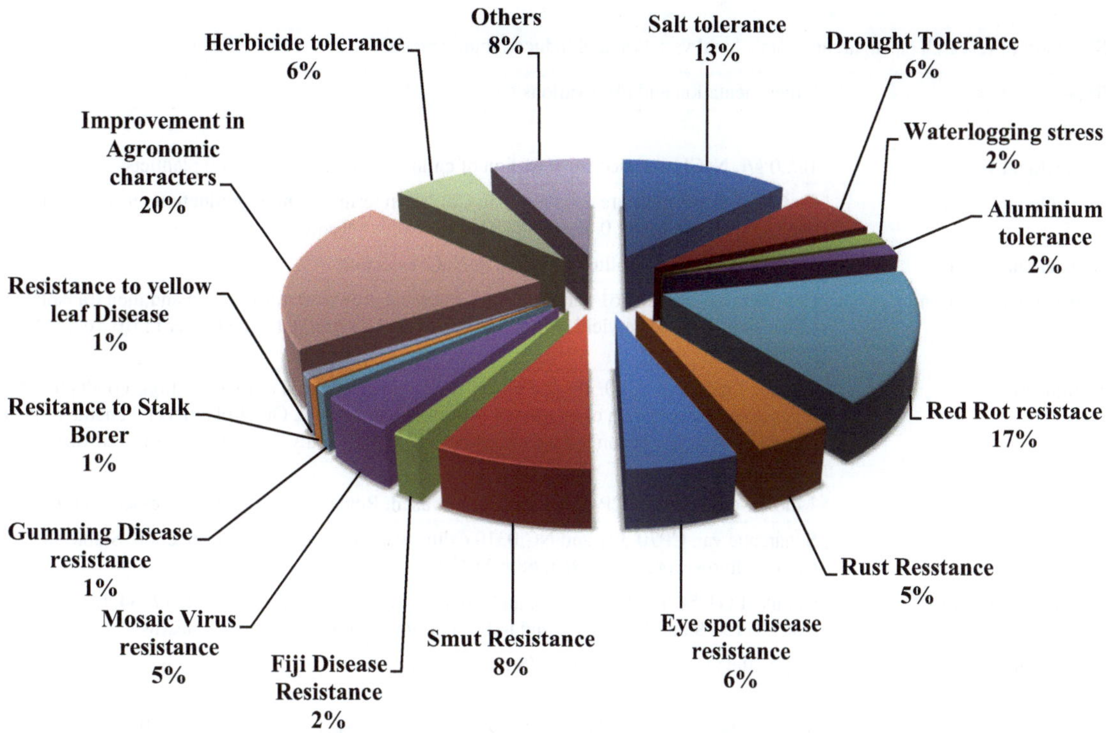

Fig. 3 Trends in sugarcane research applying in vitro variation methods for crop improvement

rust caused by *Puccinia melanocephala*, pokka boeng caused by *Fusarium moniliforme*, sett rot caused by *Ceratocystis paradoxa,* brown spot disease caused by *Cercospora longipes*, and yellow spot disease caused by *Mycovellosiella koepkei.* Similarly, sugarcane is also affected by various bacterial diseases such as ratoon stunting disease caused by *Leifsonia xyli* subsp. xyli, leaf scald disease caused by *Xanthomonas albilineans*, red strip disease caused by *Acidovorax avenae* subsp. avenae, and gumming disease caused by *Xanthomonas vasculorum.* There are several viral diseases of sugarcane reported in India and the world. Sugarcane mosaic disease is caused by *Sugarcane mosaic virus and Sorghum mosaic virus*, streak mosaic disease is caused by *Sugarcane streak mosaic virus,* sugarcane yellow leaf disease is caused by *Sugarcane yellow leaf virus,* and sugarcane leaf fleck disease is caused by *Sugarcane bacilliform virus.* Fiji disease is caused by *Fiji disease virus.* Due to biotic stress including diseases sugar production in India is reported to be reduced by 10% annually (Viswanathan and Rao 2011). Some of the viral diseases like bacilliform having infection rate as high as 100%. Numerous researches have been done to employ somaclonal variation and induced mutagenesis for the improvement in sugarcane (Table 2). Resistance against fungal and viral diseases was observed to be predominantly focused areas for improvement through somaclonal variation or induced mutagenesis. The details of some of those experiments are discussed below.

5.2.1 Fungal Disease Resistance

Red rot is caused by the fungus *Colletotrichum falcatum.* The control of red rot is important as it can cause profound effects in the form of epidemics. Rao et al. (1966) obtained resistant 2 canes from 500 and 3000 r gamma radiation on eye buds (third bud from the bottom of the cane) of Co 449 and inoculation of strain D of *Glomerella tucumanensis.* Ali et al. (2007) obtained 8 canes resistant to red rot by subjecting the callus of CP77400 variety to gamma radiation or sodium azide and selected on media containing partially purified toxins of *Colletotrichum falcatum* (0.05–0.5%). Kumar et al. (2012) reported two somaclones of var. CoS 88230 to be resistant on inoculation against four strains of *Colletotrichum falcatum.*

Somaclones from variant Pindar were developed for disease resistance against Fiji disease caused by virus and Downy mildew caused by *Sclerospora sacchari.* The somaclones with similar cane yield and quality with disease resistance were selected. The variety "Ono" was released for commercial propagation.

Sugarcane eyespot disease is caused by *Helminthosporium sacchari*/*Bipolaris sacchari.* Many experiments were done with the addition of toxins into the media along with incorporation of chemical or physical mutagen to increase the spectrum and efficiency of selection. Larkin and Scowcroft (1983) developed somaclones from Australian cultivar Q101. Three varieties CCe-183, CCe-283, and CCe-483 devel-

Table 2 Summary of research on sugarcane somaclonal variation and induced mutagenesis

Sr. no.	Type	Experimentation and observations
Salt tolerance		
1	Somaclonal variation	18/20 g/L NaCl was used for selection of callus as well as regenerated plantlets
2	Somaclonal variation	Co 62175 observed more resistant than Co 740 in terms of shoot production on NaCl-containing media. NaCl—0.2, 0.4, 0.8, and 1% used for callus selection
3	Somaclonal variation	Sugarcane var. F177 callus subjected to NaCl selection
4	Somaclonal variation	CP43/33, BF-162, IM-61 varieties used. Plantlets grown on normal field and then transplanted on saline sodic soil. Cane yield height, girth, and BRIX observed to be higher in BF-162, IM-61 somaclones
5	Gamma radiation	Gamma irradiation 5, 20, 40, and 60 Gy used saline field trial at four locations conducted. No plantlet regenerated on selection of irradiated callus on NaCl media. Other plantlets regenerated on control sets showed higher boron uptake on saline soil. RAPD analysis of mutants showed variation at molecular level
6	Somaclonal variation	1 month-old callus of CP-43/33 variety was used. Regenerated plantlets assessed on the field
7	Somaclonal variation	Sugarcane var. CP70-321 and NC 0310 callus was subjected to NaCl selection. Stable salt-tolerant callus cultures were obtained at 68 mM NaCl
8	Somaclonal variation	Callus of GT 54-9(C9), NCo310, and Co 413 varieties was selected on 1, 17, 34, 68, and 102 mM NaCl. GT54-9 and NCo310 found to be superior as showed less callus necrosis
9	Ethyl methane sulfonate (EMS)	15 plantlets were selected on EMS mutagenesis
10	Gamma radiation	Variety CoC 671, Co 86032, and Co 94012 were exposed to 10, 20, 30, 40, 50 Gy and NaCl selections were used for callus selection. 10–20 Gy exposed callus survived on 128.3 and 171.1 mM NaCl
11	Somaclonal variation	Variety CP48-103 was used, NaCl selection of 0, 33, 66, 99, and 132 mM was used for selection in vitro and under hydroponics condition. Increase in Ca^{2+} concentration and K/Na ratio was observed in shoots and roots of tolerant somaclones
12	Gamma radiation and ethyl methane sulfonate	Gamma radiation and EMS used for the two mutant lines out of six selected lines found to be salt tolerant
13	Gamma radiation	Co 86032 was used for the experiment, 10, 20, 30, 40, 50, 60, 70, and 80 Gy gamma radiation on callus was used for optimization of radiation dose. (0, 50, 100, 150, 200, and 250 mmol L^{-1} of NaCl was used for selection.) Out of 138 tolerant lines regenerated 18 selected for saline field trial was observed good in yield parameters
14	Gamma radiation	Co 740 was used for the experiment. 10, 20, 30, 40, 50, 60, 70, and 80 Gy of gamma radiation was used for optimization of dose on embryogenic callus. Out of 214 tolerant lines obtained in vitro, 24 plantlets selected for saline field trial based on agronomic performance in normal soil
15	Gamma radiation	PS862, PSJT941 varieties were used for the experiment. 5, 10, 15 20, 25, 30, and 35 Gy were used on callus. 122 mutant plantlets originating from the PS862 variety and 66 mutant plantlets originating from PSJT941
16	Gamma radiation	CoM 0265 embryogenic callus was subjected to different doses of gamma radiation. The callus and regenerated plantlets were further selected on NaCl (50, 100, 150, and 200 mM NaCl). In field trial on saline patches mutant showed higher NMC, Yield, and CCS and some showed higher sucrose percent and juice purity. Significant correlation between antioxidant enzyme activity with TWC, chlorophyll content, yield, and CCS was reported
Drought tolerance		
17	Sodium nitrite	Q77N1232, Co6519, Cadmus varieties used for the experiment. Sodium nitrile 5 mg/L was used for mutagenesis and PG 8000 (100 g/L) used for selection. Regenerates grown from stress-selected callus showed similar drought tolerance index as regenerates from tolerant variety indicating drought tolerance
18	Somaclonal variation	Isd 28. PEG-5, 7.5 and 10% used for selection. 5 out of 20 somaclones studied in vivo showed high tolerance
19	Somaclonal variation	Three varieties were used in experiment. Isd 20, Isd 35, Isd 36, Isd 37, and Isd 38 0%, 5%, 7.5%, and 10% PEG in medium was used for selection. Effect of PEG on plant regeneration, root numbers, and root length was studied
20	Somaclonal variation	20% PEG-6000 used as selection agent on callus. Effect of PEG on POX, SOD, Cat, ascorbate, and proline content was studied

(continued)

Table 2 (continued)

Sr. no.	Type	Experimentation and observations
21	Gamma radiation	Gamma radiation (0, 10, 20, 30, 40, and 50 Gy) was used on callus, 0 and 10% PEG used for selection of drought tolerance. 17 plantlets out of 42 showed high drought tolerance in vitro
22	EMS	NCo310 sugarcane varieties were exposed to 16 mM EMS and 20% (w/v) PEG-6000 used as selection agent, 18 plantlets survived
23	Somaclonal variation	Somaclones produced from CP48-103, CP69-1062, and CP57-614 varieties. 0, 25, 50, and 75 g/L of polyethylene glycol was used for selection. Tolerant somaclones showed more proline content upon stress implementation
Tolerance to water logging		
24	Gamma radiation	"Isd-2/54," "Nagarbari," and "Latarijaba," 1-291/87, 1-525/85 variety callus subjected to 20, 30, and 40 Gy gamma radiation. Three variants selected for waterlogging at MV6, 4 variants were selected for delayed flowering at MV3
Aluminum tolerance		
25	EMS	VMC 7616, PS 862 variety callus subjected to 0.1%, 0.3%, and 0.5% EMS for 30, 60, and 120 min. Further selection was carried out on 0, 100, 200, 300, 400, and 500 ppm $AlCl_3{\cdot}6H_2O$ and Low pH (4.0)
26		Sugarcane variety N51 callus was subjected to 5-azaC (100 μM) and Al (1.5–50 mM) treatment. About 8 plants/gm callus were regenerated in in vitro selection as compared to 105 plants in control. One mutant with higher green leaf area, longer roots under stress conditions of 60 mM Al stress
Resistance to fungus Colletotrichum falcatum (red rot)		
27	Gamma radiation	Five canes obtained from one clump (out of 1600) irradiated with 500 r showed high resistance to the pathogen
28	Gamma radiation	Co 1148 variety setts were subjected to 2, 4, and 6 Kr gamma radiations. Out of 3600 plantlets planted, 2254 survived under field conditions
29	EMS, diethyl-sulfonate	Sugarcane eye buds were subjected to different concentrations of EMS, diethyl sulfonate and nitrosomethyl urea but no red rot-resistant mutant obtained
30	Gamma radiation (EMS)	Variety CoJ 64 eye buds subjected to 0.8% EMS, 6 Kr gamma rays (both LD_{50}). 4 kr gave high number of resistant mutants
31	Gamma radiation, EMS	Single eye bud setts of CoS 687 variety of sugarcane with wide adaptability but susceptible to red rot was exposed to gamma radiation (30, 40, 50, 60 Gy) and EMS (0.2 and 0.4% EMS in 0.1 M, pH 7-phosphate buffer for 6 h)
32	Somaclonal variation	500 resistant plants to red rot and six clones tolerant to water-logged conditions were obtained by somaclonal variation through callus
33	Somaclonal variation	Variety CoS 91279 callus selected on fungal toxin in media. Only 3 somaclones out of 42 were found moderately resistant against 4 isolates of *Colletotrichum falcatum* by plug method and 2 somaclones were found resistant by bud inoculation method
34	Somaclonal variation	Three somaclones exhibited moderate resistance to red rot and three clones found to be tolerant to water logging
35	Gamma radiation	Varieties "Isd-2/54," "Nagarbari" and "Latarijaba" callus were subjected to 20, 30, and 40 Gy gamma radiation treatment. 37 out of 2954 plantlets found to be resistant to red rot on exposure to red rot spore suspension
36	Gamma radiation, sodium azide	Variety CP77400 was subjected to 10–50 Gy gamma rays, 1.0–5.0 mg/L sodium azide. And selection of the callus was carried out on partially purified toxin of *Colletotrichum falcatum* (0.05–0.5%) using ethyl acetate and Na_2SO_4. Out of 164 in vitro selected lines 8 found resistant to red rot on field analysis
37	Somaclonal variation	Three somaclones of CoJ 88 variety were found to be moderately resistant to two pathogenic strains of *C. falcatum*
38	Somaclonal variation	Somaclones generated from varieties CoJ88, CoJ64 through in vitro selection exhibited higher disease resistance under field condition
39	Somaclonal variation	Somaclones generated from S97US297. Variation through RAPD and SSR markers was observed in red rot-resistant somaclones
40	Somaclonal variation	Somaclones generated from BF-162. Variation through RAPD and SSR markers was observed in red rot-resistant somaclones
41	Somaclonal variation	50 somaclones of var. CoS 88230 were inoculated with mixture of four strains of *Colletotrichum falcatum* of which 2 observed to be resistant

(continued)

Table 2 (continued)

Sr. no.	Type	Experimentation and observations
42	Somaclonal variation	Variation through RAPD and SSR markers in red rot-resistant somaclones of variety Isd 37 was observed
43	Gamma radiation	Single eye buds of cv CoJ64 were different gamma radiation doses (10, 20, 25, and 30 Gy). MIVI generation of the clones showed improvement in agronomical characters and red rot tolerance
44	Somaclonal variation	Out of 581 somaclones generated, 121 survived after transplantation. Of which 10% showed resistance
Resistance to fungus Puccinia melanocephala (brown rust)		
45	Somaclonal variation	Callus of B4362 variety was used for somaclonal variation
46	Somaclonal variation	261 somaclones of var. CoA7601 were produced and field planted. Co 92007 variety was obtained
47	Somaclonal variation	Plantlets derived through callus of var. B4362 showed higher incidences of disease than control
48	Gamma radiation, NaN_3	Out of 22928 regenerated plants—1 mutant obtained from NaN_3 (0.01%) and 2 mutants from gamma radiation (10–40 Gy) found rust resistant
49	Gamma radiation, NaN_3	NaN_3 produced high number of resistant against rust. Five brown rust resistant mutants were obtained from 11167 regenerated plantlets
Disease resistance to fungus Bipolaris sacchari (eye spot disease)		
50	Somaclonal variation	480 plantlets were obtained from Q101 variety by selection of the callus on media containing fungal toxin. Plantlets exposed to purified toxins for selection. In round 2 of selection, 6 somaclones were found resistant to the toxin
51	Somaclonal variation	About 2000 somaclones obtained by in vitro selection on purified toxin were subjected to field analysis. On 10 year analysis, four somaclones resistant to eye spot were obtained. Two somaclones showed difference in restriction digestion pattern of mitochondrial DNA
52	Somaclonal variation	Sugarcane variety Co 419 was selected on media containing purified toxin of *Helminthosporium sacchari*
53	Somaclonal variation	c-DNA AFLP analysis of resistant somaclones to either smut or eye spot was carried out by 15 with Eco RI + 1/Mse I + 1 combination and 15 with Eco RI + 2/Mse I + 2 combinations. 62 differential genes were identified of which 19 showed similarities to the known sequence. Most of these proteins were related to plant defense or signaling
54	Ethyl methane sulfonate (EMS)	Embryogenic callus of sugarcane variety NCo376 was subjected to 32 mM EMS solution for 4 h and selected on media with culture filtrate (CF). ISSR analysis showed presence of polymorphic bands in mutants
Disease resistance to fungus Ustilago scitaminea (smut disease)		
55	Somaclonal variation	Out of nine somaclones of variety F177 tested for smut resistance only one had lower (43.8%) infection than donor (88.2%). Change in isoenzyme pattern was observed
56	Gamma radiation	23 putative smut-resistant mutants of variety Co 547 regenerated by 2 Kr gamma radiation were tested for 2 years which lead to identification of 15 stable smut-resistant mutant
57	Somaclonal variation	Toxin of Ustilago used for selection of smut-resistant mutant
58	Somaclonal variation	Six varieties Co 740, Co 7704, Co85-261, Co 6907, G 86-20, and CoC 671 subjected to selection of somaclonal variation. Somaclones from all except Co 740 resistant/mildly resistant to smut were obtained
59	Gamma radiation	3 mutant clones obtained by gamma radiation treatment to setts of CoC 671 var. were found smut resistant with similar in yield characteristics as parent
60	Gamma radiation	Callus of sugarcane variety My5450, Ja-60-5 was subjected to gamma radiation (10–30 Gy). Out of 15,500 regenerated plantlets none selected for resistance
61	Gamma radiation	Sugarcane variety setts of B47419, BJ6502, Co62175, and Co957 were subjected to 30–50 kGy gamma radiation. Eight clones resistant to smut were observed in field analysis
62	EMS and somaclonal variation	Callus of CoC 671 was subjected to 0.8 M EMS and 13.57 M 2–4D. From 310 regenerated plantlets, 17 were selected for rod row trial and smut resistance. ITS (internal transcribed spaces) markers of *Sporisorium scitaminae* were used for the detection of uninfected plantlets by PCR which showed 405 bp amplicon in parent and susceptible variety Co740 but not in some of the mutant clones
63	Ethyl methane sulfonate (EMS)	Variety CoC 671 callus was subjected to EMS mutagenesis (0.8 M EMS and 0.125 M PEG). About 300 plantlets regenerated, 11 selected and checked on field trial. The mutant also showed polymorphism on molecular analysis by RAPD primers

(continued)

Table 2 (continued)

Sr. no.	Type	Experimentation and observations
Resistance to Fiji disease		
64	Somaclonal variation	Fiji disease-resistant variety "Ono" was developed from susceptible sugarcane variety "Pindar"
Resistance to mosaic virus		
65	Gamma radiation	Un-germinated eye bud setts were exposed to gamma radiation. More than 500 subclones tested on field with Nco 310 infected with strain H of mosaic gave only one resistant mutant from variety CP 65-357. But inferior juice quality was observed
66	Somaclonal variation	RAPD markers used for the variant analysis of SCMV-resistant somaclones
67	Gamma radiation	Sugarcane variety B6749 was subjected to 2 krad of gamma radiation. 15 resistant clones were screened for RAPD polymorphism
68	Somaclonal variation	SSR markers used for detection of polymorphism of somaclonal variants resistant to mosaic virus
69	Somaclonal variation	Out of 581 somaclones generated, 121 survived after transplantation. Of which 10% showed resistance
Sugarcane yellow leaf disease resistance		
70	Sodium azide, sodium nitrile, ethyl methane sulfonate	Sodium azide (SA) at 0.5, 1.0, 1.5 mg L^{-1}; sodium nitrite (SN) at 3, 5, 7 mg L^{-1}; ethyl methane sulfonate (EMS) at 0.6, 0.8, 1.0 L ML^{-1}, and 2,4-D at 4, 5, 6 mg L^{-1} were used for optimization. Four out of 207 asymptomatic plants (16T22, 16T23, 16T29, and 16T31) were devoid of virus coat protein band and were considered to be YLD resistant
Change in agronomic characteristics		
71	X-rays	Early flowering was detected in some mutant clones
72	Pyrogallol, pyrocatechol, quinol, acenaphthene	Shortened internodes, glabrous leaf sheath was observed in come mutant clones
73	Gamma radiation	Mutant clones showed reduction in flowering
74	Gamma radiation	Mutant selected for glabrous leaf sheath
75	Somaclonal variation	Morphological, cytological, and enzymatic variation was observed in somaclones obtained from saccharum hybrids
76	Gamma radiation	Glabrous leaf sheath, non-flowering, and early maturity mutant formed by gamma radiation was selected. In another experiment mutants resistant to red rot were also evaluated
77	Gamma radiation	Fast growing vigorous, 60 day earlier maturing clone of var. Co 527 was obtained through gamma radiation. Mutant showed 2–3 lesser chromosome than parental variety
78	X-rays	Drastic mutants of Co 419 had chromosome $2n = 80–85$ than parent $2n = 110–115$
79	Gamma radiation	Out of 207 somaclones screened 8 plants showed glabrous leaf sheath, Co 91017 selected after multi-location trial
80	Somaclonal variation	Morphological variation and higher sucrose content and yield was observed in CoC671 somaclone variants
81	Gamma radiation	Out of 5000 regenerated plantlets 8 somaclones were found with flowering elimination or reduction, which also showed increase in yield by 20% due to increase in stalk height
82	Gamma radiation	0.5 Kr regenerated plants showed improvement in plant height, number of tillers/plant, cane thickness
83	Somaclonal variation	Study of somaclones in two ratoon crop season—somaclone no. 7 surpasses on stalk height, stalk number, cane yield. Somaclone 4 surpasses on BRIX, purity, POL and sugar recovery
84	Somaclonal variation	39 hybrids of crosses between *Saccharum officinarum, Saccharum spontaneum,* and *Erianthus arundinaceus* were used as material for generation on callus. The regenerated plants from callus were further selected for agronomic characters
85	Somaclonal variation	21 somaclones of var. BL were selected out of 324 callus regenerated plantlets. Of which 7 clones showed higher sucrose/plant on field analysis
86	Ethyl methane sulfonate	At field evaluation the mutants of Co 99004, Co 94012 formed by EMS mutagenesis (0.5% for 2 h) and NaCl selection (100 mM) were subjected to field selection. Mutants found to be superior in tillering and other yield characters in both varieties
87	Gamma radiation	CoC 671 callus was subjected to gamma radiation. Nine high sucrose variants were studied for sucrose-related enzyme distribution
88	EMS and somaclonal variation	Isd 37, Isd 39, and Isd 40 variety callus was subjected to EMS (0, 0.01, 0.02, and 0.03 M) for 24 h. High number of mutants generated from ISD 37 variety which showed variation in tillering, number of canes, leaf sheath color

(continued)

Table 2 (continued)

Sr. no.	Type	Experimentation and observations
89	Gamma radiation	Variety CoJ 85 shy in tillering was used for the experiment. Setts were subjected to gamma irradiation (15, 20, 25, and 30 Gy). Many mutant clones with high tillering were obtained. This further showed difference in lignin content. Also on analysis of COMT gene sequence variation in SNP was reported
Effect on callus regeneration		
90	Gamma radiation	Callus of verities B.O. 91, Co S 8408 subjected to gamma radiation mutagenesis. Increase in peroxidase activity in plantlets was observed with increase in radiation
91	Gamma radiation	Use of RAPD markers for variability analysis in mutant regenerated by gamma radiation
92	Gamma radiation	Partial desiccation improves plant regeneration from gamma-irradiated callus
93	Gamma radiation	Partial desiccation of CoC 671 variety callus for 6 h on sterile silica gel under desiccator improves plant regeneration from callus
94	Gamma radiation	Chlorophyll content, root length, and plantlet regeneration were observed to be affected due to radiation
95	Ethyl methane sulfonate	CRP48-103, CP57-614 varieties subjected to EMS (4, 8, …, 48 mM). Effect of EMS dose on plant regeneration was studied
	(EMS)	SSR markers were used for diversity analysis in regenerated plantlets which showed variation in amplification pattern
96	Ethyl methane sulfonate (EMS) and sodium nitrile	Combination of 3 mg L^{-1} of sodium nitrile and 0.6 μM EMS was found effective for callus induction
Cytological changes		
97	Colchicine-induced variation	Reported first time chromosome doubling in commercial sugarcane variety. Colchicine-treated cells resulted in polyploid, mixoploid, and diploid plants
98	Colchicine-induced variation	Callus of H69-9092 subjected to colchicine. Aneuploid cells obtained in callus culture. Stalk length, internode length, stalk number, volume, and diameter also, leaf area showed negative correlation with number of chromosomes. Whereas positive correlation was obtained between chromosome number and stomata length
99	Somaclonal variates from hybrid	About 1000 somaclones regenerated from inter-generic hybrid clone H83-9998 (2n = ca. 94) involving H62-4671 (Saccharum spp. hybrid, 2n = ca. 115) × 28NG7 (*Erianthus arundinaceum*, 2n = 60) as parents. 126 somaclones were selected based on phenotypic variation which was subjected to field selection and isoenzyme and cytological analysis. Change in isoenzyme pattern of *peroxidase*, *phosphoglucomutase*, and *phosphoglucose isomerase* was recorded. Also chromosome number in somaclone observed to vary from 2n = 66 to 2n = 109
Herbicide tolerance		
100	Somaclonal variation	Cv. V64-10 was checked for ametryn tolerance. Tolerant calli were able to regenerate on 6.25 higher ametryn as compared to the sensitive calli. The plantlets regenerated from those calli were also reported to tolerate 5.35-fold higher herbicide concentrations than sensitive plants
101	Somaclonal variation	Cell suspension culture of three varieties (PR62258, V64-10, and V71-51) was subjected to 0.8 mM glyphosate in liquid medium. Plantlets regenerated from the tolerant cell suspension were able to withstand sixfold higher glyphosate as compared to sensitive calli. On RAPD analysis, tolerant lines showed 564 bp polymorphic fragments by OPA-07 primer
102	Somaclonal variation	0, 0.156, 0.312, and 0.625 kg ha^{-1} Imazapyr application in field level in split plot design. 27 different clones selected of which 6 found to be resistant to Imazapyr dose 0.625 kg ha^{-1}
103	Ethyl methane sulfonate (EMS)	An EMS used for mutagenesis on N12 sugarcane variety. *Acetohydroxyacid synthase* activity of tolerant line was observed to be 2.8–4 times higher than sensitive variety
104	Somaclonal variation	Variation plantlets regenerated from 5-azacytidine-treated callus through methylation-sensitive restriction endonuclease iso-schizomers Hpa II and MSp I. Both these enzymes recognize and cut the same sequence (CCGG). However, Hpa II can cut only when outer cytosine of this recognition sequence in hemi-methylated (mCCGG) and Msp I can cut only when internal cytosine is fully methylated (CmCGG), where increase in hemi-methylated cytosine was found to increase under mutagenesis

oped from C87-51 through induced mutagenesis are registered for release in 1993 (FAO-IAEA database). Two varieties Co85017 and Co85035 developed from Co 740 by using 150 Gy gamma rays were registered in 1985 (FAO-IAEA database for mutant variety released). Sreenivasan et al. (1987) reported obtaining somaclones resistant to smut disease caused by fungal pathogen *Ustilago scitaminiae.*

In sugarcane, brown rust disease is caused by the fungus *Puccinia melanocephala.* In variety CoA 7601, out of 261 plants regenerated, somaclone Co 92007 resistant to rust disease has been registered for release in 1987 (Sreenivasan et al. 1987). Perez Ponce et al. (1998) also, obtained one resistant clone from NaN_3 mutation and two clones from gamma radiation mutagenesis from 22,928 plantlets tested. Oloriz et al. (2011) have used both mutagenic treatments (gamma radiation and NaN_3), wherein higher frequency of resistant mutants from NaN_3 treatment.

5.2.2 Development of Sugarcane Resistant to Viral Diseases

Sugarcane-growing regions around the world report occurrences of diseases caused by various viruses. Mosaic disease in sugarcane can be caused by many viruses such as *Sugarcane mosaic virus, Sorghum mosaic virus,* and *Sugarcane streak mosaic virus.* The yellow leaf disease in sugarcane is caused by *Sugarcane yellow leaf virus* and in minor incidences by *Maize yellow mosaic virus.* The viral diseases are reported to reduce rate of sett germination, and also reduce yield by hampering photosynthesis efficiency leading to 30–80% losses in yield (Vamsi Krishna et al. 2023).

Resistance to Mosaic Virus

Researcher used 2.5 Kr, 5 Kr, and 7.5 Kr gamma radiations on non-germinated eye buds of nine sugarcane varieties susceptible to strain H of mosaic virus. Out of more than 500 plants tested on the field along with heavily mosaic infected N: Co310 lead to the identification of only one resistant clone with inferior juice quality. Oropeza et al. (1995) developed two somaclonal variants resistant to mosaic virus, namely, AT626 and BT627.

Resistance to Yellow Leaf Disease

Yellow leaf disease in sugarcane is caused by yellow leaf disease mosaic virus. Kona et al. (2019) used chemical-induced mutagenesis on Co 86032 variety. Out of 207 plantlets four asymptomatic plants (16T22, 16T23, 16T29, and 16T31) showed absence of virus coat protein band in PCR analysis and were considered to be YLD resistant. But these clones showed inferior agronomic traits.

5.3 Changes in Agronomic Traits

Many researchers have reported somaclonal variation or mutagenesis-induced change in agronomic character such as flowering reduction, increase in cane sugar content, increase in height, increase in tillering, glabrous leaf sheath, improvement in stem girth, and late in maturity. The performance of the variants at the field is a benchmark for their release for commercial propagation. The details of commercially released sugarcane varieties formed through induced mutagenesis or somaclonal variation are provided in Table 3.

Altogether, thirteen mutant varieties have been registered in FAO-IAEA database with improved disease resistance or improved agronomic performance: "Co 8153" with improved juice quality and higher yield was developed from a hybrid of Co 6304 × Co 6806 by gamma radiation (450 Gy on setts), "Nanei" with longer, thicker stalk, better tillering, higher yield of cane and sugar was developed from var. "Ni" by 150 Gy gamma rays treatment. Both these varieties are released in 1981 (FAO-IAEA database, 2025, Table 3). "CCe-10582" developed from "C 87-51" in 1990 shows an increase in yield. "Yuetangfu 83-5" developed from "Yuetang 71-210" was also released in 1992. Whereas the variety "Guifu 80-29" developed from "Guitang 72-28" by 80 Gy gamma radiation in 2005 shows late maturity, high sugar, and low stem height. The First Indian variety developed from CoC 671 by somaclonal variation "Co 94012 (Phule Savitri)" was released in 2006 (Jalaja et al. 2006). Co 94012 was selected for early maturity, high sucrose content, and moderate resistance to diseases red rot and smut. In the same year, variety "Guitang 22" developed from "Xintaitang 1" by 80 Gy gamma radiation was released. "Guifu 98-296" developed by mutation breeding from "Guitang 72-28" was released in 2011. It was selected for high yield and high sugar content. In 2016, the "VSI 434" sugarcane variety developed through somaclonal variation was released for early maturity and high sugar content. There is also a report of the release of one variety developed from CP 52-43 by 10-30 Gy gamma radiation on callus selected for flowering reduction and two varieties developed from "POJ-28 78" by (5–20 Gy) gamma radiation selected for higher sugar content from Institute of Plant Biotechnology, Cuba (Perez Ponce et al. 1998). However, the name of those varieties could not be revealed from the literature.

Table 3 Commercially released sugarcane somaclonal variant and mutant varieties

Sugarcane variety	Year of release	Characteristic	Parent	Mutagen and dose	Country
VSI 434	2016	Early maturing, high Sucrose	CoC 671	No mutagen	India
Co 94-012 (Phule Savitri)	2006	Early maturing, high sucrose	CoC 671	No mutagen	India
Guitang 22	2005	Higher biomass, energy cane	Xintaitang No. 1	Gamma rays 80 Gy	China
CCe-183, CCe-283, CCe-48	1993	Resistance to sugarcane eye spot disease	C 87-51	–	Cuba
Yuetangfu 83-5	1992	Sugar content	Yuetang 71-210	–	China
CCe-10582	1990	Higher yield	C 87-51	–	Cuba
Guifu 80-29	1989	Early maturity, high yield, high sucrose, medium to large stem	Guitang 72-28	Gamma rays 80 Gy	China
Co 85035	1985	Earliness	Co 740	Gamma rays 150 Gy	India
Co 85017	1985	Adaptability, resistance to smut, higher yield, higher sucrose (%)	Co 740	Gamma rays 150 Gy	India
Nanei	1981	Size and thickness of stalk, improved tillering	Ni 1	Gamma rays 420 Gy	Japan
Co 8153	1981	Improved juice quality	Co 6304 × Co 6806	Gamma rays 150 Gy	India
Ono	1974	Improved disease resistance and yield	Pindar	No mutagen	India
Co997 mutant	1967	Resistance to red rot	Co 997	Gamma rays 30–50 Gy	India
Co 6608	1966	Resistance to red rot	Co 449	–	India

6 Advantages of Somaclonal Variation/Induced Mutation

Sugarcane crosses between different varieties are difficult to carry out due to multiple reasons such as differences in the time of flowering in different genotypes differences in fertilization efficiencies which can result in the production of small seeds which germinate and survive only under favorable conditions are strictly maintained. The somaclonal variation/induced mutation protocols can be optimized even for recalcitrant varieties. Sugarcane is a genetically complex crop with an average of 10–15 years need for development through conventional breeding, selection cycle, and vegetative propagation of resulting cultivars. The selection process is challenging due to the complex polyploidy nature of sugarcane, which can mask the effect of desirable genes throught the presence of similar alleles across multiple chromosome sets. The time for development by somaclonal variation/induced mutation is lower. A large number of plantlets can be generated, maintained, and selected in vitro, thus requirement of relatively smaller area till preliminary screening. Mutagenesis can be useful in identification of SNP responsible for resistance/tolerance in a crop. In barley, mutant raised from sodium azide mutagenesis was selected for Imidazoline herbicides. The tolerant mutant showed single amino acid substitution of serine 653 to asparagine amino acid in gene *acetohydroxy acid synthase* (Lee et al. 2011).

Both somaclonal variation and induced mutagenesis can be applied to change specific characters in promising variety without altering its other characteristics for improving disease resistance or abiotic stress tolerance in commercially superior variety. POJ-2878 variety of high commercial use was severely prone to viral diseases. A clonal selection followed by mutation induction has led to the development of commercial cultivar development with increased sugar content (Perez Ponce 1998). Lower expenses are involved as compared to the genomic transformation procedures. Approval for the release of the improved varieties produced through somaclonal variation or induced mutagenesis is legally acceptable in most countries. Hybridization is an essential and fundamental source of variability. The induced mutation can also be employed along with hybridization where inter-specific or inter-generic hybrids generated with high mortality can be treated with mutagenic treatment for increasing number of survival. These populations can be used for backcrossing programs.

7 Possible Constraints in Sugarcane Variation Research

Dominant or recessive mutations can occur by induced mutations or somaclonal variation. However, the chances of occurrence of recessive mutations are higher, thus phenotypic expression of changed genotype is not possible unless two recessive homozygous genes come together. In sugarcane being polyploidy, the chances of phenotypic expression further decrease due to the presence of multiple copies of the gene.

Many phenotypic changes can be epigenic in nature, i.e., temporary changes that are not an effect of changes at the genomic level and only be as a response to particular environmental conditions thus not inherited to the next progeny. For example, the eyespot-resistant mutant did not show the resistance after over 10 years (Smith and Drew 1990). Infection by more virulent strains or different from those against which resistance was developed was also predicted as possible reason.

The genomic locations and frequency of the mutation are unpredictable. Mutation can also occur in genomic locations that control other agronomically important characteristics. Thus, careful analyses of the traits are required in agronomic trials. Many mutant clones/somaclonal variants selected for disease resistance/abiotic stress tolerance are often rejected in field trials due to undesired phenotypes such as thinner canes which reduce overall yield.

The phenotypic expression may arise by epigenic variation. Thus, the stability of the expression pattern must be done for at least M1V4 generation. If this time for selection is taken into the consideration for the release of variety, the advocated advantage of lesser time in vegetatively propagated plants release becomes questionable.

Some cells which are more exposed to mutagen may get more damaged than those that are sub-lethally exposed in the same sample. This phenomenon is known as the "diplontic effect." particularly when multicellular tissue such as eye buds, are used for induced mutagenesis.

8 Opportunity for Targeted Mutagenesis Through Genome Editing Technology in Sugarcane

In the era of genome editing technology, precise genome modification has become possible. As discussed in previous parts of the chapter, several limitations occurred by complex genome with multiple copies of the gene, most preferred methods of induced mutagenesis so far are randomized mutagenesis imposing long screening required for selection of the mutant possessing desired trait and further to evaluate stability of the trait to reject epigenic effect. This has created requirement of mutation system with higher precision. Advancement of applied technology in biological systems such as genomics, proteomics, metabolomics, and transcriptomics has opened up the plethora of information and understanding of complex interaction at cellular, tissue, and environmental interaction. With knowledge of pathways controlling stress, resistance protein, genes responsible for metabolite production, effect of different metabolites, and factors responsible for pathogenicity, identifying targets for the precision mutagenesis has become easy. Availability of sugarcane genomes with precise annotation of putative genes and their regulatory region obtained through high-throughput sequencing technology has created strong platform for target decision, screening, and allied processes such as evaluation of the crops after genome editing. Genome editing technologies like Zink finger nuclease (ZFN), clustered regularly inter-spaced short palindromic sequence repeats (CRISPER)/Crisper-associated protein-9 (Cas 9), and transcriptional activator like effector molecule (TALEN) have become a cornerstone in crop improvement.

9 Conclusion

It was postulated that, if mutagens are to be used in crop improvement programs with the level of specificity expected by breeders, they should either targate replicating or transcribing genes specifically or act as an internal sieveto favor or eliminate specific type of mutations. In sugarcane, such specificity has not been achieved. Very few varieties formed by somaclonal variation or induced mutagenesis have been registered for commercial use in sugarcane despite many attempts as seen in research publications for crop improvement. Still today the exact mechanism underlying somaclonal variation and mutagenesis is not fully known. New approaches such as tilling can be employed to uncover the changes at a genomic level. Some attempts of transcriptomic data analysis are providing light on the mechanism of improvement and functioning at the genomic level. These studies are important and can be employed in the future where a single or limited gene of function is involved in providing tolerance; for example, EMS or sodium azide-based mutagenesis to look for single amino acid substitution at acetolactate synthase (ALS) for increasing tolerance to Imidazoline herbicides. Such targeted approaches will increase the rate of success in obtaining desired variants. Alternatively, a more targeted type of mutagenesis by genome editing technology, zinc finger nuclease meganuclease which can be tailored for recognition and binding with specific gene locations is getting developed. There underlies a vast potential for understanding defense mechanism against biotic, abiotic stresses, understanding effect of differences in cell wall compositions through mutagenesis. Mutant possesses unique genomic alterations which could be tapped for designing future biotechnological strategies through genome editing. Furthermore, addressing issue of metal toxicity such as aluminum and chromium toxicity has been attempted at very less extent through mutagenesis. The incidences of metal toxicity are increasing at an alarming rate. Induce mutagenesis and somaclonal variations combined with recent technological advancements are poised to play a vital role in future sugarcane development.

References

Ali A, Naz S, Alam SS, Iqbal J (2007) In vitro induced mutation for screening of red rot (Colletotrichum falcatum) resistance in sugarcane (Saccharum officinarum). Pak J Bot 39:1979–1994

Ahmed MS, Gardezi DA, Batley J, Iqbal MZ, Javid A (2019) Somaclonal variations for red rot and sugarcane mosaic virus resistance and candidate genes integrity assessment in somaclones of selected sugarcane varieties (Saccharum officinarum L). Pak J of Agri Sci 56(1):15–27. https://doi.org/10.21162/PAKJAS/19.6578

Borras-Hidalgo O, Thomma Bart PHJ, Carmona E, Borroto CJ, Pujol M, Arencibia A, Lopez J (2005) Identification of sugarcane genes induced in disease resistant somaclones upon inoculation with *Ustilago scitaminea* or *Bipolaris sacchari*. Plant Physiol Biochem 43:1115–1121. https://doi.org/10.1016/j.plaphy.2005.07.011

Grant-Downton RT, Dickinson HG (2005) Epigenetics and its implications for plant biology. Ann Bot 96:1143–1164. https://doi.org/10.1093/aob/mci273

FAO IAEA Mutant Variety Database (2025) https://nucleus.iaea.org/sites/mvd/SitePages/Search.aspx

Jalaja NC, Sreenivasan TV, Pawar SM et al (2006) Co 94012—a new sugarcane variety through somaclonal variation. Sugar Tech 8:132–136. https://doi.org/10.1007/BF02943647

Koch AC, Ramgareeb S, Rutherford RS, Snyman SJ, Watt MP (2012) An in vitro mutagenesis protocol for the production of sugarcane tolerant to the herbicide imazapyr. In Vitro Cellular and Developmental Biology – Plant. https://doi.org/10.1007/s11627-012-9448-x

Khan SJ, Khan MA, Ahmed HK, Khan RD, Yusuf Z (2004) Somaclonal variation in sugarcane through tissue culture and subsequent screening for salt tolerance. Asian J Plant Sci. 3(3):330–334. https://doi.org/10.3923/ajps.2004.330.334

Koetle M, Osborn C, Snyman S, Rutherford R (2023) The 5-azacytidine-induced epimutagenesis of sugarcane (Saccharum spp. hybrids) for aluminium tolerance. Biochimica et Biophysica Acta (BBA) - General Subjects 1867(12):130491. https://doi.org/10.1016/j.bbagen.2023.130491

Kona P, Kumar MH, Reddy K et al (2019) Regeneration and evaluation of somaclones of sugarcane variety Co86032 for yellow leaf disease resistance and yield traits. J Biosci 44(29). https://doi.org/10.1007/s12038-019-9846-x

Kumar P, Agarwal A, Tiwari AK, Lal M, Jabri MLA (2012) Possibilities of development of red rot resistance in sugarcane through somaclonal variation. Sugar Tech 14:192–194. https://doi.org/10.1007/s12355-012-0138-0

Larkin PJ, Scowcroft WR (1983) Somaclonal variation and eyespot toxin tolerance in sugarcane. Plant Cell Tissue Organ Cult 2:111–121. https://doi.org/10.1007/BF00043356

Lee H, Rustgi S, Kumar N, Burke I, Yenish JP, Gill KS, Dv W, Ullrich SE (2011) Single nucleotide mutation in the barley acetohydroxy acid synthase (AHAS) gene confers resistance to imidazolinone herbicides. PNAS 108(21):8909–8913. https://doi.org/10.1073/pnas.1105612108

Majid MA, Shamsuzzaman KM, Howlider MAR, Islam MM (2001) Development of sugarcane mutants with resistance to red rot, waterlogging and delayed for non-flowering through induced mutations. IAEA Tec Doc 1227:31–44. https://eurekamag.com/research/003/710/003710070.php

Masoabi M, Lloyd J, Kossmann J, van der Vyver C (2018) Ethyl methane sulfonate mutagenesis and in vitro polyethylene glycol selection for drought tolerance in sugarcane (Saccharum spp.). Sugar Tech 20:50–59. https://doi.org/10.1007/s12355-017-0524-8

Mutant Variety Database (2018) Officially released mutant varieties. FAO-IAEA database. http://mvgs.iaea.org/AboutMutantVarieties.aspx

Nagai C, Ahloowalia BS, Tew TL (1991) Somaclonal variants from an intergeneric hybrid: Saccharum spp. hybrid × *Erianthus arundinaceum*. Euphytica 53:193–199. https://doi.org/10.1007/BF00023271

Negi P, Pandey M, Dorn KM et al (2020) Transcriptional reprogramming and enhanced photosynthesis drive inducible salt tolerance in sugarcane mutant line M4209. J Exp Bot 71(19):6159–6173. https://doi.org/10.1093/jxb/eraa339

Oloriz MI, Gil V, Rojas L, Veitía N, Höfte M, Jiméne E (2011) Selection and characterization of sugarcane mutants with improved resistance to brown rust obtained by induced mutation. Crop Pasture Sci 62:1037–1044. https://doi.org/10.1071/CP11180

Oropeza M, Guevara P, de García E et al (1995) Identification of somaclonal variants of sugarcane (Saccharum spp.) resistant to sugarcane mosaic virus via RAPD markers. Plant Mol Biol Rep 13:182–191. https://doi.org/10.1007/BF02668790

Perez Ponce J, Jimenez EG, Gomez RK (1998) Field performances of selected sugarcane (Saccharum spp. hybrid) mutants. In: Jain SM, Brar DS, Ahloowalia BS (eds) Somaclonal variation and induced mutations in crop improvement. Kluwer Academic Publishers, Dordrecht, Netherlands, pp 447–462. https://doi.org/10.1007/978-94-015-9125-6_22

Piperidis G, D'Hont A (2001) Chromosome composition analysis of various Saccharum interspecific hybrids by genomic in situ hybridization (GISH). In: Proceedings of the international society of sugar cane technologists vol 24. pp 565–566. https://doi.org/10.1139/G10-010

Phillips RL, Kaeppler SM, Olhoft P (1994) Genetic instability of plant tissue cultures: breakdown of normal controls. Proc Natl Acad Sci U S A 91:5222–5226. https://doi.org/10.1073/pnas.91.12.5222

Rao TJ, Srinivasan KV, Alexander KG (1966) A Red rot resistant mutant of sugarcane induced by gamma irradiation. Proc Indian Acad Sci 64:402–407. https://www.ias.ac.in/article/fulltext/secb/064/04/0224-0230

Rutherford RS, Snyman SJ, Watt MP (2014) In vitro studies on somaclonal variation and induced mutagenesis: progress and prospects in sugarcane (Saccharum spp.)—a review. J Hortic Sci Technol 89(1):1–16. https://doi.org/10.1080/14620316.2014.11513041

Shabbir R, Javed T, Afzal I et al (2021) Modern biotechnologies: innovative and sustainable approaches for the improvement of sugarcane tolerance to environmental stresses. Agronomy 11(6). https://doi.org/10.3390/agronomy11061042

Sharma S, Malhotra PK, Goyal M, Sharma V, Mittal A, Yadav IS, Sanghera GS, Chhuneja P (2023) Characterization of sugarcane mutants developed through gamma irradiations for their lignin content and caffeic acid-O-methyl transferase (COMT) gene mutations. Int J Radiat Biol 100(4):619–626. https://doi.org/10.1080/09553002.2023.2295962

Sreenivasan J, Sreenivasan TV, Alexander KC (1987) Somaclonal variation for smut disease (Ustilago scitaminea Syd.) resistance in sugarcane. In: Reddy GM (ed) Proc symp plant cell and tissue culture of economically important plants, Hyderabad, pp 259–264

Smith MK, Drew RA (1990) Current applications of tissue culture in plant propagation and improvement. Aust J Plant Physiol 17:267–289. https://era.dpi.qld.gov.au/id/eprint/12383/1/PP9900267.pdf

Vamsi Krishna G, Manoj Kumar V, Kishore Varma P, Bhavani B, Vijaya Kumar G (2023) Identification of resistance to sugarcane mosaic virus, sugarcane streak mosaic virus, and sugarcane bacilliform virus in new elite sugarcane accessions in India. Front Microbiol. https://doi.org/10.3389/fmicb.2023.1276932

Viswanathan R, Rao GP (2011) Disease scenario and management of major sugarcane diseases in India. Sugar Tech 13:336–353. https://doi.org/10.1007/s12355-011-0102-4

MIX
Papier aus verantwortungsvollen Quellen
Paper from responsible sources
FSC® C105338

If you have any concerns about our products,
you can contact us on
ProductSafety@springernature.com

In case Publisher is established outside the EU,
the EU authorized representative is:
Springer Nature Customer Service Center GmbH
Europaplatz 3, 69115 Heidelberg, Germany

Printed by Libri Plureos GmbH
in Hamburg, Germany